Alcohol-Induced Oxidative Stress in Health and Disease

Alcohol-Induced Oxidative Stress in Health and Disease

Guest Editor

Marco Fiore

MDPI

Basel • Beijing • Wuhan • Barcelona • Belgrade • Novi Sad • Cluj • Manchester

Guest Editor
Marco Fiore
Institute of Biochemistry and
Cellular Biology IBBC-CNR
Sapienza Universitá di Roma
Rome
Italy

Editorial Office
MDPI AG
Grosspeteranlage 5
4052 Basel, Switzerland

This is a reprint of the Special Issue, published open access by the journal *Antioxidants* (ISSN 2076-3921), freely accessible at: www.mdpi.com/journal/antioxidants/special_issues/alcohol.

For citation purposes, cite each article independently as indicated on the article page online and using the guide below:

Lastname, A.A.; Lastname, B.B. Article Title. *Journal Name* **Year**, *Volume Number*, Page Range.

ISBN 978-3-7258-3648-2 (Hbk)
ISBN 978-3-7258-3647-5 (PDF)
https://doi.org/10.3390/books978-3-7258-3647-5

Contents

About the Editor

Marco Fiore

The research focuses on understanding the neurobiological mechanisms underlying Alcohol Use Disorders (AUDs) and their interplay with pediatric syndromes, including Fetal Alcohol Spectrum Disorders (FASDs) and other neurodevelopmental conditions. The lab employs a multidisciplinary approach, integrating molecular biology and behavioral science to uncover how alcohol exposure alters brain development and function across the lifespan, as well as the study of neurotrophins such as NGF and BDNF.

Key investigations target how chronic alcohol use disrupts neural circuits involved in reward, decision making, and emotional regulation. Using advanced techniques, the research maps changes in neurotransmitter systems, synaptic plasticity, and epigenetic regulation resulting from alcohol exposure. Complementary studies in pediatric populations examine how prenatal alcohol exposure impairs cognitive and social development, with a focus on identifying early biomarkers for FASD.

The lab, located in the Policlinico Umberto I of the Sapienza University of Rome, also explores potential therapeutic interventions, including pharmacological agents and neuromodulatory techniques, to mitigate the neurodevelopmental deficits associated with pediatric syndromes and reduce relapse rates in AUD patients. Collaborative efforts with clinicians and public health specialists aim to translate findings into practical strategies for prevention and treatment. This research holds promise for advancing our understanding of alcohol's impact on the brain and improving outcomes for vulnerable populations.

Preface

Alcohol abuse is a global public health concern, with its impact extending beyond social and behavioral consequences to great physiological effects. Among these, oxidative stress plays a critical role in alcohol-related pathophysiology, influencing a wide range of health situations, from liver disorder to neurodegeneration. This Special Issue, "Alcohol-Induced Oxidative Stress in Health and Disease", published in *Antioxidants*, brings together pioneering research that investigates the mechanisms of alcohol-induced oxidative damage, its implications for human health, and possible therapeutic interventions.

The aim of this collection was extensive, embracing both basic and clinical research. Contributions from important experts provided insights into how alcohol abuse disrupts redox homeostasis, the role of antioxidant defense mechanisms, and novel approaches to mitigate oxidative damage. The studies compiled herein highlighted the importance of understanding oxidative stress as a central process in alcohol-related diseases. They offered perspectives on how antioxidant therapies may serve as potential therapeutic options.

The motivation for presenting this Special Issue derived from the growing recognition of oxidative stress as a crucial mediator of alcohol-induced tissue injury. Despite significant advancements, gaps remain in our comprehension of the precise molecular pathways involved and the development of applicable interventions. By assembling assorted research articles, we aimed to provide a comprehensive reference that could aid researchers, clinicians, and healthcare professionals in advancing knowledge and improving patient outcomes.

This work is intended for a wide audience, including biomedical researchers, clinicians, pharmacologists, toxicologists, and public health professionals attracted by the connection between alcohol metabolism, oxidative stress, and disease evolution. Moreover, this reprint could be a precious tool for students and early-career scientists to extend their understanding of this evolving scenario.

The realization of this Special Issue would not have been conceivable without the dedication and expertise of the contributors, whose work has significantly enriched our understanding of alcohol-induced oxidative stress. I sincerely thank all researchers, reviewers, and editorial staff at *Antioxidants* (MDPI) for their priceless contributions. Their commitment to scientific merit has safeguarded the high quality of this publication.

Further, I hope that this collection of studies will inspire more research and promote new strategies for alleviating the adverse health effects of alcohol-induced oxidative stress.

Marco Fiore
Guest Editor

antioxidants

MDPI

Editorial

Oxidative Stress in Alcohol Abuse: An Unfortunately Still Open Question

Marco Fiore

Institute of Biochemistry and Cell Biology, IBBC-CNR, c/o Campus Internazionale, Via E. Ramarini, 32, 00015 Rome, Italy; marco.fiore@cnr.it

As the guest editor of this Special Issue "Alcohol-Induced Oxidative Stress in Health and Disease" of *Antioxidants* (https://www.mdpi.com/journal/antioxidants/special_issues/alcohol accessed on 29 July 2024), I am pleased to introduce this collection of papers exploring the multidimensional and complex relationship between alcohol consumption and oxidative stress, alongside the consequent health effects. Alcohol, a substance embedded in human culture for millennia, causes both instants of conviviality and severe health risks. This dichotomy makes alcohol an undeniable topic of interest for scientific studies.

Alcohol consumption is prevalent, and while moderate consumption is often socially tolerable and sometimes even associated with certain health benefits, the negative effects of alcohol abuse are intense. These outcomes range from instantaneous impacts such as neurocognitive damage, nausea, and hangovers, to long-term effects including dependence, liver and brain injury, and an elevated risk of various tumors. Fundamental to many of these detrimental consequences is alcohol-induced oxidative stress, an issue that this Special Issue aims to disclose.

Oxidative stress results from a disproportion between the production of reactive oxygen species (ROS) and reactive nitrogen species (RNS), and the body's capability to cleanse these reactive intermediates or restore the resulting injury. Ethanol metabolism plays a crucial role in this imbalance. Indeed, alcohol metabolism mostly occurs in the liver through enzymatic paths involving alcohol dehydrogenase (ADH), cytochrome P450 2E1 (CYP2E1), and catalase. These routes produce ROS as by-products, which lead to oxidative stress if not effectively neutralized by the endogenous antioxidant systems.

The function of mitochondria and the microsomal ethanol-oxidizing system (MEOS) in this activity is central. Mitochondrial dysfunction caused by chronic alcohol abuse intensifies ROS production, leading to cellular loss. Furthermore, alcohol-induced activation of CYP2E1 potentiates ROS and RNS production, further moving the redox balance towards a pro-oxidant level. This oxidative stress negatively disrupts several organs, particularly the brain and liver, leading to disorders such as neurodegeneration and alcoholic liver disease (ALD).

This Special Issue collects novel research and review papers that expand our knowledge of the processes by which alcohol-induced oxidative stress participates in health and disease. The contributions focus on both clinical and preclinical outcomes, offering a fascinating view of current understanding and forthcoming research directions (Contributions 1–5).

One of the recurrent topics in this Special Issue is the sophisticated relationship between alcohol-induced oxidative stress and the body's antioxidant defense systems. Some papers discuss how chronic alcohol abuse harms the functionality of the main antioxidant enzymes such as catalase, glutathione peroxidase (GPx), and superoxide dismutase (SOD). These changes potentiate oxidative impairment, eliciting the progression of disorders such as alcoholic cardiomyopathy and ALD (Contributions 2–3).

check for updates

Citation: Fiore, M. Oxidative Stress in Alcohol Abuse: An Unfortunately Still Open Question. *Antioxidants* **2024**, *13*, 934. https://doi.org/10.3390/antiox13080934

Received: 29 July 2024
Accepted: 30 July 2024
Published: 1 August 2024

Another topic of this Issue regards the toxic effect of alcohol on the fetus if consumed during pregnancy. Likewise, alterations in oxidative stress induced by alcohol may also severely affect the newborn through epigenetic mechanisms (Contributions 6–7).

Another important topic to focus on is the possible therapeutic role of antioxidants in counteracting alcohol-induced oxidative injury. Studies included in this Issue explore various antioxidant compounds, including synthetic antioxidants and natural antioxidants like polyphenols (Contributions 8–9). These reports offer promising evidence that increasing the body's antioxidant ability can reduce some of the harmful effects of alcohol abuse. However, the intricacy of alcohol-induced oxidative stress requires a more direct approach to antioxidant treatment, considering aspects such as timing, dosage, and the definite paths involved (Contributions 10–11).

While noteworthy progress has been made in the comprehension of alcohol-induced oxidative stress, several questions persist. Future research should be conducted to unravel the detailed molecular mechanisms by which alcohol affects redox homeostasis. There is a crucial necessity for more comprehensive studies on the relationship between several ROS and RNS species and their specific cellular targets (Contributions 1, 3).

Additionally, translating preclinical findings into efficient clinical therapies remains a challenge. More accurate clinical trials are needed to assess the efficacy and safety of antioxidant treatments in humans. It should be noted that personalized medicine methods, considering individual differences in antioxidant capacity and alcohol metabolism, could also improve the efficacy of these interventions.

In conclusion, this Special Issue of *Antioxidants* underlines the subtle role of oxidative stress in alcohol-related health topics and focuses on potential possibilities for treatments. The combined efforts of investigators contributing to this Issue may expand our knowledge and provide novel expectations for mitigating alcohol's negative effects through targeted antioxidant approaches.

I extend my sincere gratitude to all the contributors, reviewers, and the editorial team for their priceless roles in the foundation of this Special Issue. I do believe that the insights disclosed from these studies will stimulate further studies to improve clinical practices intended to decrease the burden of alcohol-induced disorders.

Conflicts of Interest: The author declares no conflicts of interest.

List of Contributions

1. Tsermpini, E.; Goričar, K.; Kores Plesničar, B.; Plemenitaš Ilješ, A.; Dolžan, V. The Disease Model of Addiction: The Impact of Genetic Variability in the Oxidative Stress and Inflammation Pathways on Alcohol Dependence and Comorbid Psychosymptomatology. *Antioxidants* **2024**, *13*, 20. https://doi.org/10.3390/antiox13010020.
2. Gauthier, T.; Ping, X.; Harris, F.; Brown, L. Liposomal Glutathione Augments Immune Defenses against Respiratory Syncytial Virus in Neonatal Mice Exposed in Utero to Ethanol. *Antioxidants* **2024**, *13*, 137. https://doi.org/10.3390/antiox13020137.
3. Chapp, A.; Shan, Z.; Chen, Q. Acetic Acid: An Underestimated Metabolite in Ethanol-Induced Changes in Regulating Cardiovascular Function. *Antioxidants* **2024**, *13*, 139. https://doi.org/10.3390/antiox13020139.
4. Idelfonso-García, O.; Alarcón-Sánchez, B.; Guerrero-Escalera, D.; López-Hernández, N.; Pérez-Hernández, J.; Pacheco-Rivera, R.; Serrano-Luna, J.; Resendis-Antonio, O.; Muciño-Olmos, E.; Aparicio-Bautista, D.; et al. Nucleoredoxin Redox Interactions Are Sensitized by Aging and Potentiated by Chronic Alcohol Consumption in the Mouse Liver. *Antioxidants* **2024**, *13*, 257. https://doi.org/10.3390/antiox13030257.
5. Mudyanselage, A.; Wijamunige, B.; Kocoń, A.; Turner, R.; McLean, D.; Morentin, B.; Callado, L.; Carter, W. Alcohol Triggers the Accumulation of Oxidatively Damaged Proteins in Neuronal Cells and Tissues. *Antioxidants* **2024**, *13*, 580. https://doi.org/10.3390/antiox13050580.
6. Derme, M.; Piccioni, M.; Brunelli, R.; Crognale, A.; Denotti, M.; Ciolli, P.; Scomparin, D.; Tarani, L.; Paparella, R.; Terrin, G.; et al. Oxidative Stress in a Mother Consuming Alcohol during Pregnancy and in Her Newborn: A Case Report. *Antioxidants* **2023**, *12*, 1216. https://doi.org/10.3390/antiox12061216.

7. Terracina, S.; Tarani, L.; Ceccanti, M.; Vitali, M.; Francati, S.; Lucarelli, M.; Venditti, S.; Verdone, L.; Ferraguti, G.; Fiore, M. The Impact of Oxidative Stress on the Epigenetics of Fetal Alcohol Spectrum Disorders. *Antioxidants* **2024**, *13*, 410. https://doi.org/10.3390/antiox13040410.
8. Zhou, H.; Khan, D.; Gerdes, N.; Hagenbeck, C.; Rana, M.; Cornelius, J.; Muhammad, S. Colchicine Protects against Ethanol-Induced Senescence and Senescence-Associated Secretory Phenotype in Endothelial Cells. *Antioxidants* **2023**, *12*, 960. https://doi.org/10.3390/antiox12040960.
9. Jegal, K.; Park, H.; Choi, B.; Kim, J.; Ku, S. Synergistic Protective Effect of Fermented Schizandrae Fructus Pomace and Hoveniae Semen cum Fructus Extracts Mixture in the Ethanol-Induced Hepatotoxicity. *Antioxidants* **2023**, *12*, 1602. https://doi.org/10.3390/antiox12081602.
10. Ibáñez, C.; Acuña, T.; Quintanilla, M.; Pérez-Reytor, D.; Morales, P.; Karahanian, E. Fenofibrate Decreases Ethanol-Induced Neuroinflammation and Oxidative Stress and Reduces Alcohol Relapse in Rats by a PPAR-α-Dependent Mechanism. *Antioxidants* **2023**, *12*, 1758. https://doi.org/10.3390/antiox12091758.
11. Santilli, A.; Shapiro, D.; Han, Y.; Sangwan, N.; Cresci, G. Tributyrin Supplementation Rescues Chronic–Binge Ethanol-Induced Oxidative Stress in the Gut–Lung Axis in Mice. *Antioxidants* **2024**, *13*, 472. https://doi.org/10.3390/antiox13040472.

antioxidants

MDPI

Article

Alcohol Triggers the Accumulation of Oxidatively Damaged Proteins in Neuronal Cells and Tissues

Anusha W. Mudyanselage [1,2], Buddhika C. Wijamunige [1,2], Artur Kocoń [1], Ricky Turner [1], Denise McLean [3], Benito Morentin [4], Luis F. Callado [5,6] and Wayne G. Carter [1,*]

1 Clinical Toxicology Research Group, School of Medicine, University of Nottingham, Royal Derby Hospital Centre, Uttoxeter Road, Derby DE22 3DT, UK; wijesekara@agri.sab.ac.lk (A.W.M.); buddhikawijamunige@agri.sab.ac.lk (B.C.W.); artek.1993@googlemail.com (A.K.); ricky.turner@nhs.net (R.T.)
2 Department of Export Agriculture, Faculty of Agricultural Sciences, Sabaragamuwa University of Sri Lanka, Belihuloya 70140, Sri Lanka
3 School of Life Sciences, University of Nottingham, Nottingham NG7 2UH, UK; denise.mclean@nottingham.ac.uk
4 Section of Forensic Pathology, Basque Institute of Legal Medicine, E-48001 Bilbao, Spain; morentin.b@justizia.eus
5 Department of Pharmacology, University of the Basque Country-UPV/EHU, E-48940 Leioa, Spain; lf.callado@ehu.eus
6 Centro de Investigación Biomédica en Red de Salud Mental (CIBERSAM), Spain
* Correspondence: wayne.carter@nottingham.ac.uk; Tel.: +44-(0)1332-724-738

Abstract: Alcohol is toxic to neurons and can trigger alcohol-related brain damage, neuronal loss, and cognitive decline. Neuronal cells may be vulnerable to alcohol toxicity and damage from oxidative stress after differentiation. To consider this further, the toxicity of alcohol to undifferentiated SH-SY5Y cells was compared with that of cells that had been acutely differentiated. Cells were exposed to alcohol over a concentration range of 0–200 mM for up to 24 h and alcohol effects on cell viability were evaluated via MTT and LDH assays. Effects on mitochondrial morphology were examined via transmission electron microscopy, and mitochondrial functionality was examined using measurements of ATP and the production of reactive oxygen species (ROS). Alcohol reduced cell viability and depleted ATP levels in a concentration- and exposure duration-dependent manner, with undifferentiated cells more vulnerable to toxicity. Alcohol exposure resulted in neurite retraction, altered mitochondrial morphology, and increased the levels of ROS in proportion to alcohol concentration; these peaked after 3 and 6 h exposures and were significantly higher in differentiated cells. Protein carbonyl content (PCC) lagged behind ROS production and peaked after 12 and 24 h, increasing in proportion to alcohol concentration, with higher levels in differentiated cells. Carbonylated proteins were characterised by their denatured molecular weights and overlapped with those from adult post-mortem brain tissue, with levels of PCC higher in alcoholic subjects than matched controls. Hence, alcohol can potentially trigger cell and tissue damage from oxidative stress and the accumulation of oxidatively damaged proteins.

Keywords: alcohol; alcohol-related brain damage; developmental neurotoxicity; oxidative stress; protein carbonylation; reactive oxygen species

check for updates

Citation: Mudyanselage, A.W.; Wijamunige, B.C.; Kocoń, A.; Turner, R.; McLean, D.; Morentin, B.; Callado, L.F.; Carter, W.G. Alcohol Triggers the Accumulation of Oxidatively Damaged Proteins in Neuronal Cells and Tissues. *Antioxidants* **2024**, *13*, 580. https://doi.org/10.3390/antiox13050580

Academic Editor: Marco Fiore

Received: 9 April 2024
Revised: 30 April 2024
Accepted: 4 May 2024
Published: 8 May 2024

1. Introduction

Ethyl alcohol (ethanol) is the most widely imbibed, licit, psychoactive drug. Although drinking alcohol is an element of the social fabric of many cultures, there are serious health concerns and consequences that can arise from excessive alcohol intake [1–3]. The relationship between alcohol and human harm is complex and multidimensional but does increase monotonically with increased consumption [3]. The number of global deaths attributed to the harmful use of alcohol was over 3 million in 2016, constituting 1 in

20 deaths [4]. In terms of disability-adjusted life years (DALYs), over 5% of the global burden of disease is causally linked to alcohol usage [4,5].

The impact of alcohol on health relates to both the volume of alcohol consumed and the pattern of drinking, including the number of heavy drinking sessions [6]. Epidemiological analyses have established an association between alcohol usage and over 200 somatic diseases [7]. For some of these diseases, such as liver cirrhosis, a relative-risk dose response exists [8] but the relationship between alcohol intake and risk of disease is not uniformly dose-dependent in all tissues. For some tissues, a curvilinear relationship such as a J- or U-shaped curve may exist such that low to moderate drinkers have a reduced health risk compared with certain cohorts of abstainers. Although still a moot point, some epidemiological studies have suggested a protective benefit of low-level alcohol consumption for reduced risk of diabetes mellitus, ischemic heart disease, and dementia [7,9,10]. Nevertheless, the evidence base for long-term cognitive damage to alcoholics is considerable. Some epidemiological studies have suggested a reduced risk of development of dementia for certain minimal and light drinking cohorts when compared with abstainers, but many studies have concluded that heavy drinking is associated with an increased risk of dementia and cognitive decline [11–17].

In support of an association between excessive alcohol drinking and dementia, brain atrophy, damage, and neuronal loss have all been detected in many but not all post-mortem studies of brains of alcoholics [18–25]. Likewise, brain shrinkage of white and/or grey matter in response to longitudinal alcohol exposure has been detected using a range of in vivo imaging techniques [26–34]. Furthermore, specific localised volumetric reductions of subcortical structures including the prefrontal cortex and hippocampal regions have also been detected in alcoholics [30,32], and correlate with cognitive decline [32].

Adolescence is a period of notable vulnerability to the neurotoxic effects of alcohol, with binge drinking associated with reduced grey matter and detrimental effects on attention and cognition [35,36]. The elderly may also be more responsive to the toxic effects of alcohol [36], and there is a decline in brain structure with age that mirrors that observed in alcoholic patients [25]. Alcohol also has teratogenic effects, such that excessive maternal alcohol consumption during pregnancy impacts the neurodevelopment of the foetus and results in foetal alcohol spectrum disorders (FASD), and negative effects on cognition [36–40]. FASD is recognised by the presence of a range of impairments to growth, dysmorphia, and central nervous system (CNS) dysfunction, including deficits in cognition and neurobehavioural abnormalities as a consequence of brain damage [36–40]. Reduced grey and white matter contribute to the collective reduction in brain size for babies with FASD [39,40]. Alcohol may therefore be particularly neurotoxic during periods of neurodevelopment and in the elderly, and this could be mediated by mechanisms including cellular redox stress and induction of apoptosis [40–43].

Excessive alcohol exposure can result in a depletion of numbers of neurons (cell death), but alcohol also has a broad impact on neurocircuitry and plasticity [40,44] and these can diminish the functionality of surviving neurons [39,40,45]. Hence, to gain more insight into the effects of alcohol on newly differentiated neuronal cells, and the potential impact of oxidative stress, the toxicity of alcohol was directly compared between undifferentiated and differentiated SH-SY5Y cells. Neurotoxicity was assessed via quantitation of alcohol effects on cell viability, mitochondrial morphology and functionality, the induction of reactive oxygen species (ROS), and the accumulation of oxidatively damaged proteins. Studies were also undertaken to consider whether the oxidative damage observed in cells after alcohol exposure was mirrored by that present within human post-mortem brain tissue from alcoholics.

2. Materials and Methods

2.1. Cell Culture and Cell Image Capture

The SH-SY5Y human neuroblastoma cell line was purchased from the European Collection of Authenticated Cell Culture (ECACC) (ECACC-94030304). Experiments were conducted with cells from passages 13–14. SH-SY5Y cells were grown in the following culture medium: 43.5% Eagle's Minimum Essential Medium (EMEM) (M4655, Sigma, Poole, UK) supplemented with 43.5% Ham's F12 nut mix (217665-029, Gibco, Waltham, MA, USA), 10% heat-inactivated fetal bovine serum (FBS) (F9665, Sigma, Poole, UK), 1% MEM Non-Essential Amino Acid Solution (NEAA) (RNBF3937, Sigma, Poole, UK), 2 mM glutamine, and 1% penicillin–streptomycin solution containing 10,000 IU penicillium and 10 mg/mL streptomycin (p/s) (P4333, Sigma, Poole, UK) in 25 or 75 cm^2 flasks (Thermo Fisher Scientific, Rochester, UK) at 37 °C with an atmosphere of 5% CO_2 and 95% humidity, as previously described [46]. Cells were observed daily and grown until the cells reached approximately 80% confluence, after which the culture medium was refreshed every other day.

For differentiation, SH-SY5Y cells were seeded on either poly-D-lysine (PDL) hydrobromide (5 mg/mL) (P6407, Sigma, Poole, UK) coated 25 cm^2 flasks (T25, 130189, Thermo Fisher Scientific, Rochester, UK) or in 96-well microtiter plates (6005649, Perkin Elmer, Groningen, The Netherlands) with 10% FBS media. After the cells had settled, they were grown to 60% confluency. The following day, the cells were treated with differentiation medium (10 μM all-trans retinoic acid (RA) (R2625, Sigma, Poole, UK) in low-serum SH-SY5Y medium (1% FBS) for 6 days and then treated with 20 ng/mL brain-derived neurotrophic factor (BDNF) (B3795, Sigma, Poole, UK) with low-serum medium containing RA for 2 more days, after which the cells displayed a fully differentiated morphology [46,47].

Cells treated with alcohol (10–200 mM) were monitored with an inverted microscope with phase-contrast optics (Olympus, DP70, London, UK) to compare the general morphological changes with untreated controls for both undifferentiated and differentiated cells at the end of the treatment period. Cells that were cultured in 12-well PDL-coated plates were used to study the neurite length changes in differentiated cells in response to 0–200 mM alcohol treatments. Cells were considered to be differentiated if each neuronal cell contained at least one component that was longer than its cell body [48]. The neurite lengths from 200 randomly chosen cells were measured in 5 selected quadrants per well using the neurite tracer tool from Image J (Image J 1.49k, National Institute of Health, Bethesda, MD, USA), in three independent wells for each treatment [49].

Untreated cells and those incubated with alcohol for 24 h were prepared for transmission electron microscopy (TEM) according to the methods described in [50]. In brief, after a 24 h incubation, the medium was removed and cells were washed with medium containing fixative (3% glutaraldehyde in 0.1 M cacodylate buffer). The media–fixative solution (1:1 (v/v)) was then replaced with fixative alone, before the cells were fixed in an incubator for 1 h at 37 °C. Cells were scraped into the fixative, collected by centrifugation, and then further fixed at 4 °C for 1 h. Cells were then washed in a 0.1 M cacodylate buffer and transferred to flat-bed embedding capsules, before incubation with 1% osmium tetroxide in 0.1 M cacodylate buffer for 1 h. Cells were water-washed and then dehydrated using a series of ethanol solutions: 50, 70, 90, and 100% ethanol and a transitional solution, 100% propylene oxide. Cells were then infiltrated with an epoxy resin–propylene oxide mix (1:1) overnight. The following day, the samples were infiltrated with epoxy resin for 3 × 2 h and then embedded and polymerized by heating at 60 °C for 48 h. Ultra-thin samples (80 nm) of the cells were sectioned with a diamond knife on a Leica EM UC6 ultramicrotome, mounted on 200 mesh copper grids, and then analysed using a Tecnai G2 BioTWIN TEM (FEI company, Eindhoven, The Netherlands).

2.2. Thiazolyl Blue Tetrazolium Bromide (MTT) Assays

Cell metabolic activity and cell viability were determined using a Thiazolyl Blue Tetrazolium Bromide (MTT) (M5655, Sigma, Poole, UK) assay, as described previously [51]. SHSY-5Y cells were seeded at 3×10^4 cells/well in 96-well plates with growth medium (10% FBS). After 24 h, undifferentiated cells were exposed to ethanol (0–200 mM) diluted in growth media (10% FBS). Differentiated cells were prepared as described above and then treated with ethanol (0–200 mM) diluted in differentiation medium supplemented with

20 ng/mL BDNF. After incubation, spent medium was removed and then replaced with medium containing 10% 5 mg/mL MTT and incubated for 4 h. Plate wells which only received 10% MTT and respective growth medium served as background controls. The generated formazan crystals were suspended in a 1:1 dimethyl sulphoxide (DMSO, D8418, Sigma, Poole, UK)–isopropanol (279544, Sigma, Poole, UK) solution. The absorbance of wells was then read at 570 nm using a spectrophotometer (Multiskan Spectrum, Thermo Electron Corporation, Waltham, MA, USA). An average value was calculated from experiments performed in triplicate after the subtraction of blank (negative control) values. Cell viability was expressed as a percentage of survival compared with that from mock-treated cells. The inhibitor concentrations producing 50% loss of viability of cells (IC_{50} values) were obtained from the concentration–response curves and expressed as mean ± standard deviation (SD).

2.3. Lactate Dehydrogenase (LDH) Assays

Undifferentiated or differentiated SHSY-5Y cells were prepared as described above for the MTT assay and similarly treated with ethanol. After ethanol treatment, 50 µL of spent medium was removed and LDH activity determined using an assay kit (ab102526, Abcam, Cambridge, UK) according to the manufacturer's guidelines. NADH standards were prepared according to the manufacturer's protocol and were transferred into the same assay plate. Assays were performed at 450 nm using a spectrophotometer (Multiskan Spectrum, Thermo Electron Corporation, Waltham, MA, USA) in kinetic mode, with readings every 2 min at 37 °C, protected from light, for a total of 60 min. A NADH standard curve was generated and LDH activity measurements interpolated from the NADH standard curve. An average value was calculated from experiments performed in triplicate after the subtraction of blank (negative control) values.

2.4. Adenosine 5′-Triphosphate (ATP) Assays

Undifferentiated SH-SY5Y cells were seeded in 6-well plates (CC7682-7506, STARLAB International GmbH, Hamburg, Germany) at a density of 1×10^6 cells/well for analysis. For differentiated cells, cells were seeded at 5×10^4 cells/well in PDL-coated 6-well plates, with the differentiation protocol followed for 7 days, as described above. Cells were treated with ethanol, as before, and ATP levels were quantified using an ATP luminescence assay kit (ATP Bioluminescence Assay Kit CLS II (11 699 695 001, Roche, Germany), as per the manufacturer's protocol. The ATP content in control and ethanol-treated samples was interpolated from an ATP standard curve, as described previously [52]. Average values were calculated from experiments performed in triplicate after the subtraction of blank (negative control) values.

2.5. Measurements of Reactive Oxygen Species

The generation of reactive oxygen species (ROS) was quantified using a 2′,7′-dichlorof luorescin diacetate (DCFDA) (D6883, Sigma, Poole, UK) assay. SHSY-5Y cells were seeded at 3×10^4 cells/well in clear-bottom black 96-well plates (165305, Thermo Fisher Scientific, Rochester, UK) with growth medium (at 10% FBS). After 24 h, undifferentiated cells were exposed to ethanol (0–200 mM) diluted in growth medium (10% FBS) and differentiated cells were prepared as described above and then treated with ethanol (0–200 mM) diluted in differentiation medium supplemented with 20 ng/mL BDNF and 10 µM RA. Cells were treated with ethanol for 3, 6, 12, or 24 h, with 50 µM DCFDA included for the experiment duration. Cells were washed twice with ice-cold PBS and then their fluorescence was quantified using a Varioskan™ LUX multimode microplate reader (Thermo Fisher Scientific, Waltham, MA, USA) at excitation and emission spectra of 495 nm and 529 nm, respectively. Hydrogen peroxide (0.5 mM) was used as a positive control for ROS, set as 100% fluorescence [46,53]. Three to six replicate assays were performed for all data points, from which an average was calculated.

2.6. Cell Lysis

After ethanol or vehicle treatment of undifferentiated or differentiated SH-SY5Y cells, cells were washed with cold phosphate-buffered saline (PBS) (10010015, Life Technologies, Paisley, UK) before addition of 0.5 mL of radioimmunoprecipitation assay (RIPA, 20-188, Millipore, Burlington, MA, USA) buffer containing protease inhibitors (04693124001, mini-protease inhibitor cocktail, Sigma, Poole, UK) and a phosphatase inhibitor cocktail (P0044, Sigma, Poole, UK), with flask agitation on ice for 5 min. Cells were then scraped into the RIPA buffer, vortexed thoroughly, and then homogenized by passage through a 28 g needle 25 times. Homogenates were stored at −20 °C until required.

2.7. Protein Quantification

The quantitation of protein concentration was performed based on the Lowry assay [54]. Bovine serum albumin (BSA) protein was used as a protein standard. The modified Lowry assay was performed in 96-well plates using protein standard amounts of 1.25, 2.5, 5, 7.5, and 10 µg of protein. For a volume of 40 µL of cell lysates or protein standards, 20 µL of Reagent A was added followed by 160 µL of Reagent B. After 15 min, spectrophotometric measurements were taken at 740 nm using a Spectramax plate reader (Multiskan Spectrum, Thermo Electron Corporation, Waltham, MA, USA). Protein amounts of unknowns were interpolated from the BSA standard curve.

2.8. Determination of Protein Carbonyl Content

Undifferentiated or differentiated SH-SY5Y cells were grown to 80% confluence and then treated with ethanol for 3, 6, 12, or 24 h, as described above. After alcohol treatment, cells were washed with ice-cold PBS three times and then solubilized and lysed with RIPA buffer containing protease and phosphatase inhibitors (according to Section 2.5). Samples were vortexed for 30 s and then sonicated for 15 min on ice-cold water. Samples were then spun at 500× g for 10 min at 4 °C, and the supernatant was retained and centrifuged at 23,100× g for 40 min at 4 °C to prepare a crude cytosolic fraction [24]. Protein concentration was determined using a modified Lowry assay (according to Section 2.6) and then adjusted to 1 mg/mL for cells or brain tissue homogenates (refer to Section 2.9). An equivalent volume of 10 mM 2,4-dinitrophenylhydrazine (DNPH) (Sigma-Aldrich, Poole, UK) prepared in 2 N HCL (231-5957, Scientific Laboratory Suppliers, Nottingham, UK) was added to samples or blanks and vortex mixed, and then, samples were left in the dark for 1 h at room temperature, with vortex mixing every 10 min. Protein precipitation was initiated by the addition of an equivalent volume of ice-cold 20% (w/v) trichloroacetic acid (TCA) (Sigma-Aldrich, Poole, UK) and the samples were retained on ice for 15 min. The precipitate was washed according to a previously published method [46], before solubilization in 6 M guanidine hydrochloride (50950, Fluka Chemie AG, Buchs, Switzerland) in 50 mM phosphate buffer, pH 2.3, with incubation at 37 °C for 30 min and with vortex mixing. The protein carbonyl content (PCC) was then determined spectrophotometrically (Thermo Fisher Scientific, Fluoroskan Ascent FC, Waltham, MA, USA) at 366 nm using a molar absorption coefficient of 22,000 $M^{-1}cm^{-1}$ after subtraction of blanks. Data points were generated from assays performed in triplicate, from which an average was calculated.

2.9. Western Oxy-Blotting

Immuno-blotting for reactive carbonyl groups (oxidatively damaged proteins) was undertaken using an OxyBlot Protein Oxidation Detection Kit (S7150, Millipore, Burlington, MA, USA) as recommended by the manufacturer. Cytosolic protein concentrations were quantified as detailed above using a modified Lowry assay (Section 2.6). Proteins were then prepared to a concentration of 2 mg/mL via the addition of 12% sodium dodecyl sulphate (SDS) and 2,4-dinitriophenylhydrazine (DNPH) solution, and carbonyl groups were derivatized by incubation at room temperature for 15 min. Neutralization solution and then β-mercaptoethanol were added to the sample mixture, and then the proteins were resolved using Novex NuPAGE 10% Bis–Tris gel (Thermofisher Scientific, Rochester, UK) in an Xcell surelock mini-cell system with (3-N-morpholino)propanesulphonic acid (MOPS) running buffer (Thermofisher scientific, Rochester, UK), as described previously [55]. Gel proteins were transferred in a BioRad mini trans-blot cell to polyvinylene difluoride (PVDF) (Millipore, USA) membranes and probed with a rabbit anti-DNP primary antibody, followed by a goat anti-rabbit IgG (horseradish peroxidase (HRP)-conjugated) secondary antibody, as described previously [46]. Immunoreactivity was detected using a ChemiDoc MP imager (BioRad, Hertfordshire, UK), with light captured with an autoexposure setting to ensure signal linearity.

2.10. Human Brain Samples

The human brain samples used in this study were used in accordance with the Human Tissue Act (2004) (UK) and were supplied by the Neuropsychopharmacology Research Group from the Department of Pharmacology of the University of the Basque Country (UPV/EHU) (https://www.ehu.eus/en/web/neuropsicofarmacologia/home, accessed on 8 April 2024). Brain tissue collection was conducted in compliance with the research policies and ethical review boards for post-mortem brain studies (Basque Institute of Legal Medicine, Bilbao, Spain) and registered in the National Biobank Register of the Spanish Health Department with the study number C.0000035 (https://biobancos.isciii.es/ListadoColecciones.aspx, accessed on 8 April 2024). Diagnosis of alcoholism was carried out according to the *Diagnostic and Statistical Manual of Mental Disorders* (DSM-III-R, DSM-IV, or DSM-IV-TR; American Psychiatric Association) or International Classification of Diseases criteria (ICD-10; World Health Organization). All diagnoses were established by clinicians in charge of the patients prior to death. Six control brain samples were used, matched by age and sex to 6 alcoholic subjects, as detailed in previous studies [24,25] (Supplementary Table S1). Toxicological screening of the blood (quantitative assays for antidepressants, antipsychotics, other psychotropic drugs, and ethanol) was performed at the National Institute of Toxicology, Madrid, Spain. The brain samples used were all from the prefrontal cortex (Brodmann's area 9) (BA 9), macroscopically dissected at the time of autopsy and stored at -80 °C until required.

2.11. Statistical Analysis

Data for cell viability and ATP assays are presented as means ± standard error of the mean (SEM). Statistical analysis was performed using GraphPad Prism 9.2.0 (GraphPad Prism, San Diego, CA, USA). Concentration–response curves were plotted using a non-linear regression curve fit model as line of best fit. To assess differences between control and treatment groups, one-way analysis of variance (ANOVA) or two-way ANOVA with Dennett's multiple comparison test and Tukey's multiple comparisons, respectively, were performed. Results were considered significant at a p-value below 0.05.

3. Results

3.1. Alcohol Effects on Cell Viability

Undifferentiated and differentiated SH-SY5Y cells were exposed to alcohol at concentrations of 0–200 mM for 3, 6, 12, or 24 h and cell metabolic activity and viability were quantified using an MTT assay (Figure 1A–D). Alcohol reduced cell metabolic activity and viability in a concentration- and exposure duration-dependent manner from a threshold of ≥20 mM for both undifferentiated and differentiated SH-SY5Y cells (Figure 1A–D and Supplementary Table S2).

After 3 or 6 h alcohol exposure, cell metabolic activity for both undifferentiated or differentiated SH-SY5Y cells was similar and inversely proportional to alcohol concentration, such that there was an approximately linear decline in cell viability with increasing alcohol concentration (Figure 1A,B). After 12 or 24 h incubation with alcohol, the inhibitor–response curves showed significant reduction in cell viability at 50 mM alcohol ($p < 0.0001$) (Figure 1C,D). Differentiated cells were more resistant to alcohol toxicity than undifferentiated cells, with higher concentrations required to induce 50% inhibition of cell viability

(IC_{50}) (Figure 1A–D, Table 1, and Supplementary Table S3). The lowest concentration of alcohol examined (10 mM) increased cell metabolic activity, although non-significantly ($p = 0.113$), by 6–11% in differentiated cells and 1–10% in undifferentiated cells ($p = 0.08$) (Figure 1A–D).

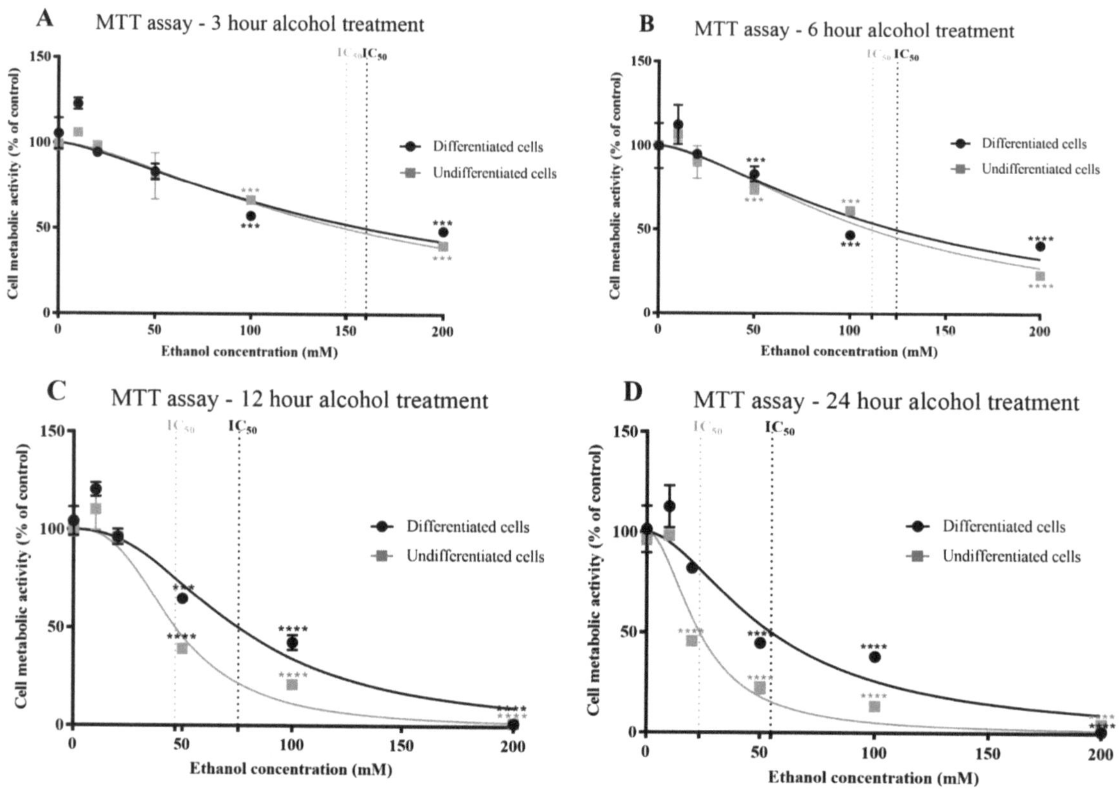

Figure 1. Effects of alcohol on cellular metabolic activity and viability, determined using an MTT assay. Undifferentiated or differentiated SH-SY5Y cells were exposed to alcohol (0–200 mM) for durations of 3 (**A**), 6 (**B**), 12 (**C**), and 24 (**D**) h and the levels of metabolic activity and cell viability were quantified using an MTT assay. Each data point represents the mean of at least 5 individual experiments. Marked significance: *** = *p*-value < 0.001, **** = *p*-value < 0.0001.

Since MTT assays provide insight into cell metabolic activity and this may not always correlate with cell viability, the liberation of extracellular LDH was used as an independent method for the determination of cell viability in response to alcohol. Similar to the MTT assays, undifferentiated and differentiated cell viability decreased in proportion to the alcohol concentration and length of exposure time (Figure 2A–D and Supplementary Table S4). The threshold for a significant reduction of cell viability was a concentration of alcohol of $\geq$20 mM for 6 h exposure time ($p < 0.001$ for undifferentiated cells and $p < 0.0001$ for differentiated cells) (Figure 2B). Non-linear regression analysis showed that undifferentiated cells were more sensitive to the toxic effects of alcohol, with lower IC_{50} concentrations, in keeping with the MTT data (Table 1 and Supplementary Table S2).

Table 1. Toxicity of alcohol to undifferentiated and differentiated SHSY-5Y cells.

Cell Type	Treatment Duration (Hours)	MTT Assay		LDH Assay		ATP Assay	
		IC_{50}	R^2	IC_{50}	R^2	IC_{50}	R^2
Undifferentiated	3	149.8 ± 18.6	0.8800	110.6 ± 3.1	0.9878	158.5 ± 17.3	0.9149
Differentiated		160.5 ± 25.8	0.7969	172.5 ± 3.4	0.9863	179.4 ± 26.3	0.7732
Undifferentiated	6	111.5 ± 7.6	0.9453	107.2 ± 4.6	0.9722	124.4 ± 10.6	0.9430
Differentiated		124.4 ± 14.7	0.8573	136.1 ± 5.9	0.9507	158.2 ± 29.2	0.7196
Undifferentiated	12	46.7 ± 3.1	0.9648	46.9 ± 3.1	0.9743	42.2 ± 3.9	0.9445
Differentiated		75.25 ± 7.0	0.9268	75.8 ± 4.5	0.9551	74.4 ± 3.7	0.9745
Undifferentiated	24	23.3 ± 2.1	0.9372	24.9 ± 2.374	0.9371	36.0 ± 5.1	0.8476
Differentiated		54.83 ± 6.5	0.9016	59.10 ± 2.3	0.9391	48.08 ± 3.4	0.9551

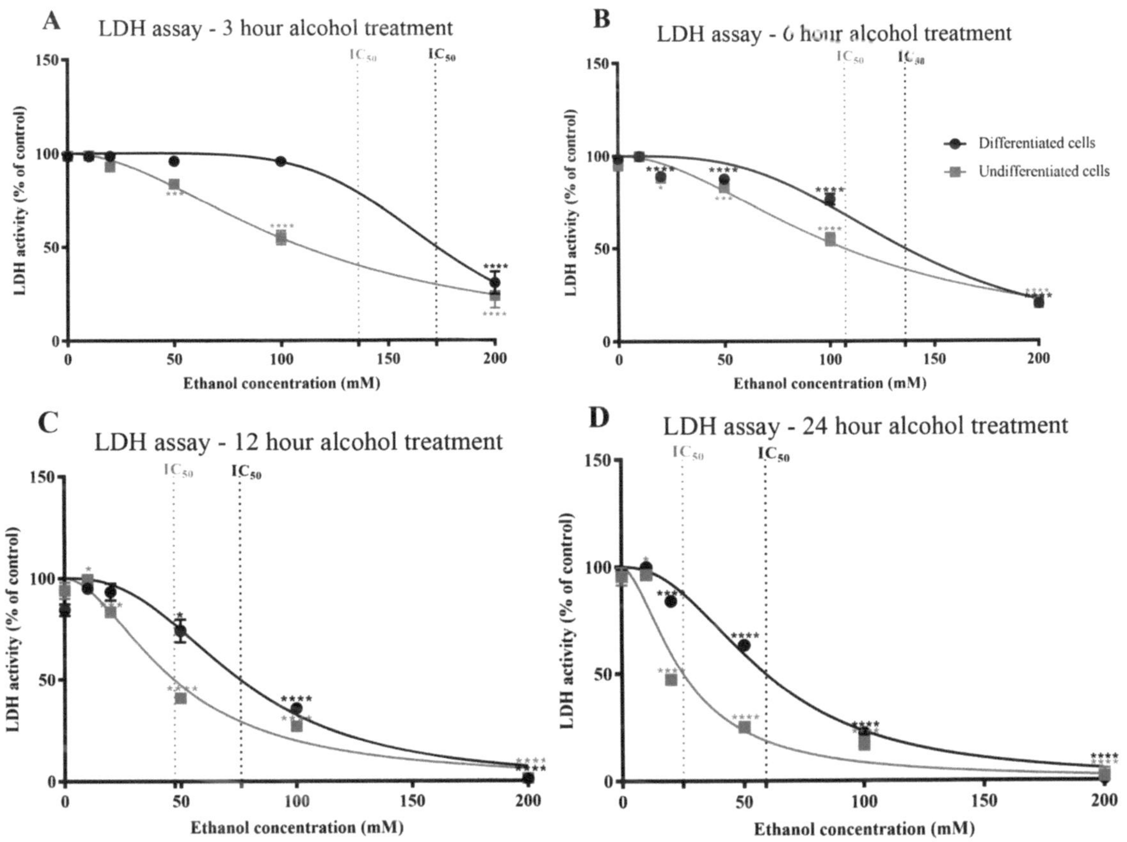

Figure 2. Alcohol effects on cell viability determined using an LDH activity assay. Undifferentiated or differentiated SH-SY5Y cells were exposed to alcohol (0–200 mM) for durations of 3 (**A**), 6 (**B**), 12 (**C**), and 24 (**D**) h, and the activity of extracellular LDH was quantified. Each data point represents the mean of at least 5 individual experiments. Marked significance: * = p-value < 0.05, *** = p-value < 0.001, **** = p-value < 0.0001.

Additionally, the alcohol-induced reduction in cell viability and influence on neuritic projections (Figure 3) were assessed via direct observation of the cells and photographic image capture (Supplementary Figure S1 and Supplementary Table S5). Alcohol triggered a significant reduction in neuritic arborization from a threshold concentration of 50 mM for 6 ($p < 0.001$), 12 ($p < 0.001$), and 24 h ($p < 0.001$) exposures (Figure 3 and Supplementary Table S5).

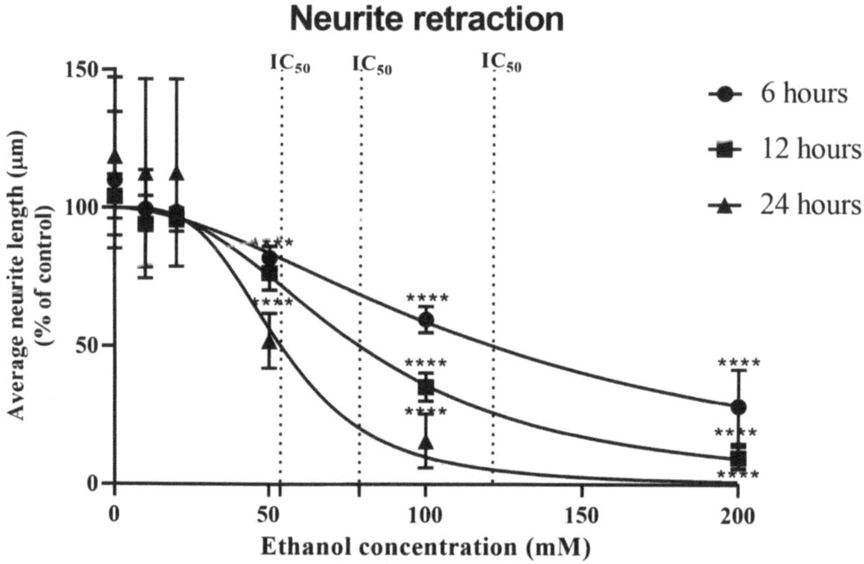

Figure 3. Neurite retraction in response to alcohol treatment. Differentiated SHSY-5Y cells were treated with alcohol over a concentration range of 0–200 mM for 6–24 h and the length of neuritic projections was quantified. Experiments were conducted in triplicate and each data point represents the mean of at least 5 individual experiments (±SD), with vehicle control experiments set at 100%. Significant reductions in neuritic projections were observed at 50 mM alcohol for all time points. Marked significance: **** = p-value < 0.0001.

3.2. Alcohol Effects on Cellular Bioenergetics and the Liberation of Reactive Oxygen Species

Direct effects on mitochondrial morphology were examined using transmission electron microscopy (TEM) (Figure 4A–D). Alcohol at concentrations of ≥50 mM resulted in more translucent mitochondria (less electron dense) and some vacuoles were present within cells, which may reflect mitophagy.

The effect of alcohol on cellular bioenergetic capacity was determined via quantitation of ATP levels. An alcohol-induced decline in ATP levels was observed which correlated with alcohol concentration and exposure duration and mirrored the MTT alcohol response curves for both undifferentiated and differentiated SH-SY5Y cells (Figure 5A–D, and Supplementary Table S6), with similar IC_{50} values (refer to Table 1 and Supplementary Table S2). A significant reduction in ATP levels was evident from an exposure concentration of ≥20 mM and 3 h exposure for undifferentiated cells ($p < 0.0001$). Interestingly, the induction of cell metabolic activity (MTT assay results) observed after 10 mM alcohol exposure was reiterated for ATP production. The resistance of differentiated cells to alcohol was also evident from measurements of ATP levels, such that a significant reduction in ATP was observed with alcohol concentrations of ≥50 mM and application of at least 6 h ($p = 0.0294$).

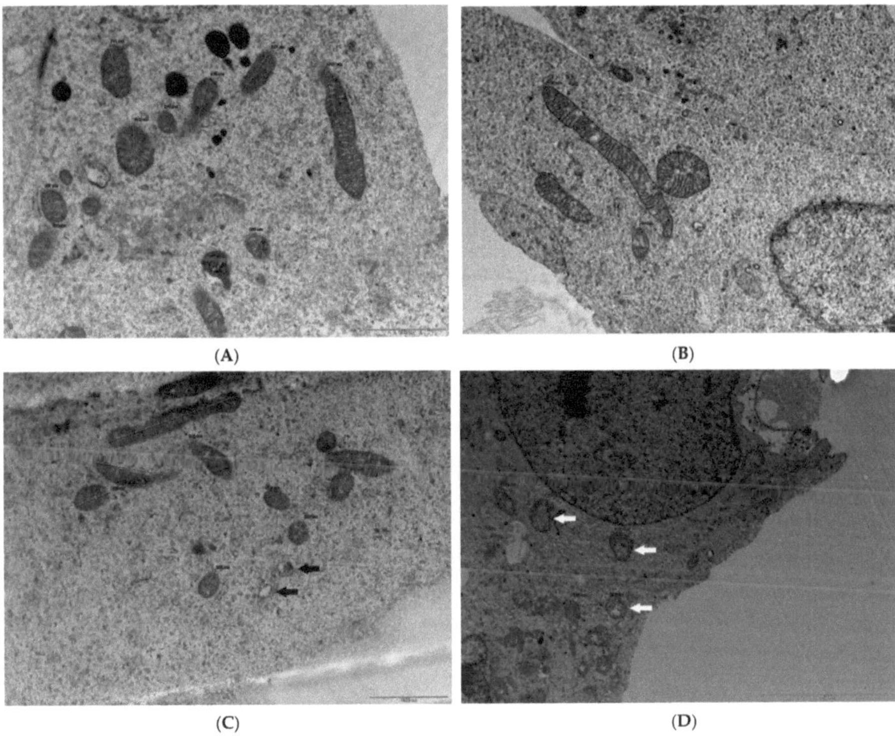

Figure 4. TEM images of control and alcohol-treated cells. (**A**) ×16,500 magnification of control (untreated) cells. Mitochondria are clear with well-visible cristae, some of which have been sized (in red) for reference. Scale bar: 1000 nm. (**B**) ×9900 magnification of cells treated with 50 mM ethanol for 24 h. Mitochondria are distinguishable with visible cristae that are patchy in places, and some elongated mitochondria can be observed. Mitochondrion measurements have been included (in red) for reference. Scale bar: 2000 nm. (**C**) ×16,500 magnification of cells treated with 100 mM ethanol for 24 h. Mitochondria are distinguishable with visible cristae that are patchy in places, some elongated mitochondria are visible, as well as some vacuolar regions perhaps generated from mitophagy (examples indicated with black arrows). Mitochondrion measurements (in red) have been included for reference. Scale bar: 1000 nm. (**D**) ×6000 magnification of cells treated with 200 mM ethanol for 24 h. Mitochondria are distinguishable with some visible cristae but clear regions within mitochondria (examples indicated with white arrows) and some vacuolar regions presumed to be generated from mitophagy. Mitochondrion measurements (in red) have been included for reference. Scale bar: 5000 nm. For TEM transverse section images, up to 19 fields of view were analysed, with random unbiased selection. Images were captured using a MegaView SIS camera, with representative images included.

The production of reactive oxygen species (ROS) was monitored over the 3–24 h time course through measuring the oxidation of 2′,7′-dichlorodihydrofluorescein in a DCFDA assay. ROS levels increased in proportion to alcohol concentrations at all time points, with ROS levels that peaked at 3 and 6 h (Figure 6A–D and Supplementary Table S7). Differentiated cells were notably more potent producers of ROS than undifferentiated cells, with significantly higher levels of ROS liberated after 20 and 50 mM alcohol exposures at the 3 and 6 h time points ($p < 0.0001$) (Figure 6A–D and Supplementary Table S3).

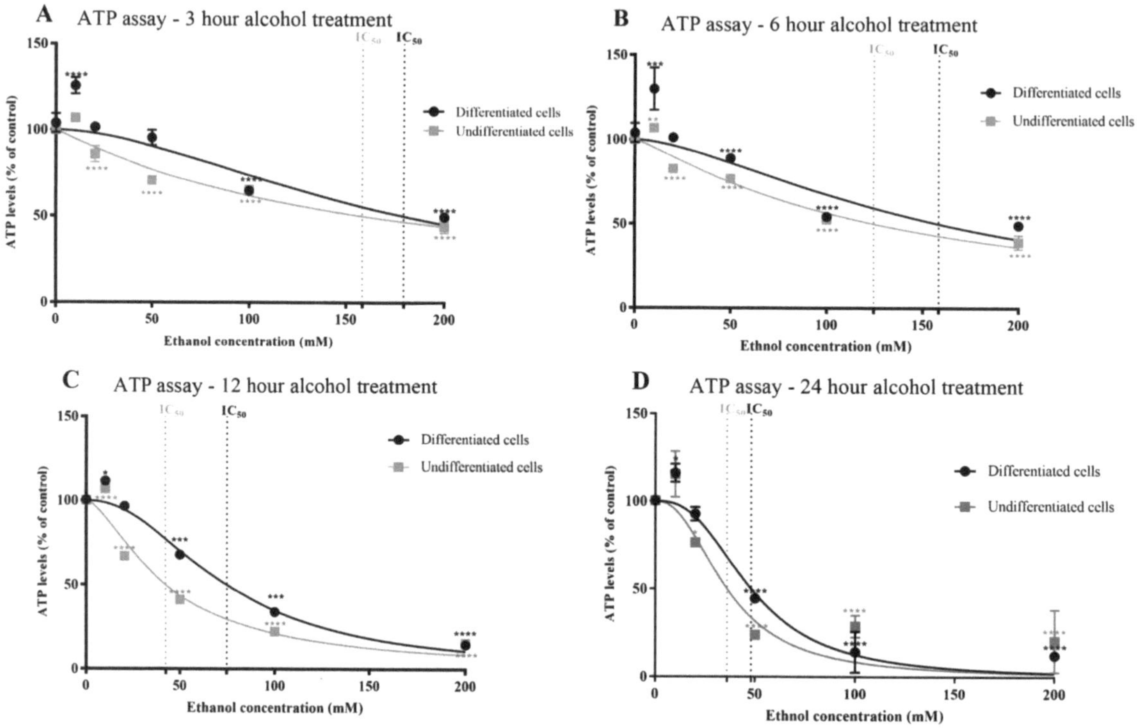

Figure 5. Effects of alcohol on cellular ATP levels determined using an ATP bioluminescence assay. Undifferentiated or differentiated SH-SY5Y cells were exposed to alcohol (0–200 mM) for durations of 3 (**A**), 6 (**B**), 12 (**C**), and 24 (**D**) h, and the level of cellular ATP was quantified using an ATP bioluminescence assay. Each data point represents the mean of at least 5 individual experiments. Marked significance: * = p-value < 0.1, ** = p-value < 0.01, *** = p-value < 0.001, **** = p-value < 0.0001.

In line with the production of cellular ROS, the level and time course of production of oxidatively damaged proteins was quantified via determination of total protein carbonyl content (PCC). PCC increased in undifferentiated and differentiated SH-SY5Y cells in accordance with the concentration of alcohol; significant levels were detected from 10 mM alcohol, the lowest concentration examined ($p < 0.0001$) (Figure 7A,B). However, there was a delay in the accumulation of PCC, with significant increases above baseline ($\approx$1 nmol/mg of protein) detected after 12 or 24 h, and this had a higher positive correlation with ROS levels (Supplementary Table S8). PCC profiles were similar for undifferentiated cells and differentiated cells but with statistically higher levels in differentiated cells from a threshold concentration of 50 mM and 12 h alcohol exposure ($p < 0.0001$) (Supplementary Table S3).

In order to characterise the carbonylated proteins, oxy-blotting was performed. Carbonylation (protein oxidation) was detected in several proteins, at denatured molecular weights of 120, 110, 90, and 50 kDa and with levels that increased in accordance with alcohol concentration (Figure 8). The profile of carbonylated proteins was similar to that detected in the brains of alcoholic subjects and, to a lesser extent, age- and sex-matched controls (Figure 8). Total PCC was increased in alcoholic brains compared with those of control subjects, with levels of approximately 4–8 nmols/mg of protein in the alcoholic brain samples, similar to those detected after the highest acute alcohol treatment of cells (Figure 8).

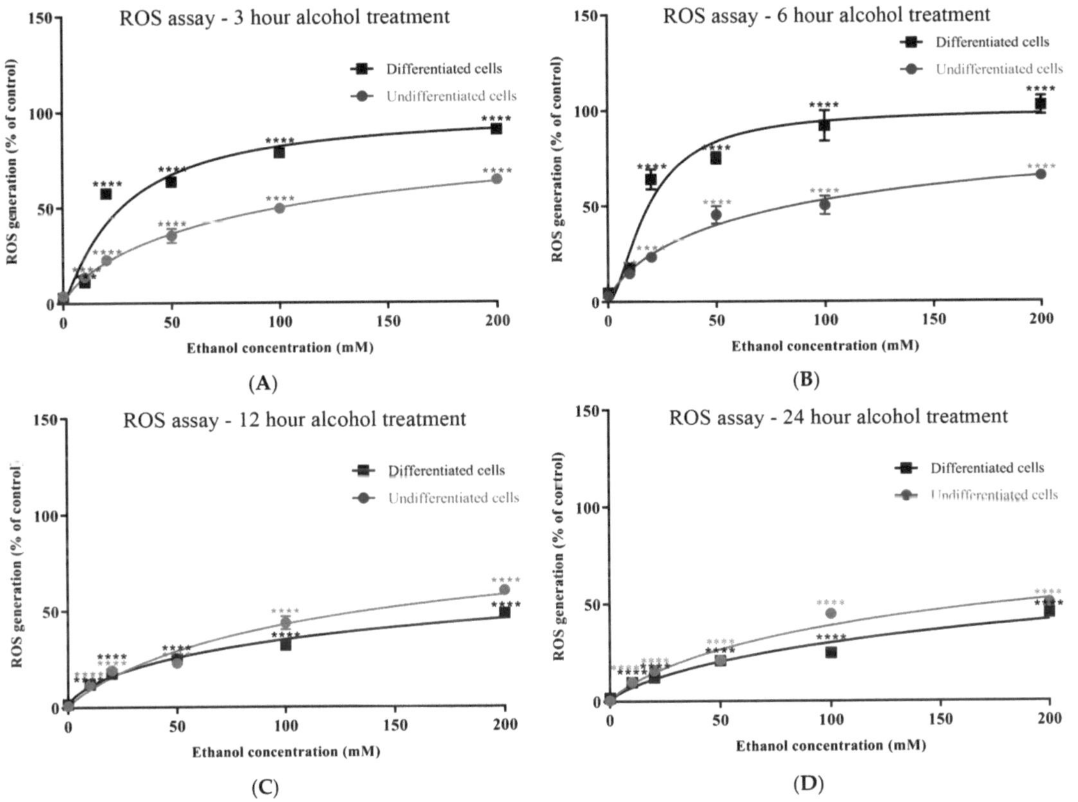

Figure 6. Alcohol induction of ROS levels determined using a DCFDA assay. Undifferentiated or differentiated SH-SY5Y cells were exposed to alcohol (0–200 mM) for durations of 3 (**A**), 6 (**B**), 12 (**C**), and 24 (**D**) h, and the levels of cellular ROS were quantified relative to that induced by H_2O_2. Each data point represents the mean of at least 5 individual experiments. Marked significance: ** = p-value < 0.01, *** = p-value < 0.001, **** = p-value < 0.0001.

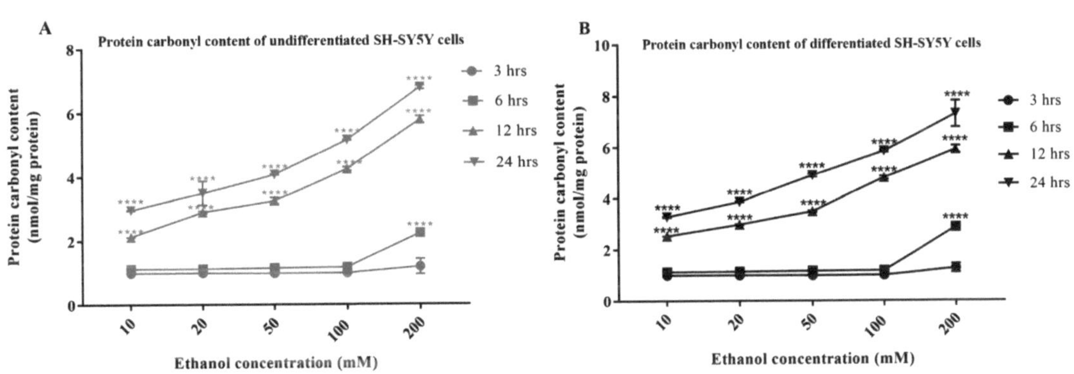

Figure 7. Alcohol induction of protein carbonyl content. Undifferentiated (**A**) or differentiated SH-SY5Y cells (**B**) were exposed to alcohol (0–200 mM) for durations of 3, 6, 12, and 24 h, and the levels of protein carbonyl content (PCC) were quantified via spectrophotometry. Each data point represents the mean of at least five individual experiments. For marked significance: **** = p-value < 0.0001.

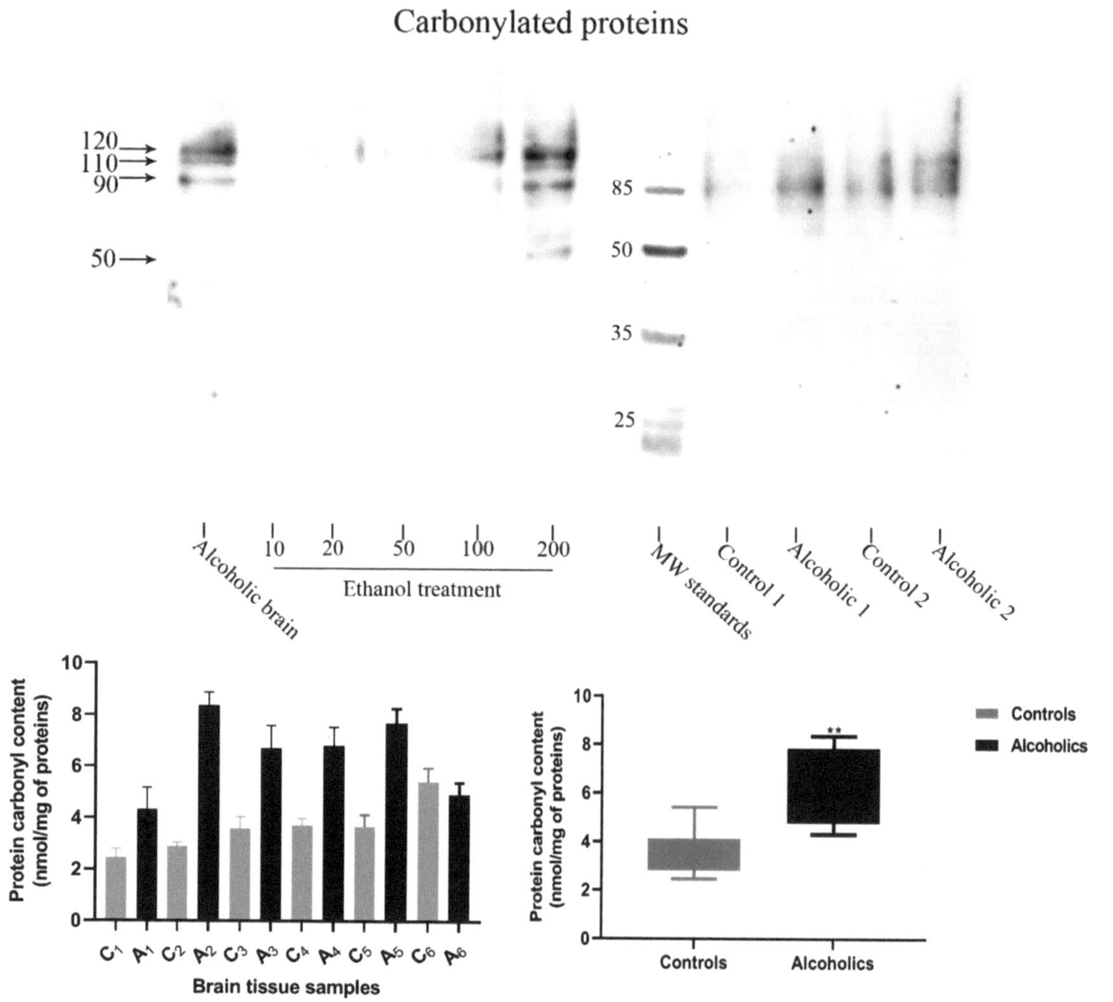

Figure 8. Quantitation and characterization of carbonylated proteins. Differentiated SH-SY5Y cells were exposed to alcohol (10–200 mM) for 24 h and carbonylated proteins were detected via oxy-blotting. Major carbonylated proteins were detected at 120, 110, 90, and 50 kDa in cells and control or alcoholic brain tissue (upper panel). Protein carbonylated content of proteins from six control and six matched alcoholic brain tissue samples were quantified via spectrophotometry (lower panels). Each data point or blotting image is a representation of at least 3 individual experiments. Significance: ** = p-value < 0.01.

4. Discussion

Alcohol has toxic effects on the brain that may be particularly detrimental during periods of neurogenesis and differentiation, such as those experienced during neurodevelopment. To consider this further, including the potential involvement of redox stress, a comparison of alcohol neurotoxicity was undertaken between undifferentiated neuroblastoma cells and those that had been acutely differentiated into a neuronal phenotype. Cytotoxicity assessment using MTT and LDH assays showed that differentiation rendered cells more resistant to alcohol, with higher alcohol concentrations required to reduce cell viability. However, somewhat in contrast to alcohol's effects on cell viability, the levels of ROS and corresponding production of carbonylated (oxidatively damaged) proteins were more extensive in differentiated cells. The characterization of carbonylated proteins revealed proteins with denatured molecular weights that overlapped with those present within the brains of alcoholic subjects, and further, PCC increased in alcoholics compared with matched controls. Hence, cell differentiation may promote resistance to alcohol-induced death but render cells more susceptible to the accumulation of oxidatively damaged proteins.

We chose to model alcohol neurotoxicity using SH-SY5Y cells due to their human origin, broad application for neurotoxicity studies, and potential for manipulation to cell cycle-synchronized, trophic-dependent, differentiated cells that display morphology, neuritic arborization, and protein expression indicative of neurons [46,47,56–59] (Supplementary Figure S1). We assessed the cytotoxicity of alcohol using MTT and LDH assays, as well as through visual inspection of cells to confirm reduced viability (Figures 1–3, and Supplementary Figure S1). Cell viability using MTT assays primarily relies on the activity of oxidoreductase and dehydrogenase enzymes in healthy (metabolically active) cells [60]. However, relatively low concentrations of agents such as phytochemicals can induce cell metabolic activity, with optical density readings that exceed those of control values [52], and this was observed after incubation with 10 mM alcohol (Figure 1). We therefore undertook another independent method for the quantification of changes in cell viability, using the liberation of extracellular LDH due to loss of membrane integrity [61]. Both methods generated similar IC_{50} values for undifferentiated or differentiated cells to those from MTT assays. Surprisingly, IC_{50} values were higher for differentiated cells (Table 1), indicative that differentiation was protective against alcohol. This contrasts with the effects of some toxic agents, such as organophosphate and carbamate pesticides, which are more toxic to differentiated SH-SY5Y cells [46] but not to other neurotoxicants, such as 1-methyl-4-phenyl-1,2,3,6-tetrahydropyridine (MPTP) [62], an agent that can induce Parkinsonian phenotypes in animals [63,64].

Cell viability experiments were undertaken across a broad concentration range of 10–200 mM alcohol for 3–24 h. This starting point for cell toxicity assays reflected blood alcohol concentrations (BACs) of 0.04–0.05% (≈9–11 mM) that represent a central nervous system (CNS) threshold for impact on psychomotor tasks [65]. The exposure to 20 to 50 mM alcohol corresponds to BACs that can arise from the consumption of several alcoholic beverages in a period of a few hours, concentrations consistent with intoxication for susceptible individuals [65]. The very high concentrations of 100 and 200 mM alcohol that were assessed can induce loss of consciousness, coma, or even death, although patients with alcohol use disorders (AUDs) often develop tolerance to alcohol's CNS effects as well as displaying heightened alcohol metabolism enabling them to withstand such high systemic BACs (>100 mM) and still may not display signs of intoxication [66].

Alcohol can damage and alter the morphology of mitochondria and promote the liberation of ROS [67,68]. We therefore investigated the ability of alcohol to affect cellular ATP levels and the production of ROS. The time course of ATP decline in response to alcohol mirrored the concentration–response curves observed from MTT and LDH assays, consistent with a shutdown of ATP production and loss of cell viability [69]. The opacity of the inner mitochondrial regions was reduced after exposure to the higher alcohol concentrations, with mitochondria observed as less electron-dense within the cristae (Figure 4), and this may correlate with a lowered ability to synthesize ATP [70]. Additionally, the capacity to produce ATP is reduced if sufficient mitochondria are damaged to trigger mitophagy and their removal, and at higher concentrations of alcohol exposure, more vacuoles were evident, which may have arisen from ongoing mitophagy (Figure 4).

Alcohol exposure induced ROS and increased the levels of oxidatively damaged proteins. Relatively low levels of ROS can impact cellular signalling pathways and may be functional, but there is a threshold at which ROS levels are detrimental to the cell and induce apoptosis [71]. Our studies show that alcohol induced ROS production and increased protein carbonyl content at the lowest levels of alcohol exposure examined (10 mM)

(Figure 6), and for these exposures, there was no reduction in cell viability (Figures 1 and 2). By contrast, higher alcohol concentrations (and exposure durations) increased ROS production and reduced cell viability, in keeping with the ability of alcohol to induce apoptotic cell death [40–43]. From alcohol exposures of 20 mM, immuno-blotting provided a means to characterise the major proteins that were oxidatively damaged; it was noteworthy that the proteins that accumulated oxidative damage after exposure to alcohol in vitro mirrored those observed in brain tissue from control and alcoholic patients. This suggests that there is a subset of cellular proteins that are particularly vulnerable to oxidative damage.

The endogenous levels of oxidative damage in alcoholic brains were similar to those from the highest induction of cellular toxicity in vitro (100–200 mM alcohol exposure) and were higher than those from age- and sex-matched control subjects (Figure 8). The molecular weights of these proteins (120, 100, 90, and 50 kDa) were found to be similar to those that accumulated in SH-SY5Y cells in response to exposures to organophosphate and carbamate pesticides [46], and this presumably reflects their relative abundance and vulnerability to oxidation. We have postulated that these protein bands may include MAP-tau and tubulin (90 and 50 kDa, respectively) due to their increased expression during differentiation [46], but the identity of these proteins, and how oxidative damage could influence protein function, will need to be addressed in future studies.

5. Conclusions

Our results show that newly differentiated neuronal cells are, surprisingly, more resistant to cell death from alcohol than undifferentiated cells. However, for similar levels of alcohol exposure, alcohol induced higher levels of ROS and the formation of oxidatively damaged proteins in newly differentiated cells. Neuritic arborization was blunted and neuronal cells were killed after 6 and 12 h exposure to $\geq$50 mM alcohol. Such levels of alcohol would correspond to exposure likely to be experienced only through sustained excessive drinking. Importantly, our experiments were limited since we could not take into account reduced alcohol concentrations due to metabolism. Our in vitro study was also limited in its capacity to reproduce the complexity of the multiple interacting cell types in vivo, since only a single population of neuronal cells was examined. Furthermore, brain tissue exhibits regional damage to alcohol [29,31–33] which may reflect differences in vulnerability between cell types, and our model may not be representative of other cell types. Nevertheless, a benefit of our approach is that the cells employed were homogenous, facilitating the generation of controlled experiments and reproducible and robust experimental data.

Since the lowest concentrations of alcohol examined (10 mM) can still induce the production of ROS and increase the levels of carbonylated proteins, depending on the turnover of these proteins, they could persist and impact neuronal cell function. Hence, the reduced cognitive capacity that arises in FASD [38–40] or that experienced by chronic heavy drinkers [9,11,14] could reflect both a reduction in numbers of neurons and the cellular damage to and limited functional capacity of surviving neurons. This raises the possibility that countering the induction of oxidative stress, such as through enhancement of the cellular antioxidant capacity, could have benefits for acute and possibly chronic alcohol exposure through reducing the potential for neuronal loss and accrued oxidative damage.

Supplementary Materials: The following supporting information can be downloaded at https://www.mdpi.com/article/10.3390/antiox13050580/s1, Supplementary Table S1: Demographics of the human brain samples. Supplementary Table S2: Alcohol toxicity in undifferentiated and differentiated SH-SY5Y cells measured via MTT assay. Supplementary Table S3: A comparison of the effects of alcohol on undifferentiated vs. differentiated SHSY-5Y cells. Supplementary Table S4: Alcohol toxicity to undifferentiated and differentiated SH-SY5Y cells measured via LDH assay. Supplementary Table S5: Neurite reduction in response to alcohol treatment. Supplementary Table S6: Alcohol toxicity in undifferentiated and differentiated SH-SY5Y cells measured via LDH assay. Supplementary Table S7: Alcohol induction of ROS in undifferentiated and differentiated SH-SY5Y cells measured via DCFDA assay. Supplementary Table S8: Correlation between the levels of ROS and protein carbonyl content in response to alcohol exposure. Supplementary Figure S1: Treatment of undifferentiated and differentiated SHSY-5Y cells with alcohol. Figure S2: Samples of original Western oxy-blots.

Author Contributions: Conceptualization, A.W.M. and W.G.C.; methodology, A.W.M., B.C.W., A.K., R.T. and D.M.; validation, A.W.M., B.C.W. and W.G.C.; formal analysis, A.W.M., B.C.W., A.K. and W.G.C.; investigation, A.W.M., B.C.W., A.K., R.T., D.M. and W.G.C.; resources, B.M., L.F.C. and W.G.C.; writing—original draft preparation, A.W.M. and W.G.C.; writing—review and editing, A.W.M., B.C.W., A.K., R.T., B.M., L.F.C. and W.G.C.; supervision, W.G.C.; project administration, W.G.C.; funding acquisition, A.W.M., W.G.C. and L.F.C. All authors have read and agreed to the published version of the manuscript.

Funding: This research was funded by a UK Foreign, Commonwealth and Development Office (FCDO) Commonwealth Scholarship Commission (UK) PhD award (grant number LKCS-2016-678) to A.W.M. This research was also supported by the European Foundation for Alcohol Research (ERAB) (EA 18 19 to L.F.C.) and the Basque Government (grant number IT1512/22).

Institutional Review Board Statement: The human brain samples used in this study were used in accordance with the Human Tissue Act (2004) (UK) and were supplied by the Neuropsychopharmacology Research Group of the Department of Pharmacology of the University of the Basque Country (UPV/EHU) (https://www.ehu.eus/en/web/neuropsicofarmacologia/home). Brain tissue collection was conducted in compliance with the research policies and ethical review boards for post-mortem brain studies (Basque Institute of Legal Medicine, Bilbao, Spain) and registered in the National Biobank Register of the Spanish Health Department with the study number C.0000035 (https://biobancos.isciii.es/ListadoColecciones.aspx).

Informed Consent Statement: Human tissue was from post-mortem patients and donated prior to autopsy.

Data Availability Statement: Additional data that supports this work are available as Supplementary Data files.

Acknowledgments: The authors would like to thank the staff members of the Basque Institute of Legal Medicine, Spain for their assistance with this study.

Conflicts of Interest: The authors declare no conflicts of interest.

References

1. Rehm, J.; Room, R.; Monteiro, M.; Gmel, G.; Graham, K.; Rehn, N.; Sempos, C.T.; Jernigan, D. Alcohol as a risk factor for global burden of disease. *Eur. Addict. Res.* **2003**, *9*, 157–164. [CrossRef] [PubMed]
2. Rehm, J.; Mathers, C.; Popova, S.; Thavorncharoensap, M.; Teerawattananon, Y.; Patra, J. Global burden of disease and injury and economic cost attributable to alcohol use and alcohol-use disorders. *Lancet* **2009**, *373*, 2223–2233. [CrossRef] [PubMed]
3. Global Burden of Disease (GBD) 2016 Alcohol Collaborators. Alcohol use and burden for 195 countries and territories, 1990–2016: A systematic analysis for the Global Burden of Disease Study 2016. *Lancet* **2018**, *392*, 1015–1035. [CrossRef] [PubMed]
4. *Global Status Report on Alcohol and Health 2018*; World Health Organization: Geneva, Switzerland, 2018. Available online: https://www.who.int/publications/i/item/9789241565639 (accessed on 31 March 2024).
5. World Health Organization. Alcohol Factsheet. 2022. Available online: https://www.who.int/news-room/fact-sheets/detail/alcohol (accessed on 31 March 2024).
6. Rehm, J.; Room, R.; Graham, K.; Monteiro, M.; Gmel, G.; Sempos, C.T. The relationship of average volume of alcohol consumption and patterns of drinking to burden of disease: An overview. *Addiction* **2003**, *98*, 1209–1228. [CrossRef] [PubMed]
7. Rehm, J.; Gmel Sr, G.E.; Gmel, G.; Hasan, O.S.M.; Imtiaz, S.; Popova, S.; Probst, C.; Roerecke, M.; Room, R.; Samokhvalov, A.V.; et al. The relationship between different dimensions of alcohol use and the burden of disease-an update. *Addiction* **2017**, *112*, 968–1001. [CrossRef] [PubMed]
8. Rehm, J.; Taylor, B.; Mohapatra, S.; Irving, H.; Baliunas, D.; Patra, J.; Roerecke, M. Alcohol as a risk factor for liver cirrhosis: A systematic review and meta-analysis. *Drug Alcohol Rev.* **2010**, *29*, 437–445. [CrossRef] [PubMed]
9. Anstey, K.J.; Mack, H.A.; Cherbuin, N. Alcohol consumption as a risk factor for dementia and cognitive decline: Meta-analysis of prospective studies. *Am. J. Geriatr. Psychiatry* **2009**, *17*, 542–555. [CrossRef] [PubMed]
10. Ronksley, P.E.; Brien, S.E.; Turner, B.J.; Mukamal, K.J.; Ghali, W.A. Association of alcohol consumption with selected cardiovascular disease outcomes: A systematic review and meta-analysis. *BMJ* **2011**, *342*, d671. [CrossRef] [PubMed]
11. Sabia, S.; Elbaz, A.; Britton, A.; Bell, S.; Dugravot, A.; Shipley, M.; Kivimaki, M.; Singh-Manoux, A. Alcohol consumption and cognitive decline in early old age. *Neurology* **2014**, *82*, 332–339. [CrossRef] [PubMed]

12. Holst, C.; Tolstrup, J.S.; Sorensen, H.J.; Becker, U. Alcohol dependence and risk of somatic diseases and mortality: A cohort study in 19 002 men and women attending alcohol treatment. *Addiction* **2017**, *112*, 1358–1366. [CrossRef] [PubMed]
13. Xu, W.; Wang, H.; Wan, Y.; Tan, C.; Li, J.; Tan, L.; Yu, J.T. Alcohol consumption and dementia risk: A dose-response meta-analysis of prospective studies. *Eur. J. Epidemiol.* **2017**, *32*, 31–42. [CrossRef] [PubMed]
14. Topiwala, A.; Ebmeier, K.P. Effects of drinking on late-life brain and cognition. *Evid. Based Ment. Health* **2018**, *21*, 12–15. [CrossRef]
15. Schwarzinger, M.; Pollock, B.G.; Hasan, O.S.M.; Dufouil, C.; Rehm, J.; QalyDays Study Group. Contribution of alcohol use disorders to the burden of dementia in France 2008-13: A nationwide retrospective cohort study. *Lancet Public Health* **2018**, *3*, e124–e132. [CrossRef] [PubMed]
16. Wang, G.; Li, D.Y.; Vance, D.E.; Li, W. Alcohol Use Disorder as a Risk Factor for Cognitive Impairment. *J. Alzheimers Dis.* **2023**, *94*, 899–907. [CrossRef] [PubMed]
17. Kilian, C.; Klinger, S.; Rehm, J.; Manthey, J. Alcohol use, dementia risk, and sex: A systematic review and assessment of alcohol-attributable dementia cases in Europe. *BMC Geriatr.* **2023**, *23*, 246. [CrossRef] [PubMed]
18. Jensen, G.B.; Pakkenberg, B. Do alcoholics drink their neurons away? *Lancet* **1993**, *342*, 1201–1204. [PubMed]
19. Harper, C. The neuropathology of alcohol-specific brain damage, or does alcohol damage the brain? *J. Neuropathol. Exp. Neurol.* **1998**, *57*, 101–110. [CrossRef] [PubMed]
20. Kril, J.J.; Halliday, G.M. Brain shrinkage in alcoholics: A decade on and what have we learned? *Prog. Neurobiol.* **1999**, *58*, 381–387. [CrossRef] [PubMed]
21. Zahr, N.M.; Kaufman, K.L.; Harper, C.G. Clinical and pathological features of alcohol-related brain damage. *Nat. Rev. Neurol.* **2011**, *7*, 284–294. [CrossRef] [PubMed]
22. Skuja, S.; Groma, V.; Smane, L. Alcoholism and cellular vulnerability in different brain regions. *Ultrastruct. Pathol.* **2012**, *36*, 40–47. [CrossRef]
23. Whittom, A.; Villarreal, A.; Soni, M.; Owusu-Duku, B.; Meshram, A.; Rajkowska, G.; Stockmeier, C.A.; Miguel-Hidalgo, J.J. Markers of apoptosis induction and proliferation in the orbitofrontal cortex in alcohol dependence. *Alcohol. Clin. Exp. Res.* **2014**, *38*, 2790–2799. [CrossRef] [PubMed]
24. Erdozain, A.M.; Morentin, B.; Bedford, L.; King, E.; Tooth, D.; Brewer, C.; Wayne, D.; Johnson, L.; Gerdes, H.K.; Wigmore, P.; et al. Alcohol-related brain damage in humans. *PLoS ONE* **2014**, *9*, e93586. [CrossRef] [PubMed]
25. Labisso, W.L.; Raulin, A.C.; Nwidu, L.L.; Kocon, A.; Wayne, D.; Erdozain, A.M.; Morentin, B.; Schwendener, D.; Allen, G.; Enticott, J.; et al. The loss of alpha- and beta-tubulin proteins are a pathological hallmark of chronic alcohol consumption and natural brain ageing. *Brain Sci.* **2018**, *8*, 175. [CrossRef]
26. Monnig, M.A.; Tonigan, J.S.; Yeo, R.A.; Thoma, R.J.; McCrady, B.S. White matter volume in alcohol use disorders: A meta-analysis. *Addict. Biol.* **2013**, *18*, 581–592. [CrossRef]
27. Xiao, P.; Dai, Z.; Zhong, J.; Zhu, Y.; Shi, H.; Pan, P. Regional gray matter deficits in alcohol dependence: A meta-analysis of voxel-based morphometry studies. *Drug Alcohol Depend.* **2015**, *153*, 22–28. [CrossRef] [PubMed]
28. Yang, X.; Tian, F.; Zhang, H.; Zeng, J.; Chen, T.; Wang, S.; Jia, Z.; Gong, Q. Cortical and subcortical gray matter shrinkage in alcohol-use disorders: A voxel-based meta-analysis. *Neurosci. Biobehav. Rev.* **2016**, *66*, 92–103. [CrossRef] [PubMed]
29. Zahr, N.M.; Pfefferbaum, A. Alcohol's effects on the brain: Neuroimaging results in humans and animal models. *Alcohol Res.* **2017**, *38*, 183–206. [PubMed]
30. Topiwala, A.; Allan, C.L.; Valkanova, V.; Zsoldos, E.; Filippini, N.; Sexton, C.; Mahmood, A.; Fooks, P.; Singh-Manoux, A.; Mackay, C.E.; et al. Moderate alcohol consumption as risk factor for adverse brain outcomes and cognitive decline: Longitudinal cohort study. *BMJ* **2017**, *357*, j2353. [CrossRef] [PubMed]
31. Fritz, M.; Klawonn, A.M.; Zahr, N.M. Neuroimaging in alcohol use disorder: From mouse to man. *J. Neurosci. Res.* **2019**, *100*, 1140–1158. [CrossRef] [PubMed]
32. Shim, J.H.; Kim, Y.T.; Kim, S.; Baek, H.M. Volumetric reductions of subcortical structures and their localizations in alcohol-dependent patients. *Front. Neurol.* **2019**, *10*, 247. [CrossRef] [PubMed]
33. Daviet, R.; Aydogan, G.; Jagannathan, K.; Spilka, N.; Koellinger, P.D.; Kranzler, H.R.; Nave, G.; Wetherill, R.R. Associations between alcohol consumption and gray and white matter volumes in the UK Biobank. *Nat. Commun.* **2022**, *13*, 1175. [CrossRef] [PubMed]
34. Immonen, S.; Launes, J.; Järvinen, I.; Virta, M.; Vanninen, R.; Schiavone, N.; Lehto, E.; Tuulio-Henriksson, A.; Lipsanen, J.; Michelsson, K.; et al. Moderate alcohol use is associated with decreased brain volume in early middle age in both sexes. *Sci. Rep.* **2020**, *10*, 13998. [CrossRef] [PubMed]
35. Lees, B.; Meredith, L.R.; Kirkland, A.E.; Bryant, B.E.; Squeglia, L.M. Effect of alcohol use on the adolescent brain and behavior. *Pharmacol. Biochem. Behav.* **2020**, *192*, 172906. [CrossRef] [PubMed]
36. Squeglia, L.M.; Boissoneault, J.; Van Skike, C.E.; Nixon, S.J.; Matthews, D.B. Age-related effects of alcohol from adolescent, adult, and aged populations using human and animal models. *Alcohol. Clin. Exp. Res.* **2014**, *38*, 2509–2516. [CrossRef] [PubMed]
37. May, P.A.; Blankenship, J.; Marais, A.S.; Gossage, J.P.; Kalberg, W.O.; Joubert, B.; Cloete, M.; Barnard, R.; De Vries, M.; Hasken, J.; et al. Maternal alcohol consumption producing fetal alcohol spectrum disorders (FASD): Quantity, frequency, and timing of drinking. *Drug Alcohol Depend.* **2013**, *133*, 502–512. [CrossRef] [PubMed]

38. Flak, A.L.; Su, S.; Bertrand, J.; Denny, C.H.; Kesmodel, U.S.; Cogswell, M.E. The association of mild, moderate, and binge prenatal alcohol exposure and child neuropsychological outcomes: A meta-analysis. *Alcohol. Clin. Exp. Res.* **2014**, *38*, 214–226. [CrossRef] [PubMed]
39. Wilhelm, C.J.; Guizzetti, M. Fetal Alcohol Spectrum Disorders: An Overview from the Glia Perspective. *Front. Integr. Neurosci.* **2016**, *9*, 170319. [CrossRef] [PubMed]
40. Popova, S.; Charness, M.E.; Burd, L.; Crawford, A.; Hoyme, H.E.; Mukherjee, R.A.S.; Riley, E.P. Fetal alcohol spectrum disorders. *Nat. Rev. Dis. Primers* **2023**, *9*, 11. [CrossRef] [PubMed]
41. Haorah, J.; Ramirez, S.H.; Floreani, N.; Gorantla, S.; Morsey, B.; Persidsky, Y. Mechanism of alcohol-induced oxidative stress and neuronal injury. *Free Radic. Biol. Med.* **2008**, *45*, 1542–1550. [CrossRef] [PubMed]
42. Birková, A.; Hubková, B.; Čižmárová, B.; Bolerázska, B. Current View on the Mechanisms of Alcohol-Mediated Toxicity. *Int. J. Mol. Sci.* **2021**, *22*, 9686. [CrossRef]
43. Tsermpini, E.E.; Plemenitaš Ilješ, A.; Dolžan, V. Alcohol-Induced Oxidative Stress and the Role of Antioxidants in Alcohol Use Disorder: A Systematic Review. *Antioxidants* **2022**, *11*, 1374. [CrossRef]
44. Gimenez-Gomez, P.; Le, T.; Martin, G.E. Modulation of neuronal excitability by binge alcohol drinking. *Front. Mol. Neurosci.* **2023**, *16*, 1098211. [CrossRef]
45. Granato, A.; Dering, B. Alcohol and the Developing Brain: Why Neurons Die and How Survivors Change. *Int. J. Mol. Sci.* **2018**, *19*, 2992. [CrossRef]
46. Mudyanselage, A.W.; Wijamunige, B.C.; Kocon, A.; Carter, W.G. Differentiated Neurons Are More Vulnerable to Organophosphate and Carbamate Neurotoxicity than Undifferentiated Neurons Due to the Induction of Redox Stress and Accumulate Oxidatively-Damaged Proteins. *Brain Sci.* **2023**, *13*, 728. [CrossRef]
47. Shipley, M.M.; Mangold, C.A.; Szpara, M.L. Differentiation of the SH-SY5Y human neuroblastoma cell line. *J. Vis. Exp.* **2016**, *108*, 53193.
48. Raghunath, M.; Patti, R.; Bannerman, P.; Lee, C.M.; Baker, S.; Sutton, L.N.; Phillips, P.C.; Damodar Reddy, C. A novel kinase, AATYK induces and promotes neuronal differentiation in a human neuroblastoma (SH-SY5Y) cell line. *Brain Res. Mol. Brain Res.* **2000**, *77*, 151–162. [CrossRef]
49. Pool, M.; Thiemann, J.; Bar-Or, A.; Fournier, A.E. NeuriteTracer: A novel ImageJ plugin for automated quantification of neurite outgrowth. *J. Neurosci. Methods* **2008**, *168*, 134–139. [CrossRef]
50. Elmorsy, E.; Attalla, S.; Fikry, E.; Kocon, A.; Turner, R.; Christie, D.; Warren, A.; Nwidu, L.L.; Carter, W.G. Adverse effects of anti-tuberculosis drugs on HepG2 cell bioenergetics. *Hum. Exp. Toxicol.* **2017**, *36*, 616–625. [CrossRef] [PubMed]
51. Elmorsy, E.; Al-Ghafari, A.; Almutairi, F.M.; Aggour, A.M.; Carter, W.G. Antidepressants are cytotoxic to rat primary blood brain barrier endothelial cells at high therapeutic concentrations. *Toxicol. Vitr.* **2017**, *44*, 154–163. [CrossRef]
52. ALNasser, M.N.; AlSaadi, A.M.; Whitby, A.; Kim, D.H.; Mellor, I.R.; Carter, W.G. Acai Berry (*Euterpe* sp.) Extracts Are Neuroprotective against L-Glutamate-Induced Toxicity by Limiting Mitochondrial Dysfunction and Cellular Redox Stress. *Life* **2023**, *13*, 1019. [CrossRef]
53. El Sharazly, B.M.; Ahmed, A.; Elsheikha, H.M.; Carter, W.G. An In Silico and In Vitro Assessment of the Neurotoxicity of Mefloquine. *Biomedicines* **2024**, *12*, 505. [CrossRef] [PubMed]
54. Lowry, O.H.; Rosebrough, N.J.; Farr, A.L.; Randall, R.J. Protein measurement with the Folin phenol reagent. *J. Biol. Chem.* **1951**, *193*, 265–275. [CrossRef] [PubMed]
55. Vigneswara, V.; Lowenson, J.D.; Powell, C.D.; Thakur, M.; Bailey, K.; Clarke, S.; Ray, D.E.; Carter, W.G. Proteomic identification of novel substrates of a protein isoaspartyl methyltransferase repair enzyme. *J. Biol. Chem.* **2006**, *281*, 32619–32629. [CrossRef] [PubMed]
56. Encinas, M.; Iglesias, M.; Liu, Y.; Wang, H.; Muhaisen, A.; Ceña, V.; Gallego, C.; Comella, J.X. Sequential Treatment of SH-SY5Y Cells with Retinoic Acid and Brain-Derived Neurotrophic Factor Gives Rise to Fully Differentiated, Neurotrophic Factor Dependent, Human Neuron-Like Cells. *J. Neurochem.* **2000**, *75*, 991–1003. [CrossRef]
57. Cheung, Y.-T.; Lau, W.K.-W.; Yu, M.-S.; Lai, C.S.-W.; Yeung, S.-C.; So, K.-F.; Chang, R.C.-C. Effects of all-trans-retinoic acid on human SH-SY5Y neuroblastoma as in vitro model in neurotoxicity research. *NeuroToxicology* **2009**, *30*, 127–135. [CrossRef] [PubMed]
58. Lopez-Suarez, L.; Awabdh, S.A.; Coumoul, X.; Chauvet, C. The SH-SY5Y human neuroblastoma cell line, a relevant in vitro cell model for investigating neurotoxicology in human: Focus on organic pollutants. *Neurotoxicology* **2022**, *92*, 131–155. [CrossRef] [PubMed]
59. Kovalevich, J.; Langford, D. Considerations for the use of SH-SY5Y neuroblastoma cells in neurobiology. *Methods Mol. Biol.* **2013**, *1078*, 9–21. [PubMed]
60. Ghasemi, M.; Turnbull, T.; Sebastian, S.; Kempson, I. The MTT Assay: Utility, Limitations, Pitfalls, and Interpretation in Bulk and Single-Cell Analysis. *Int. J. Mol. Sci.* **2021**, *22*, 12827. [CrossRef] [PubMed]
61. Kaja, S.; Payne, A.J.; Naumchuk, Y.; Koulen, P. Quantification of Lactate Dehydrogenase for Cell Viability Testing Using Cell Lines and Primary Cultured Astrocytes. *Curr. Protoc. Toxicol.* **2017**, *72*, 2.26.1–2.26.10. [CrossRef] [PubMed]
62. Elmorsy, E.; Al-Ghafari, A.; Al Doghaither, H.; Hashish, S.; Salama, M.; Mudyanselage, A.W.; James, L.; Carter, W.G. Differential Effects of Paraquat, Rotenone, and MPTP on Cellular Bioenergetics of Undifferentiated and Differentiated Human Neuroblastoma Cells. *Brain Sci.* **2023**, *13*, 1717. [CrossRef] [PubMed]

63. Thirugnanam, T.; Santhakumar, K. Chemically induced models of Parkinson's disease. *Comp. Biochem. Physiol. C Toxicol. Pharmacol.* **2022**, *252*, 109213. [CrossRef] [PubMed]
64. Prakash, S.; Carter, W.G. The Neuroprotective Effects of Cannabis-Derived Phytocannabinoids and Resveratrol in Parkinson's Disease: A Systematic Literature Review of Pre-Clinical Studies. *Brain Sci.* **2021**, *11*, 1573. [CrossRef] [PubMed]
65. Eckardt, M.J.; File, S.E.; Gessa, G.L.; Grant, K.A.; Guerri, C.; Hoffman, P.L.; Kalant, H.; Koob, G.F.; Li, T.K.; Tabakoff, B. Effects of moderate alcohol consumption on the central nervous system. *Alcohol. Clin. Exp. Res.* **1998**, *22*, 998–1040. [CrossRef] [PubMed]
66. Urso, T.; Gavaler, B.S.; Van Thiel, T.H. Blood ethanol levels in sober alcohol users seen in an emergency room. *Life Sci.* **1981**, *28*, 1053–1056. [CrossRef] [PubMed]
67. Manzo-Avalos, S.; Saavedra-Molina, A. Cellular and mitochondrial effects of alcohol consumption. *Int. J. Environ. Res. Public Health* **2010**, *7*, 4281–4304. [CrossRef]
68. Shang, P.; Lindberg, D.; Starski, P.; Peyton, L.; Hong, S.I.; Choi, S.; Choi, D.S. Chronic Alcohol Exposure Induces Aberrant Mitochondrial Morphology and Inhibits Respiratory Capacity in the Medial Prefrontal Cortex of Mice. *Front. Neurosci.* **2020**, *14*, 561173. [CrossRef] [PubMed]
69. Kamiloglu, S.; Sari, G.; Ozdal, T.; Capanoglu, E. Guidelines for cell viability assays. *Food Front.* **2020**, *1*, 332–349. [CrossRef]
70. Heine, K.B.; Parry, H.A.; Hood, W.R. How does density of the inner mitochondrial membrane influence mitochondrial performance? *Am. J. Physiol. Regul. Integr. Comp. Physiol.* **2023**, *324*, R242–R248. [CrossRef] [PubMed]
71. Redza-Dutordoir, M.; Averill-Bates, D.A. Activation of apoptosis signalling pathways by reactive oxygen species. *Biochim. Biophys. Acta* **2016**, *1863*, 2977–2992. [CrossRef] [PubMed]

antioxidants

MDPI

Article

Tributyrin Supplementation Rescues Chronic–Binge Ethanol-Induced Oxidative Stress in the Gut–Lung Axis in Mice

Anthony Santilli [1], David Shapiro [1], Yingchun Han [1], Naseer Sangwan [2,3,4] and Gail A. M. Cresci [1,4,5,*]

1 Department of Inflammation and Immunity, Lerner Research Institute, Cleveland, OH 44195, USA; santila@ccf.org (A.S.)
2 Microbial Sequencing & Analytics Resource (MSAAR) Facility, Shared Laboratory Resources (SLR), Lerner Research Institute, Cleveland, OH 44195, USA; sangwan@ccf.org
3 Cardiovascular and Metabolic Sciences, Lerner Research Institute, Cleveland Clinic, Cleveland, OH 44195, USA
4 Cleveland Clinic Lerner College of Medicine, Case Western Reserve University, Cleveland, OH 44195, USA
5 Department of Gastroenterology, Hepatology and Nutrition, Digestive Disease Institute, Cleveland, OH 44195, USA
* Correspondence: crescig@ccf.org

Abstract: Excessive alcohol consumption increases the severity and worsens outcomes of pulmonary infections, often due to oxidative stress and tissue damage. While the mechanism behind this relationship is multifaceted, recent evidence suggests ethanol-induced changes to the gut microbiome impact the gut–lung axis. To assess this, a chronic–binge ethanol feeding mouse model was used to determine how ethanol altered the gut microbiome, small intestinal epithelial barrier, and immune responses, as well as neutrophil abundance and oxidative stress in the lungs, and how supporting gut health with tributyrin supplementation during chronic–binge ethanol exposure affected these responses. We found that ethanol consumption altered gut bacterial taxa and metabolic processes, distorted small intestinal immune responses, and induced both bacteria and endotoxin translocation into the lymphatic and circulatory systems. These changes were associated with increased neutrophil (Ly6G) presence and markers of oxidative stress, lipocalin-2 and myeloperoxidase, in the lungs. Importantly, tributyrin supplementation during ethanol exposure rescued gut bacterial function ($p < 0.05$), small intestinal barrier integrity, and immune responses, as well as reducing both Ly6G mRNA ($p < 0.05$) and lipocalin-2 mRNA ($p < 0.01$) in the lungs. These data suggest ethanol-associated disruption of gut homeostasis influenced the health of the lungs, and that therapeutics supporting gut health may also support lung health.

Keywords: gut microbiome; butyrate; alcohol; lung inflammation; small intestine; gut immunity

Citation: Santilli, A.; Shapiro, D.; Han, Y.; Sangwan, N.; Cresci, G.A.M. Tributyrin Supplementation Rescues Chronic–Binge Ethanol-Induced Oxidative Stress in the Gut–Lung Axis in Mice. *Antioxidants* **2024**, *13*, 472. https://doi.org/10.3390/antiox13040472

Academic Editor: Marco Fiore

Received: 22 March 2024
Revised: 12 April 2024
Accepted: 15 April 2024
Published: 17 April 2024

1. Introduction

The gut microbiome comprises trillions of microbes, and in the past several years there has been an increasing appreciation regarding the important role the gut microbiome plays in supporting health. While there is intra-individual variability (alpha-diversity) in microbiome composition amongst healthy individuals, distinct dissimilarity (beta-diversity) between healthy individuals and those with various chronic diseases is noted [1]. At the phylum level, about 90% of the gut microbiome composition of a healthy individual comprises the Bacteroidetes and Firmicutes phyla. The gut microbiome functionally supports host health by aiding in the metabolism of nondigestible food components (e.g., fiber), mitigating pathogen invasion, and modulating the immune system [2]. The gut microbiome produces many beneficial metabolites for the host, including enzymes, vitamins, hormones, and short-chain fatty acids (SCFA; e.g., acetate, propionate, and butyrate). Butyrate is known to have many important biological functions, including supporting the integrity of the gut barrier and immune function [3,4].

Chronic heavy alcohol exposure is known to disrupt the gut microbiome's composition and function, increase intestinal permeability, and activate systemic inflammatory pathways [1]. Chronic alcohol exposure reduces the abundance of Bacteroidetes and Firmicutes phyla, expands the Proteobacteria phylum, and lowers fecal SCFA levels, including that of butyrate [5]. In our prior studies in which mice exposed to chronic and chronic–binge ethanol were co-supplemented orally with tributyrin, a butyrate prodrug, or a butyrate-targeting synbiotic, supplemented mice were protected from proximal colon epithelial barrier disruption, as well as liver injury, inflammation, and oxidative stress, compared to non-supplemented ethanol-exposed mice [6–9]. Tributyrin is a structured lipid with three butyrate molecules esterified to the glycerol backbone. Upon oral ingestion, tributyrin is digested by pancreatic lipase to free up three butyrate molecules which are then absorbed throughout the intestine. The increased half-life of tributyrin compared to sodium butyrate makes it more appealing as a potential therapeutic [6].

The role of the gut microbiome in the health of other organ systems is being increasingly investigated. Among inter-organ relations, the gut–lung axis is less studied than either the gut–liver or the gut–brain axis. Chronic excessive alcohol exposure disrupts the body's innate and adaptive immune system [10], and consequentially, individuals who consume excessive alcohol are at higher risk of acute lung injury and lung infections [11]. Chronic excessive alcohol exposure disrupts alveolar epithelial barrier integrity, reduces endogenous antioxidant levels (glutathione), and causes defects in alveolar macrophage function [11]. The microbiota of the gut and the lung influence the local immune system; however, there is also a potential for crosstalk between the gut and lungs. Impairment of the intestinal epithelial barrier by overgrowth of potentially pathogenic bacteria (gut dysbiosis) enables intact bacteria, bacterial fragments, or microbiome metabolites (e.g., SCFA) to translocate from the gut lumen into systemic circulation. This can be accomplished by entry of these components into the portal vein en route to the liver and then systemic circulation, and/or entry of these components into the draining mesenteric lymphatic system, thoracic duct, and then systemic circulation [12,13]. Eventually, both routes lead to the pulmonary system.

Chronic heavy alcohol consumption is known to induce oxidative stress and impair the gut mucosal barrier through its first metabolite acetaldehyde and induction of cytochrome P450 enzyme CYP2E1 [14]. In addition to supporting the integrity of the gut barrier and immunity, butyrate has been shown to have antioxidative effects [4]. There are limited data regarding the effects of butyrate supplementation during chronic–binge ethanol exposure on the small intestine and lung axis. Since we found tributyrin-supplemented mice exposed to chronic–binge ethanol had decreased oxidative stress, inflammation, and injury in the liver, here we tested our hypothesis that tributyrin supplementation during chronic–binge ethanol exposure would reduce oxidative stress in the lungs. Our data show that due to support of the gut microbiome, small intestinal epithelial barrier, and immune responses during chronic–binge ethanol exposure, tributyrin-supplemented mice also had less neutrophil abundance and oxidative stress in the lungs.

2. Materials and Methods

2.1. Chronic–Binge Ethanol Feeding Model

C57BL/6 female mice (10–11 weeks) were purchased from Jackson Laboratory (Bar Harbor, ME, USA). Mice were randomized to receive a chronic–binge ethanol Lieber DeCarli feeding model consisting of 10 days chronic ethanol (5% v/v) exposure and $\pm$daily supplementation with saline or 5 mM tributyrin by oral gavage. Control animals were pair-fed an isocaloric diet with maltose dextrin substituted for ethanol $\pm$ 5 mM tributyrin or saline supplementation by oral gavage daily. On the 11th day, mice were administered an oral binge gavage of either 5 g/kg ethanol or maltose with either 2.5 mM tributyrin or saline as previously randomized. There were four groups in total: pair-fed saline (PF-S), pair-fed tributyrin (PF-TB), ethanol-fed saline (EF-S), and ethanol-fed tributyrin (EF-TB).

Mice were anesthetized 6 h post-gavage, blood was collected from the inferior vena cava, and tissue (small intestine, cecum, lung) was dissected and stored for future analyses.

2.2. Shotgun Metagenomics Sequencing, Bioinformatics, and Statistical Analysis

gDNA from cecal contents was extracted using a Zymo Research Quick DNA Feca/Soil Microbiome MiniPrep Kit (Zymo Research, Irvine, CA, USA) following the manufacturer's instructions. gDNA was quantified using a Nanodrop ND1000 Spectrophotometer (Thermo Fisher Scientific, Wilmington, DE, USA).

Quality control of the metagenomic reads was conducted as described previously [15]. Briefly, raw reads were processed for low-quality-based filtering using the Trimmomatic pipeline [16]. Host-derived reads were excluded by mapping the reads to the reference human genome (version GRCh38.p14) using BBMap software (sourceforge.net/projects/bbmap/ version 34.62). Quality trimmed reads were processed for taxonomic and functional profiling using Metaphlan2 [17] and Humann2 [18], respectively. Differential feature selection was performed using Fisher's exact *t*-test. We assessed the statistical significance ($p < 0.05$) throughout, and whenever necessary, we adjusted *p*-values for multiple comparisons according to the Benjamini and Hochberg method to control the False Discovery Rate [19] while performing multiple testing on taxa and pathway abundances according to sample types.

Differential abundance test benchmarking was performed using the DAtest package (https://github.com/Russel88/DAtest/wiki/usage#typical-workflow version 2.8.0). Briefly, differentially abundant methods were compared with the False Discovery Rate (FDR), Area Under the (Receiver Operator) Curve (AUC), empirical power (Power), and the False Positive Rate (FPR). Based on the DAtest's benchmarking, we selected LefSeq and analysis of variance (ANOVA) as the methods of choice to perform differential abundance analysis. We assessed the statistical significance ($p < 0.05$) throughout, and whenever necessary, we adjusted *p*-values for multiple comparisons according to the Benjamini and Hochberg method to control the False Discovery Rate [20]. Linear regression (parametric) and Wilcoxon (non-parametric) tests were performed on genera and species abundances against metadata variables using their base functions in R (version 4.1.2) [21].

2.3. Small Intestinal Isolation of Intraepithelial Lymphocytes, Lamina Propria Lymphocytes, and Intestinal Epithelial Cells

The small intestinal tract from the pylorus to the distal ileum was dissected following euthanasia. Single cells from intraepithelial lymphocytes (IELs), lamina propria lymphocytes (LPLs), and intestinal epithelial cells (IECs) were isolated and counted as described previously by Qiu et al. [22]. IELs and LPLs were stimulated with phorbol 12-myristate 13-acetate (PMA, 10 ng/mL)/ionomycin (1 μM) (MilliporeSigma, Darmstadt, Germany) for 4 h followed by 2 μM monensin (Calbiochem, San Diego, CA, USA) during the last 2 h of stimulation. The IECs were flash frozen and stored at −80 °C for future analyses.

2.4. Isolation of Lung Immune Cells

Following euthanasia, the lungs were perfused with cold PBS and the left lobe was dissected. A portion of the tissue was minced and then digested using a 1 mg/mL collagenase type 1 (Life Sciences Technologies, Grand Island, NY, USA) while incubating at 37 °C in an orbital shaker at 225 rpm for 30 min. The suspension was then filtered using a 70 μM strainer and immediately counted.

2.5. Flow Cytometry Analysis of IELs, LPLs, and Lung Immune Cells

Intestinal (IEL, LPL) and lung immune cells were washed and stained with either LIVE/DEAD Fixable Blue Dead Cell Stain (Invitrogen, Carlsbad, CA, USA) or LIVE/DEAD Fixable Near-IR Dead Cell Stain (Invitrogen, Carlsbad, CA, USA), respectively, and then fixed and permeabilized (True-Nuclear Transcription Factor Buffer Set, Biolegend, San Diego, CA, USA). Cells were washed and blocked to reduce nonspecific staining (FcR

Blocking Reagent, Miltenyi Biotec, Gaithersburg, MD, USA). IELS and LPLs were stained with fluorescence-conjugated antibodies with anti-CD45-APCFire 810, anti-CD11c-BV650, anti-IL6-PerCPeFlour710, anti-CD3-FITC, anti-CD4-BUV805, and anti-CD8a-BUV615. Lung immune cells were stained with fluorescence-conjugated antibodies with anti-CD45-AF700, anti-SiglecF-BV421, anti-CD11b-BV605, and anti-Ly6G-PeCy7. Following staining, samples were washed 3 times, and cells were acquired using a Sony ID 7000 Spectral Cell Analyzer (San Jose, CA, USA). Analysis of the flow cytometry results was performed using FloJo ver. 10.8 (BD Life Sciences, Sparks, MD, USA).

2.6. *Tissue RT-qPCR*

Tissue was stored in RNAlater immediately after its dissection. Total RNA from the lung was extracted using an RNeasy Plus Universal Mini Kit (Qiagen, Hilden, Germany) following the manufacturer's instructions. RNA was quantified using a Nanodrop ND1000 Spectrophotometer (Thermo Fisher Scientific, Wilmington, DE, USA), and 2 μg of total RNA was reverse transcribed using SuperScript IV VILIO (Invitrogen, Carlsbad, CA, USA). QuantStudio 5 (Applied Biosciences, Waltham, MA, USA) was used for Real Time PCR (RT-qPCR) amplification with a PowerUp SYBR Green Master Mix (Applied Biosciences, Waltham, MA, USA) and 1 μM primers (Table 1). The Comparative Threshold (CT) method was used to determine relative expression compared to the housekeeping gene glyceraldehyde 3-phosphate (GAPDH), and data are presented as mean ± standard deviation.

Table 1. RT-qPCR primers.

Target	Primer	Sequence
Lymphocyte Antigen 6 Complex Locus G6D (Ly6G) NM_001310438	Ly6G F	TGC GTT GCT CTG GAG ATA GA
	Ly6G R	CAG AGT AGT GGG GCA GAT GG
Lipocalin 2 NM-008491	LCN2 F	TGG CCC TGA GTG TCA TGT G
	LCN2 R	CTC TTG TAG CTC ATA GAT GGT GC

2.7. *Western Blotting: Intestinal Epithelial Cells*

Intestinal epithelial cells were suspended in lysis buffer (1% Triton X-100, 50 mM Tris-HCl pH 7.4, 150 mM NaCl, 1 mM EDTA pH 8, 0.1% Na-deoxycholate, 0.1% SDS) containing Pierce Protease and Phosphatase Inhibitor Mini Tablets (Thermo Fisher, Waltham, MA, USA) at a concentration of 8×10^6 cells/100 μL, and lysates were prepared as previously reported [8]. Membranes were probed with anti-Junctional Adhesion Molecule A 1 (JAMA-1), anti-Zonula Occludens-1 (ZO-1), anti-PhosphoStat3 (pSTAT3), and anti-TotalStat3 (TSTAT3). Anti-Heat Shock Cognate Protein 70 (HSC70) was used as a loading control. Protein expression was visualized with enhanced chemiluminescence, and intensity was measured via densitometry using ImageJ software (National Institutes of Health (NIH), Bethesda, MD, USA, ver. 1.54i).

2.8. *Plasma Endotoxin Quantification Using Limulus Amoebocyte Lysate (LAL) Assay*

Blood samples were collected, and plasma was separated using techniques to avoid contamination using endotoxin-free materials. Glass tubes for sample dilutions were incubated at 250 °C for 40 min to eliminate residual bacteria or endotoxin. Samples were prepared as previously reported [23], and endotoxin concentration was measured using the Limulus Amebocyte Lysate Assay Kinetic-QCL Kit (Lonza, Walkersville, MD, USA). A standard curve was generated using *Escherichia coli* O55:B5 Endotoxin (Lonza, Walkersville, MD, USA). Plates were read and endotoxin concentration was quantified using SpectraMax Microplate Readers and SoftMax® Pro 7.1.2 GxP Software Molecular Devices (San Jose, CA, USA).

2.9. Cecal Secretory IgA (SIgA) Enzyme-Linked Immunoassay (ELISA)

Supernatants from cecal contents were prepared and diluted as previously described [24], and IgA levels were assessed using a Mouse IgA uncoated ELISA Kit (Invitrogen, Waltham, MA, USA) according to the manufacturer's instructions. The ELISA plate was read using a BioTek Synergy H1 Microplate Reader and BioTek Gen 5 software ver. 3.12 (Agilent Technologies, Santa Clara, CA, USA).

2.10. Lipocalin 2 (LCN2) and Myeloperoxidase (MPO) ELISA in Lung Tissue

Lung tissue samples were suspended in lysis buffer as described in the Mouse MPO ELISA Kit (HycultBiotech Inc, Wayne, PA, USA) at 10 mg/200 μL, and homogenized using a Branson Digital Sonifier 450 (Branson Ultrasonics, Danbury, CT, USA). Samples were then centrifuged at 1500× *g* at 4 °C for 15 min, and the supernatant was collected. Samples were diluted 1:1000 and both MPO and LCN2 levels were assessed using either a Mouse MPO ELISA Kit (HycultBiotech inc, Wayne, PA, USA) or a Quantkine ELISA Mouse Lipocalin-2/NGAL Immunoassay Kit (R&D Systems, Minneapolis, MN, USA), respectively, according to the manufacturer's instructions. ELISA plates were read using a BioTek Synergy H1 Microplate Reader and BioTek Gen 5 software ver. 3.12 (Agilent Technologies, Santa Clara, CA, USA).

2.11. Plating of Peyer's Patches on Enterococcosel Agar

Following small intestine dissection, Peyer's patches were isolated using aseptic techniques and washed with sterile PBS. Peyer's patches were suspended in sterile 0.17% Triton in PBS and lysed using a 1.5 mL microcentrifuge tube pestle. Samples were then serially diluted and plated on BBL Enterococcosel agar (BD Life Sciences, Sparks, MD, USA). Plates were then incubated aerobically at 37 °C for 48 h and CFUs were counted.

2.12. Statistical Analysis

GraphPad Prism® 10 ver. 10.1.2 (San Diego, CA, USA) was used for statistical analysis, and data are shown as the mean ± standard error of the mean. Two-tailed *t*-tests were used to identify statistical significance between treatment groups, and a $p \leq 0.05$ was used as the threshold for statistical significance.

3. Results

3.1. Ethanol Exposure and Tributyrin Supplementation Alter the Gut Microbiome and Its Functional Pathways

Chronic alcohol consumption modifies gut microbial populations and metabolite production, impairs the intestinal barrier, and compromises immunity [25,26]. To characterize the microbiota and assess for changes to the mouse gut microbiome caused by chronic–binge ethanol treatment, we performed shotgun sequencing of bacterial gDNA isolated from cecal contents.

Measurement of diversity and evenness of bacterial species were shown to be decreased in ethanol-fed mice compared to pair-fed mice with both saline and tributyrin supplementation (Figure 1A; Simpson indexes $p < 0.041$ and $p < 0.038$, respectively). In comparing the Simpson index of each ethanol-treated group, mice supplemented with tributyrin had significantly lower diversity and evenness compared to mice supplemented with saline (Simpson index $p = 0.021$; Figure 1A). To quantify the dissimilarity between groups (beta-diversity), Bray–Curtis analysis was performed and Venn diagram and principal coordinate analysis (PCoA) plots were drawn. Depicted in the Venn diagram, a total of 107 species were shared across all groups; however, each mouse group had significant species based on treatment and supplementation, with the EF-S group having the highest number of unique species ($n = 10$), followed by PF-S ($n = 9$), PF-TB ($n = 7$), and EF-TB ($n = 4$) (Figure 1B; $p < 0.05$). PCoA analysis shows formation of three distinct clusters with significant dissimilarity between ethanol (green) and pair-fed (orange) groups (Figure 1C; $p = 0.001$). Relative abundance analysis identified bacterial species that drive diversity across treatment groups. *Bacteroides thetaiotaomicron*, a gut commensal, was the most prevalent species and was seemingly elevated with ethanol exposure, with the ethanol–tributyrin group exhibiting the highest abundance (Figure 1D). Both pair-fed groups also showed a higher abundance of several unknown species (GGB7968 SGB41545, GGB28851 SGB41518, and GGB79854 SGB41664; Figure 1D) compared to ethanol-fed mice. Five species, two of which are unknown, showed significant differences in relative abundance across the four treatment groups. *Erysipelotrichaceae bacterium* NYU-BL-F16 was induced with tributyrin supplementation ($p < 0.05$), *Lachnospiraceae bacterium* 28-4 was significantly increased in the EF-S vs. PF-S groups ($p < 0.05$) and lowered in the EF-TB group, and tributyrin lowered *Muribaculum intestinalie* in both pair- and ethanol-fed mice ($p < 0.05$) (Figure 1E).

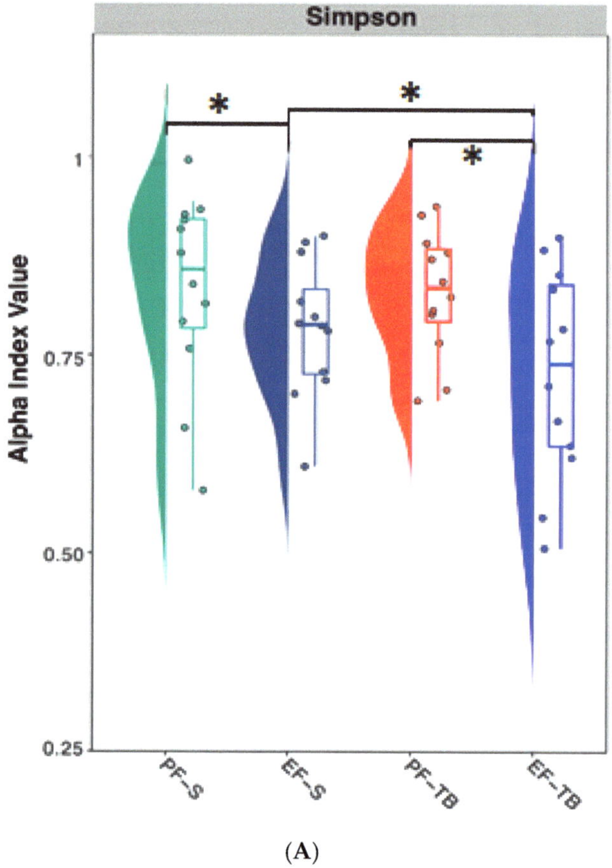

(A)

Figure 1. *Cont.*

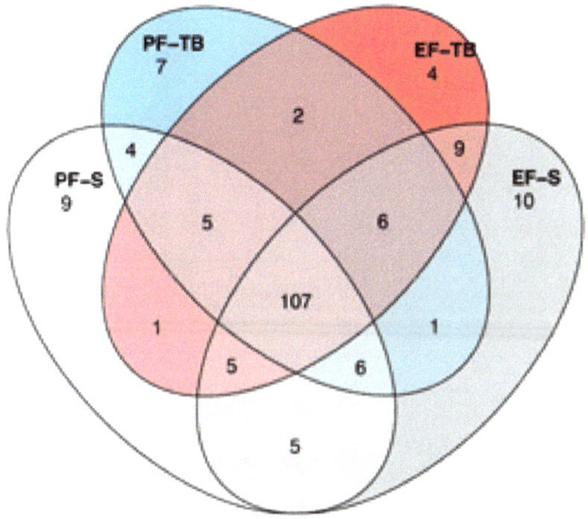

(**B**)

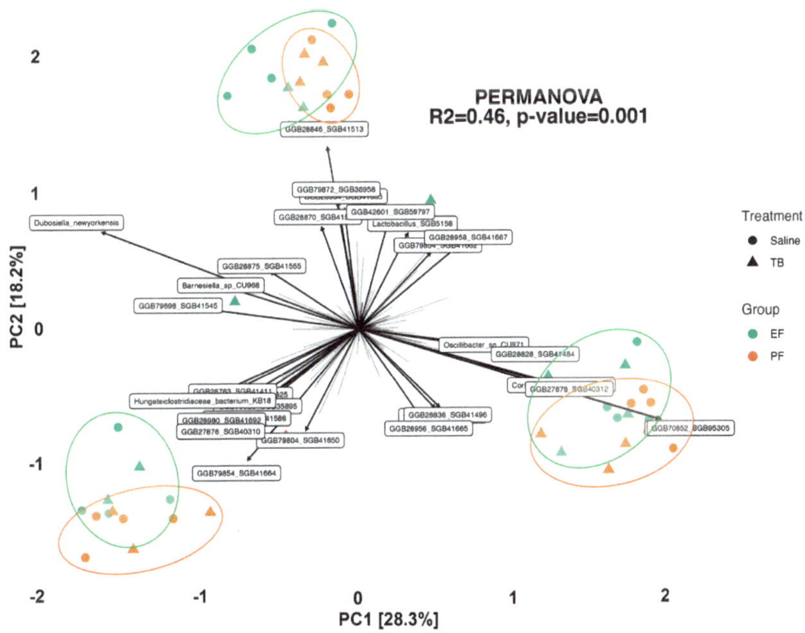

(**C**)

Figure 1. *Cont.*

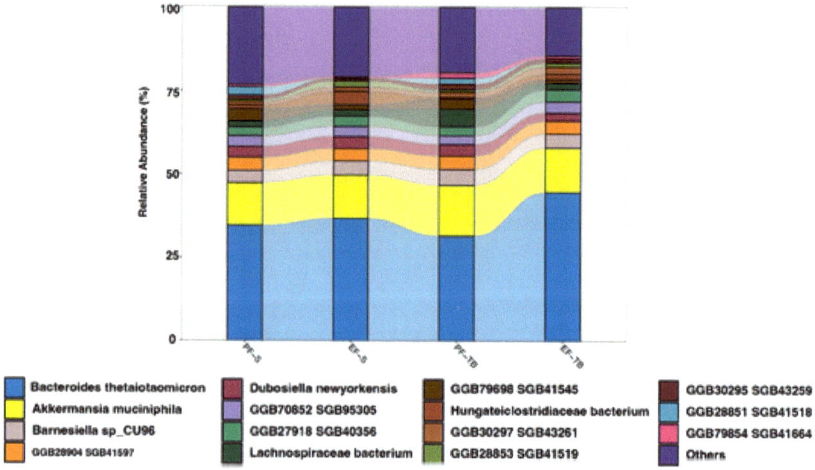

(D)

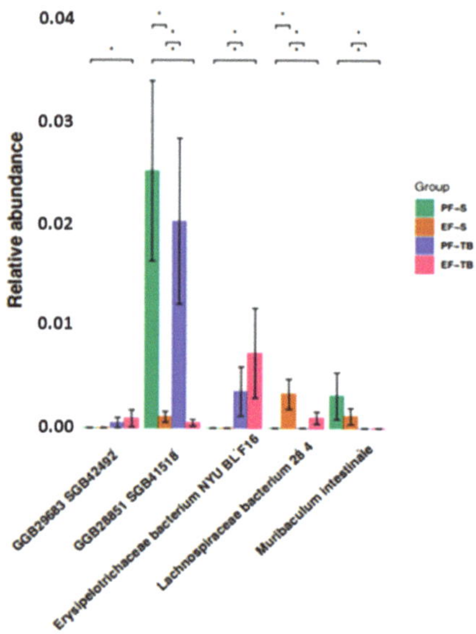

(E)

Figure 1. Effects of chronic–binge ethanol exposure ± tributyrin on the mouse gut bacteria taxonomy. (**A**) Alpha-diversity of cecal microbiome calculated using Simpson diversity index. (**B**) Venn diagram demonstrating species taxa separation and overlap among treatment groups. (**C**) Principal coordinate analysis plots generated using Bray–Curtis dissimilarity with PERMANOVA analysis. Ellipses are used to visually highlight differences between ethanol (green) and control (orange) diets and do not represent statistical analysis. (**D**) Stacked graph showing the relative abundance of top taxa at species level present in cecal contents. (**E**) Relative abundance of species showing significant differences between treatment groups with LEfSe statistical analysis. PF-S: control diet + saline gavage; EF-S: ethanol diet + saline gavage; PF-TB: control diet + tributyrin gavage; EF-TB: ethanol diet + tributyrin gavage; * $p \leq 0.05$.

Shotgun sequencing of cecal bacteria also revealed gene changes to metabolic pathways of bacteria induced by chronic–binge ethanol exposure. The Simpson alpha-diversity index of gene pathways was reduced in the EF-TB group compared to all other groups (Figure 2A). To quantify the functional dissimilarity between groups (beta-diversity), Bray–Curtis analysis was performed and Venn diagram and principal coordinate analysis (PCoA) plots were drawn. The cecal bacterial gDNA in the EF-TB group displayed the most unique metabolic pathways ($n = 16$), as well as a large overlap of 27 pathways shared with both PF groups, but distinct from the EF-S group, suggesting tributyrin supplementation during ethanol exposure restored these pathways to those of PF mice (Figure 2B). PCoA plots of Bray–Curtis dissimilarity (Figure 2C) revealed distinct gene clustering between each group, demonstrating each exposure and/or supplementation altered the function of bacteria in a unique manner. Specifically, the EF-TB group clustered separately from the other groups (Figure 2C; orange crosses highlighted in ellipse). Assessment of the relative abundance of the top ten metabolic pathways showed they were all reduced in the EF-S group, and this trend was recovered with tributyrin supplementation. Several pathways are involved in the mitigation of oxidative damage, such as Coenzyme A (CoA) production [27] (CoA and phosphopentathonate biosynthesis), tetrapyrrole biosynthesis [28–30], S-adenosyl-l-methionine (SAM) salvage [30], and L-arginine biosynthesis [31,32] (Figure 2D). Pathways for peptidoglycan synthesis and modification, such as uridine monophosphate (UMP) biosynthesis I, II, and III, uridine diphosphate N-acetylmuramyl pentapeptide synthesis, and glycolysis, were also disturbed in the same manner (Figure 2D).

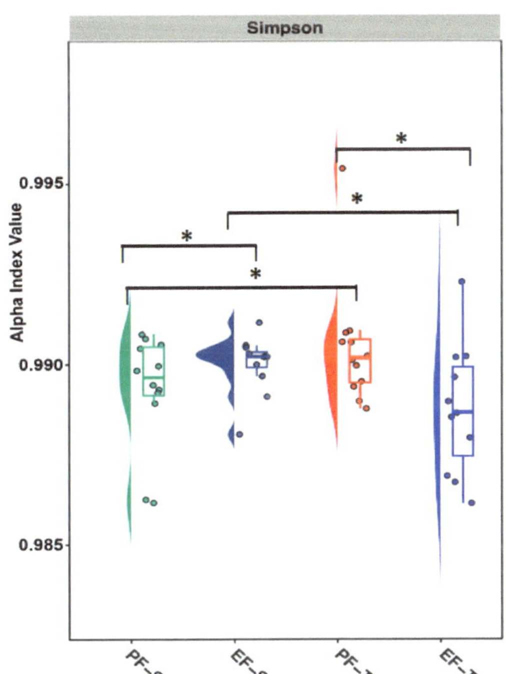

(A)

Figure 2. *Cont.*

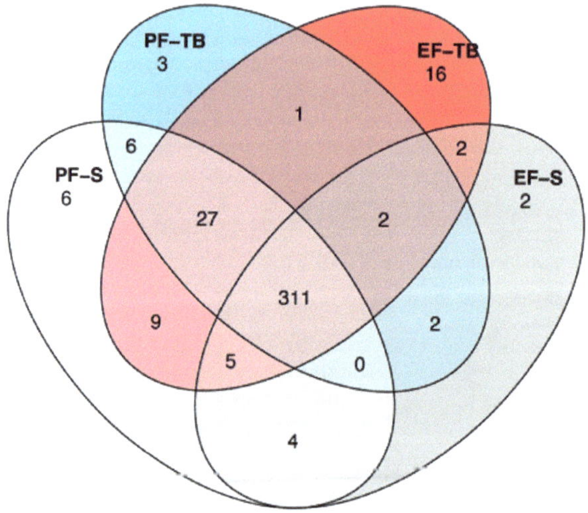

(**B**)

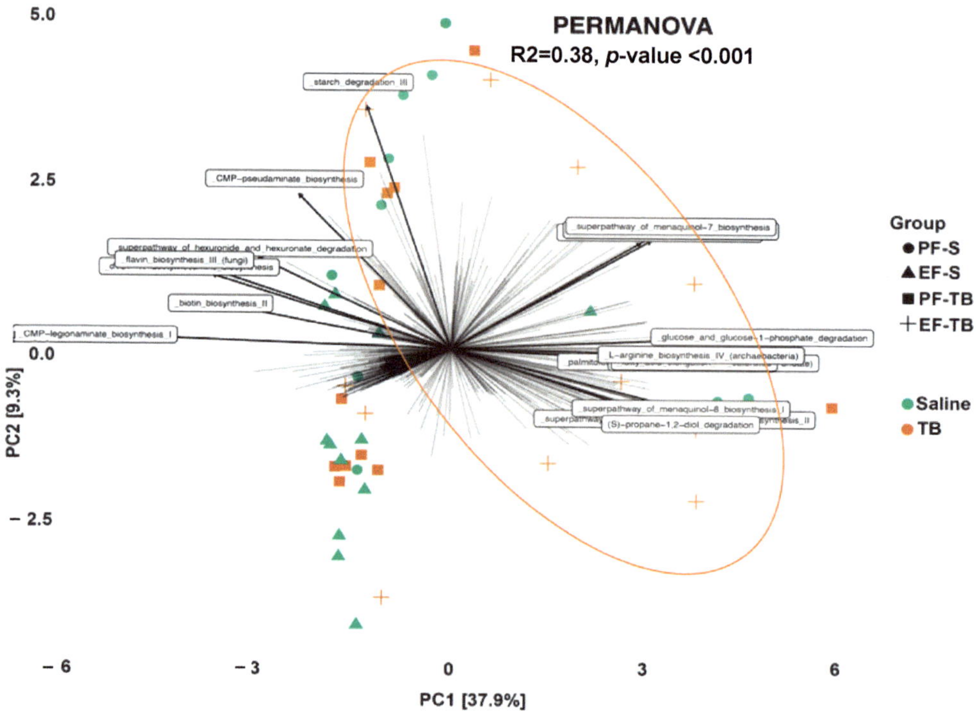

(**C**)

Figure 2. *Cont.*

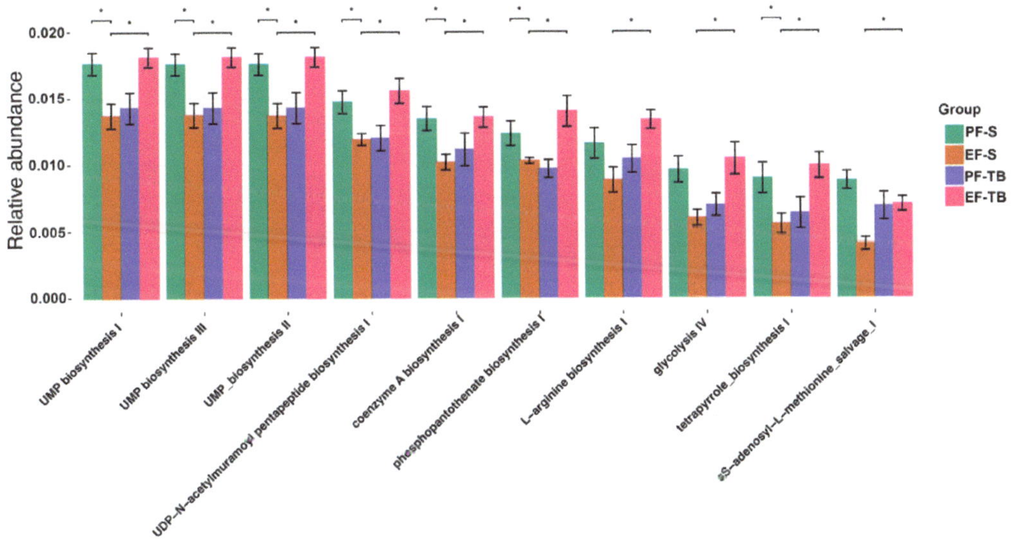

(**D**)

Figure 2. Tributyrin supplementation mitigates ethanol's effects on gut microbiome function. (**A**) Alpha-diversity of cecal microbiome metabolic pathways calculated using Simpson diversity index. (**B**) Venn diagram demonstrating cecal bacteria metabolic pathway overlap between treatment groups. (**C**) Principal coordinate analysis plots generated using Bray–Curtis dissimilarity with PERMANOVA analyses. Orange-colored ellipse visually highlights the separation of the EF-TB group but does not represent statistical analysis. (**D**) Relative abundance graphs of cecal bacteria metabolic pathways showing significant differences between treatment groups with ANOVA testing. PF-S: control diet + saline; EF-S: ethanol diet + saline; PF-TB: control diet + tributyrin; EF-TB: ethanol diet + tributyrin; * $p \leq 0.05$.

3.2. Ethanol Affects Intestinal Epithelial Immune Responses

Disruptions to the gut microbiome in mice by chronic alcohol consumption and antibiotic treatment have been reported to decrease secretory IgA (SIgA) levels in cecal contents [24] and feces [33], respectively. SIgA serves as the first line of immune defense in protecting the intestinal epithelium from enteric toxins and pathogens. SIgA is synthesized by B cells upon antigen presentation by dendritic cells, and is secreted into the intestinal lumen to prevent microbes from penetrating the mucosal barrier and interfacing with intestinal epithelial cells [34,35]. We found that mice exposed to chronic–binge ethanol and saline displayed significantly reduced cecal SIgA levels relative to pair-fed saline mice, and tributyrin supplementation mitigated this but not significantly (Figure 3A; $p = 0.006$) A deficiency in this first-line mucosal immune defense allows for luminal bacteria to translocate across the epithelial barrier. Peyer's patches are part of the gut-associated lymphoid tissue (GALT) and contain many dendritic cells, macrophages, and lymphocytes that help maintain a balanced gut microbiota and keep pathogens at bay. We found significantly increased *Enterococci* in the Peyer's patches of EF-S mice, which was resolved with tributyrin co-supplementation so that EF-TB group counts were similar to those of pair-fed mice (Figure 3B).

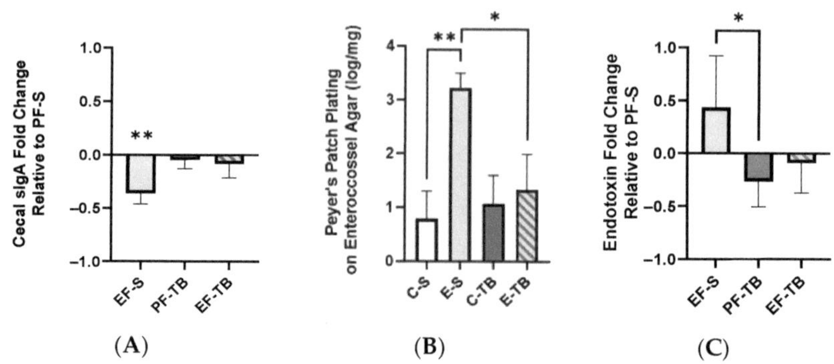

Figure 3. Ethanol exposure depletion of cecal SIgA and induction of bacterial translocation into the lymphatic and circulatory systems. ** denotes a significant decrease in the EF-S group vs. PF-S. (**A**) Concentration of SIgA in mouse cecal contents presented as fold of the control (PF-S). (**B**) *Enterococci* bacterial growth in small intestinal Peyer's patches. (**C**) Plasma endotoxin concentration. Fold of the control (PF-S). Treatment groups include 6–8 mice. PF-S: control diet + saline; EF-S: ethanol diet + saline; PF-TB: control diet + tributyrin; EF-TB: ethanol diet + tributyrin; * $p < 0.05$; ** $p < 0.01$.

Since ethanol provoked gut dysbiosis and increased the presence of bacteria in the Peyer's patches, we assessed plasma endotoxin levels as a measurement of impaired intestinal barrier and a resulting systemic proinflammatory effect. Plasma endotoxin was elevated in the EF-S group, and tributyrin supplementation normalized plasma endotoxin to levels similar to those in the pair-fed groups (Figure 3C; $p = 0.013$).

3.3. *Tributyrin Supplementation Bolsters Immune and Barrier Functions Both at the Intestinal Epithelium and within the Lamina Propria*

Proinflammatory mediators traveling from the intestinal lumen into systemic circulation must first penetrate the epithelial barrier and circumvent immune responses in both the intestinal epithelium and lamina propria. As butyrate is known to support the intestinal barrier and its immune responses, we assessed for IEL dendritic cells and found an elevation in IEL host defense via increased percentage of CD11c$^+$IL-6$^+$ dendritic cells in EF-TB compared to EF-S mice (Figure 4A). This response coincided with increased pSTAT3 in the EF-TB group vs. the PF-TB group, and a trend towards an increase from the EF-S group (Figure 4B). IL-6 at low levels has been shown to promote the phosphorylation of Stat3 in epithelial cells, which then promotes repair of the mucosal barrier by inducing B-cell differentiation needed for IgA production [36]. JAMA-1 and ZO-1 are two of many proteins that comprise the intestinal epithelial barrier complex. JAMA-1 spans the paracellular space and helps to seal the gap between epithelial cells, and ZO-1 serves as an anchor protein to multiple other junctional proteins, and when either protein is diminished the epithelial barrier's integrity may be compromised. We found an induction of JAMA-1 in the small intestinal IEC of EF-TB mice, but there were no differences in ZO-1 expression between groups (Figure 4C,D).

If intestinal epithelial defenses are breached, bacteria or bacterial byproducts can traverse through the lamina propria and its immune repertoire prior to entering circulation. Assessing for T cells within the lamina propria, we found an elevated percentage of CD3$^+$CD4$^+$ T helper/inducer cells (Figure 5A) and a reduced percentage of CD3$^+$CD8a$^+$ cytotoxic/suppressor T cells (Figure 5B) in the EF-TB group compared to the EF-S group. This resulted in an overall increased CD4:CD8 ratio in EF-TB mice compared to EF-S mice (Figure 5C), which is congruent with healthy immunity.

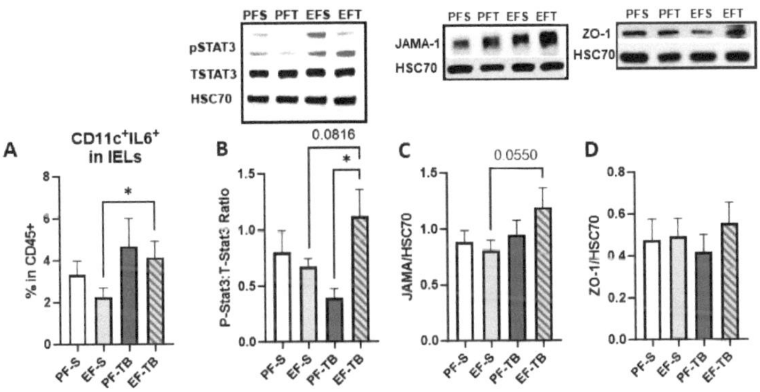

Figure 4. Tributyrin supplementation bolsters both the epithelial barrier and immunity. (**A**) %CD11c$^+$IL-6$^+$ cells isolated from IELs. Gating strategy for flow analysis is described in Figure S1. (**B**) Ratio of PhosphoStat3 to TotalStat-3 expression in IECs isolated from the small intestine. (**C**) JAMA protein expression in IECs isolated from the small intestine. (**D**) ZO-1 protein expression in IECs isolated from the small intestine. Treatment groups contain 5–8 replicates. PF-S: control diet + saline; EF-S: ethanol diet + saline; PF-TB: control diet + tributyrin; EF-TB: ethanol diet + tributyrin; * $p < 0.05$.

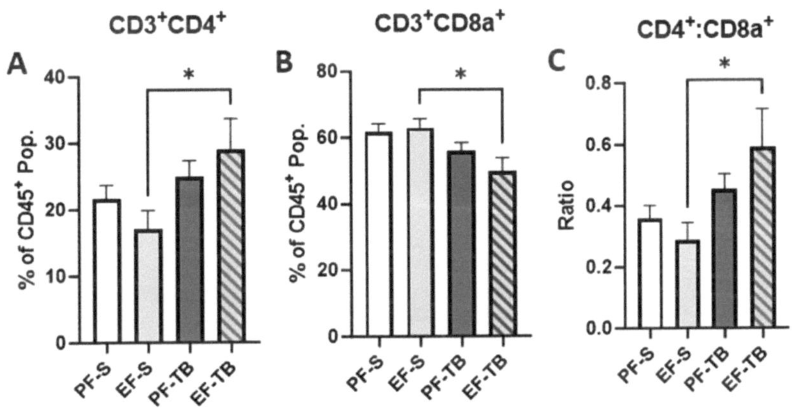

Figure 5. Percentage of CD4+ and CD8a+ T-cells in the lamina propria of the small intestine. (**A**) % CD3$^+$CD4$^+$ cells isolated from LPLs. (**B**) % CD3$^+$CD8a$^+$ cells isolated from LPLs. (**C**) Ratio of CD3$^+$CD4$^+$ to CD3$^+$CD8a$^+$ cells isolated from LPLs. Gating strategy for flow analysis is described in Figure S2. Treatment groups contain 7–8 replicates. PF-S: control diet + saline; EF-S: ethanol diet + saline; PF-TB: control diet + tributyrin; EF-TB: ethanol diet + tributyrin; * $p < 0.05$.

3.4. Tributyrin Supplementation Mitigates Ethanol-Induced Neutrophil Presence and Markers of Oxidative Stress in the Lungs

Bacteria or bacterial antigens in systemic circulation, arising from either the gut lymphatic system or vasculature, have been shown to induce immune cell infiltration and proinflammatory responses in outlying organ systems such as the hepatic, pulmonary, and nervous systems [12,13,37]. Since we found chronic–binge ethanol exposure altered the gut microbiome and impaired the gut barrier and immune responses which resulted in more bacteria in the Peyer's patches and endotoxin in circulation, we assessed how these changes affected the lungs. We hypothesized that tributyrin's protection of these effects in the gut would also impart a protective effect against immune infiltration and oxidative stress in the lungs.

We found an accumulation of the percentage of Ly6G+CD11b+ neutrophils in the EF-S mice compared to the PF-S mice by flow cytometry (Figure 6A; $p = 0.048$). Lung tissue Ly6G mRNA expression was also increased in EF-S mice, and tributyrin supplementation reduced this effect (Figure 6B; $p = 0.0001$). As activation of neutrophils can induce oxidative stress, we assessed for two markers, MPO and LCN2.

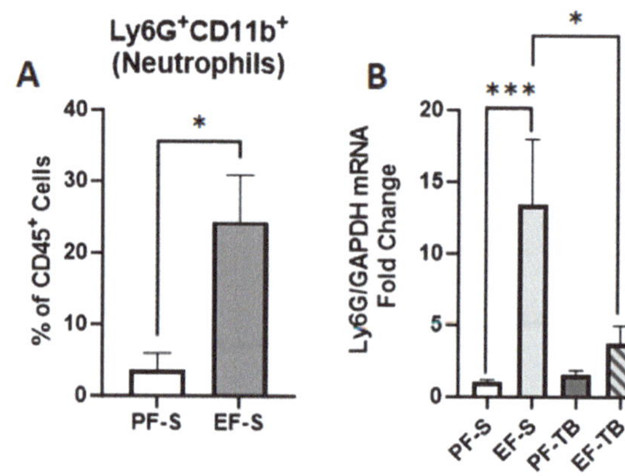

Figure 6. Neutrophil presence in the lungs. (**A**) % Ly6G$^+$CD11b$^+$ cells (neutrophils) isolated from lung homogenate. (**B**) Ly6G mRNA expression in lung tissue. Fold changes relative to PF-S group. Gating strategy for flow cytometry is described in Figure S3. Treatment groups contain 4–8 replicates. PF-S: control diet + saline gavage; EF-S: 5% ethanol diet + saline gavage; PF-TB: control diet + tributyrin gavage; EF-TB: 5% ethanol diet + tributyrin gavage; * $p < 0.05$; *** $p < 0.0001$.

Activated neutrophils can induce oxidative stress responses. As neutrophils in lungs were elevated with chronic–binge ethanol exposure, we tested for markers of oxidative stress. MPO catalyzes the production of powerful pro-oxidant species capable of damaging surrounding tissue [38]. We found that the MPO protein level was elevated in lung tissue in EF-S compared to PF-S mice (Figure 7A; $p = 0.0466$). LCN2 sequesters bacterial siderophores to starve them of iron and is involved in mediation of oxidative stress [39–41]. LCN2 mRNA and protein expression were induced in EF-S mice, and tributyrin supplementation reduced the mRNA expression of LCN2 in ethanol-exposed mice (Figure 7B,C).

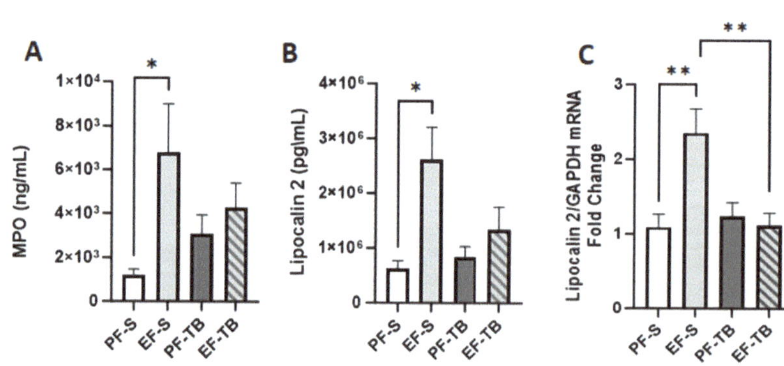

Figure 7. Markers of oxidative stress, MPO and LCN2, in lung. (**A**) MPO protein levels in lung tissue via ELISA. (**B**) LCN2 mRNA expression via qRT-PCR in lung tissue. Fold change is relative to PF-S. (**C**) LCN2 protein levels in lung tissue via ELISA. Treatment groups contain 7–8 replicates. PF-S: control diet + saline; EF-S:ethanol diet + saline; PF-TB: control diet + tributyrin; EF-TB: ethanol diet + tributyrin; * p <0.05; ** $p < 0.01$.

4. Discussion

Here, we show that mice exposed to chronic–binge ethanol have alterations in gut microbiome composition and function, decreased intestinal barrier defenses, skewed immune defenses, and increased bacteria and endotoxin in the GALT and systemic circulation, respectively, and this coincided with increased neutrophil presence and oxidative stress response in the lungs. Importantly, supplementation with tributyrin not only supported gut health disrupted by ethanol, but it also rescued the negative effects of ethanol in the lungs.

Analysis of the cecal microbiome showed that ethanol consumption significantly decreased species alpha-diversity (intra-individual variability), and this was further decreased with tributyrin supplementation. Significant dissimilarity (beta-diversity) between groups was also noted, driven by specific taxa. A portion of the microbiome was unidentified, and several unknown species of the Firmicutes phylum seemingly drove the marker of diversity assessed. Taxa clustering emerged based on both diet and treatment, indicating that while both factors are capable of shifting microbial composition, several species changed based on ethanol consumption alone. Bacteria belonging to the *Lachnospiraceae* family are known SCFA producers. The species elevated in both ethanol-exposed groups, *Lachnospiraceae* bacterium 28–4, has been predicted to possess pathways involved in butyrate synthesis [42]. Tributyrin supplementation decreased bacterial diversity and evenness, which is likely due to its alteration of specific taxa. One taxon induced by tributyrin was *B. thetaiotaomicron*, one of the most abundant commensal microbes that is involved with the metabolism of complex carbohydrates and maturation of the host immune system [43]. *Erysipelotrichaceae bacterium* NYU-BL-F16 also expanded with tributyrin supplementation [44]. While poorly characterized, its presence has been associated with a reduction in host metabolic pathways involved in diet-induced obesity [44]. In both the pair- and ethanol-fed groups, tributyrin supplementation decreased the relative abundance of *Muribaculum intestinale* (*MI*). This taxon produces a cardiolipin known to initiate an innate immune response and production of proinflammatory cytokines such as TNF-a, IL-6, and IL-23 [45]. Overall, tributyrin supplementation during ethanol exposure expanded gut microbiome species that contribute to the production of SCFA and support of the immune system compared to the results in mice only exposed to ethanol, and this may have contributed to maintenance of the immune effects noted in the hosts' cells.

Shotgun metagenomic sequencing provides information regarding the function of the bacteria through identification of gene pathways. Ethanol exposure restricted several metabolic pathways of the cecal bacteria. Importantly, all top reduced pathways by ethanol exposure were recovered with tributyrin supplementation. Several altered metabolic pathways are noteworthy for their role in mediating immunity and mitigating oxidative stress. Coenzyme A and its precursor pantothenate mitigate reactive oxygen species through the formation of disulfide bonds between the thiol functional group and sulfur-containing residues of various proteins [27]. L-arginine is involved in the production of nitric oxide, and disruption in L-arginine metabolism is associated with immune-mediated or infectious diseases [31,32]. Tetrapyrroles form the basis of a variety of metal chelating compounds that display antioxidant properties [28,29,46] and are regulated by S-adenosyl-l-methionine (SAM) enzymes [30]. Additionally, SAM and SAM enzymes regulate several processes involved with oxidative stress management, such as the production of glutathione and regulation of tetrapyrrole synthesis [30,47,48]. Taken together, the loss in each of these individual pathways with ethanol exposure alone indicates an impairment of the bacteria in mitigating oxidative damage in their local environment. Such damage not only affects microbial health but can also contribute to oxidative stress, leading to inflammation, tissue damage, and altered immune responses [49,50]. Interestingly, a loss occurred in pathways involved in the synthesis of peptidoglycan, a primary component of the gram-positive bacteria cell wall. Moreover, it was gram-positive bacteria (Enterococci) that were cultured from the Peyer's patches of EF-S mice. Peptidoglycan recognition by host immune cells is necessary to mount an immune response, and reduction and alteration to peptidoglycan structures have been suggested to allow these bacteria to circumvent such responses and contribute to their pathogenesis [51,52]. Taken together, tributyrin supplementation rescued specific gut microbiome taxa and their functions needed to combat the negative effects ethanol imposes on immunity and oxidative stress.

The changes induced in gut microbiome taxa and function by ethanol impacted immune functions in both the small intestine and lungs. Firstly, there was a depletion in cecal SIgA in the EF-S group. SIgA is found at low levels under normal conditions but is up-regulated upon presentation of antigen by dendritic cells to B cells. SIgA binds to bacteria, entrapping them so they are unable to interact with or penetrate the intestinal epithelium [34,35]. Bound bacteria can also be transported by M cells into Peyer's patches in the small intestine or lymphatic follicles in the colon as a mechanism to initiate an immune response [33–35,53]. Ethanol-induced loss of SIgA opens a route for bacteria or antigen to reach the intestinal epithelium and translocate into systemic circulation. Indeed, plasma endotoxin concentration was elevated in the EF-S group, but this was normalized with tributyrin supplementation. While the portal vein is typically thought of as the primary route for pathogens or their associated molecular patterns originating from the gut to enter circulation, the lymphatic tissue that drains the gut has also been implicated as an alternative pathway [13,54]. Bacteria, endotoxin, or activated immune components that collect in lymphatic tissue in both the small and large intestines drain into the mesenteric lymph nodes and reach the thoracic duct. The contents of the thoracic duct then enter primary circulation via the subclavian vein. Taken together, compromise in gut luminal IgA could be an initial entryway for bacteria or bacterial endotoxins to reach the lungs.

Heavy alcohol consumption has been widely observed to impact pulmonary immunity and infections. Our results corroborate prior work from Arteel et. al., who have shown that the Lieber DeCarli chronic–binge model induces neutrophil accumulation in both bronchoalveolar lavage fluid and lung tissue, leading to a mild inflammatory response [55]. We also found increased LCN2 and MPO proteins, released by neutrophils in response to bacteria and their proinflammatory compounds. We posit disrupted intestinal homeostasis by ethanol and skewed oxidative stress defenses in both the gut microbiome and host cells, and impaired mucosal barrier and immune function open a path for bacteria and endotoxins to enter the lung to drive an immune response characterized by the accumulation of neutrophils and release of MPO and LCN2. Under normal conditions, such a response is necessary for host defense, but with persistent presence of these proteins, oxidative tissue damage in the lungs and ultimately pathology could result. For instance, the release of MPO and the resulting enzymatic products displays potent bactericidal properties, but this effect is not exclusive to pathogens and, if prolonged, can damage host tissue [38,56]. Lipocalin 2 is produced by neutrophils in response to the presence of bacteria and endotoxins and sequesters bacterial siderophores, depriving them of iron necessary for their growth. Through this ability to capture and stockpile iron, LCN2 has been shown to both induce and protect against oxidative stress depending on the context [39].

As a potential therapeutic to restore and bolster gut immune and barrier function disrupted by ethanol exposure, tributyrin co-supplementation was shown to support the gut–lung axis. Tributyrin supplementation resolved the expansion of *Muribaculum intestinali* and recovered bacterial metabolic pathways important for the regulation of oxidative stress in the small intestine. Additionally, enhancement of both the mucosal and epithelial barriers and immune responses in both the epithelium and lamina propria occurred. Such protection reduced both bacterial presence in Peyer's patches and endotoxin in circulation, and in turn lowered neutrophil populations and MPO and LCN2 in the lung. Taken together, these data support our hypothesis that neutrophil activation in the lung is linked with ethanol-induced disturbances to the small intestine. These studies do have limitations in that they are associative findings and not cause and effect, necessitating futher studies. Also, we utilized a pre-clinical mouse model that replicates hepatic phenomena as human alcoholic hepatitis, with controlled environments and co-supplementation of tributyrin with ethanol exposure. Whether tributyrin protection would occur as a treatment in humans with alcoholic hepatitis and pulmonary compromise is uncertain.

5. Conclusions

Chronic–binge ethanol exposure disrupts the gut–lung axis. While the association between alcohol consumption and oxidative damage in the lungs is well documented, this work supports the notion that these processes are driven by ethanol-induced damage to the gut microbiome's composition and function, the small intestine mucosal barrier, and immunity. Importantly, tributyrin supplementation supported intestinal immune and barrier function, as well as reducing neutrophils and oxidative stress-related proteins in the lungs. These findings highlight the unique ability of intestinal health to influence the lung. Further investigations into how the gut–lung crosstalk occurs for future novel treatments of alcohol-related pulmonary conditions are warranted.

Supplementary Materials: The following supporting information can be downloaded at https://www.mdpi.com/article/10.3390/antiox13040472/s1: Figure S1. Gating strategy for CD11c$^+$IL-6$^+$ cells; Figure S2. Gating strategy for CD3$^+$CD4$^+$ and CD3$^+$CD8a$^+$ cells; Figure S3. Gating strategy for Ly6G$^+$CD11b$^+$ cells.

Author Contributions: Conceptualization, G.A.M.C.; methodology, A.S., D.S., Y.H. and G.A.M.C.; validation, A.S., D.S., Y.H. and G.A.M.C.; formal analysis, N.S. and G.A.M.C.; investigation, A.S., D.S., Y.H. and G.A.M.C.; visualization, N.S.; software, N.S.; resources, N.S. and G.A.M.C.; writing—original draft preparation, A.S.; writing—review and editing, G.A.M.C.; supervision, G.A.M.C.; funding acquisition, G.A.M.C. All authors have read and agreed to the published version of the manuscript.

Funding: This research was funded by NIH R01 grant # R01AA028043 (Cresci) NIH/NIAAA; NIH T32 trainee grant 5T32HL155005-03 (Santilli) NIH/NHLBI; and Sony ID7000 support from NIH grant S10OD02520.

Institutional Review Board Statement: All mouse procedures and experiments were approved by the Cleveland Clinic Institutional Animal Care and Use Committee (protocol #2428).

Data Availability Statement: Data supporting the reported results can be requested by writing to the corresponding author.

Conflicts of Interest: The authors declare no conflicts of interest.

References

1. Jew, M.H.; Hsu, C.L. Alcohol, the Gut Microbiome, and Liver Disease. *J. Gastroenterol. Hepatol.* **2023**, *38*, 1205–1210. [CrossRef] [PubMed]
2. Adak, A.; Khan, M.R. An Insight into Gut Microbiota and Its Functionalities. *Cell. Mol. Life Sci.* **2019**, *76*, 473–493. [CrossRef] [PubMed]
3. Liu, H.; Wang, J.; He, T.; Becker, S.; Zhang, G.; Li, D.; Ma, X. Butyrate a double edged sword for health? *Adv. Nutr.* **2018**, *9*, 21–29. [CrossRef] [PubMed]
4. Liu, P.; Wang, Y.; Yang, G.; Zhang, Q.; Meng, L.; Xin, Y.; Jiang, X. The Role of Short-Chain Fatty Acids in Intestinal Barrier Function, Inflammation, Oxidative Stress, and Colonic Carcinogenesis. *Pharmacol. Res.* **2021**, *165*, 105420. [CrossRef] [PubMed]
5. Smirnova, E.; Puri, P.; Muthiah, M.D.; Daitya, K.; Brown, R.; Chalasani, N.; Liangpunsakul, S.; Shah, V.H.; Gelow, K.; Siddiqui, M.S.; et al. Fecal Microbiome Distinguishes Alcohol Consumption from Alcoholic Hepatitis but Does Not Discriminate Disease Severity. *Hepatology* **2020**, *72*, 271–286. [CrossRef]
6. Cresci, G.A.; Bush, K.; Nagy, L.E. Tributyrin Supplementation Protects Mice from Acute Ethanol-Induced Gut Injury. *Alcohol. Clin. Exp. Res.* **2014**, *38*, 1489–1501. [CrossRef]
7. Cresci, G.A.; Glueck, B.; McMullen, M.R.; Xin, W.; Allende, D.; Nagy, L.E. Prophylactic Tributyrin Treatment Mitigates Chronic-Binge Ethanol-Induced Intestinal Barrier and Liver Injury. *J. Gastroenterol. Hepatol.* **2017**, *32*, 1587–1597. [CrossRef] [PubMed]
8. Roychowdhury, S.; Glueck, B.; Han, Y.; Mohammad, M.A.; Cresci, G.A.M. A Designer Synbiotic Attenuates Chronic-Binge Ethanol-Induced Gut-Liver Injury in Mice. *Nutrients* **2019**, *11*, 97. [CrossRef]
9. Han, Y.; Glueck, B.; Shapiro, D.; Miller, A.; Roychowdhury, S.; Cresci, G.A.M. Dietary Synbiotic Supplementation Protects Barrier Integrity of Hepatocytes and Liver Sinusoidal Endothelium in a Mouse Model of Chronic-Binge Ethanol Exposure. *Nutrients* **2020**, *12*, 373. [CrossRef]
10. Sarkar, D.; Katherine Jung, M.; Joe Wang, H.; Sarkar, D.K. Alcohol and the Immune System. *Alcohol Res.* **2015**, *37*, 153–155.
11. Simet, S.M.; Sisson, J.H. Alcohol's Effects on Lung Health Immunity. *Alcohol Res.* **2015**, *37*, 199–208. [PubMed]
12. Ma, Y.; Yang, X.; Chatterjee, V.; Wu, M.H.; Yuan, S.Y. The Gut–Lung Axis in Systemic Inflammation Role of Mesenteric Lymph as a Conduit. *Am. J. Respir. Cell Mol. Biol.* **2021**, *64*, 19–28. [CrossRef] [PubMed]
13. Enaud, R.; Prevel, R.; Ciarlo, E.; Beaufils, F.; Wieërs, G.; Guery, B.; Delhaes, L. The Gut-Lung Axis in Health and Respiratory Diseases: A Place for Inter-Organ and Inter-Kingdom Crosstalks. *Front. Cell Infect. Microbiol.* **2020**, *10*, 9. [CrossRef] [PubMed]

14. Forsyth, C.B.; Farhadi, A.; Jakate, S.M.; Tang, Y.; Shaikh, M.; Keshavarzian, A. Lactobacillus GG Treatment Ameliorates Alcohol-Induced Intestinal Oxidative Stress, Gut Leakiness, and Liver Injury in a Rat Model of Alcoholic Steatohepatitis. *Alcohol* **2009**, *43*, 163–172. [CrossRef] [PubMed]
15. Sangwan, N.; Zarraonaindia, I.; Hampton-Marcell, J.T.; Ssegane, H.; Eshoo, T.W.; Rijal, G.; Negri, M.C.; Gilbert, J.A. Differential Functional Constraints Cause Strain-Level Endemism in *Polynucleobacter* Populations. *mSystems* **2016**, *1*, 00003-16. [CrossRef] [PubMed]
16. Bolger, A.M.; Lohse, M.; Usadel, B. Trimmomatic: A Flexible Trimmer for Illumina Sequence Data. *Bioinformatics* **2014**, *30*, 2114–2120. [CrossRef] [PubMed]
17. Segata, N.; Waldron, L.; Ballarini, A.; Narasimhan, V.; Jousson, O.; Huttenhower, C. Metagenomic Microbial Community Profiling Using Unique Clade-Specific Marker Genes. *Nat. Methods* **2012**, *9*, 811–814. [CrossRef]
18. Franzosa, E.A.; McIver, L.J.; Rahnavard, G.; Thompson, L.R.; Schirmer, M.; Weingart, G.; Lipson, K.S.; Knight, R.; Caporaso, J.G.; Segata, N.; et al. Species-Level Functional Profiling of Metagenomes and Metatranscriptomes. *Nat. Methods* **2018**, *15*, 962–968. [CrossRef] [PubMed]
19. Benjamini, Y. Discovering the False Discovery Rate. *J. R. Statist. Soc. B* **2010**, *72*, 405–416. [CrossRef]
20. Benjamini, Y.; Hochberg, Y. Controlling the False Discovery Rate: A Practical and Powerful Approach to Multiple Testing. *J. R. Statist. Soc. B* **1995**, *57*, 289–300. [CrossRef]
21. R Core Team. *R: A Language and Environment for Statistical Computing*; R Foundation for Statistical Computing: Vienna, Austria, 2021; Available online: https://www.R-project.org/ (accessed on 3 November 2023).
22. Qiu, Z.; Sheridan, B.S. Isolating Lymphocytes from the Mouse Small Intestinal Immune System. *J. Vis. Exp.* **2018**, *2018*, e57281.
23. Siddiqui, M.T.; Han, Y.; Shapiro, D.; West, G.; Fiocchi, C.; Cresci, G.A.M. The Postbiotic Butyrate Mitigates Gut Mucosal Disruption Caused by Acute Ethanol Exposure. *Int. J. Mol. Sci.* **2024**, *25*, 1665. [CrossRef] [PubMed]
24. Shapiro, D.; Kapourchali, F.R.; Santilli, A.; Han, Y.; Cresci, G.A.M. Targeting the Gut Microbiota and Host Immunity with a Bacilli-Species Probiotic during Antibiotic Exposure in Mice. *Microorganisms* **2022**, *10*, 1178. [CrossRef] [PubMed]
25. Calleja-Conde, J.; Echeverry-Alzate, V.; Bühler, K.M.; Durán-González, P.; Morales-García, J.Á.; Segovia-Rodríguez, L.; Rodríguez de Fonseca, F.; Giné, E.; López-Moreno, J.A. The Immune System through the Lens of Alcohol Intake and Gut Microbiota. *Int. J. Mol. Sci.* **2021**, *22*, 7485. [CrossRef] [PubMed]
26. Bishehsari, F.; Magno, E.; Swanson, G.; Desai, V.; Voigt, R.M.; Forsyth, C.B.; Keshavarzian, A. Alcohol and Gut-Derived Inflammation. *Alcohol Res.* **2017**, *38*, 163–171.
27. Gout, I. Coenzyme A: A Protective Thiol in Bacterial Antioxidant Defence. *Biochem. Soc. Trans.* **2019**, *47*, 469–476. [CrossRef] [PubMed]
28. Perez-Ortiz, G.; Sidda, J.D.; Peate, J.; Ciccarelli, D.; Ding, Y.; Barry, S.M. Production of Copropophyrin III, Biliverdin and Bilirubin by the Rufomycin Producer, Streptomyces Atratus. *Front. Microbiol.* **2023**, *14*, 1092166. [CrossRef]
29. Wilks, A.; Ikeda-Saito, M. Heme Utilization by Pathogenic Bacteria: Not All Pathways Lead to Biliverdin. *Acc. Chem. Res.* **2014**, *47*, 2291–2298. [CrossRef] [PubMed]
30. Layer, G.; Jahn, M.; Moser, J.; Jahn, D. Radical SAM Enzymes Involved in Tetrapyrrole Biosynthesis and Insertion. *ACS Bio. Med. Chem. Au* **2022**, *2*, 196–204. [CrossRef]
31. Li, J.Y.; Guo, Y.C.; Zhou, H.F.; Yue, T.T.; Wang, F.X.; Sun, F.; Wang, W.Z. Arginine Metabolism Regulates the Pathogenesis of Inflammatory Bowel Disease. *Nutr. Rev.* **2023**, *81*, 578–586. [CrossRef]
32. Nüse, B.; Holland, T.; Rauh, M.; Gerlach, R.G.; Mattner, J. L-Arginine Metabolism as Pivotal Interface of Mutual Host–Microbe Interactions in the Gut. *Gut Microbes* **2023**, *15*, 2222961. [CrossRef] [PubMed]
33. López, M.C. Chronic Alcohol Consumption Regulates the Expression of Poly Immunoglobulin Receptor (PIgR) and Secretory IgA in the Gut. *Toxicol. Appl. Pharmacol.* **2017**, *333*, 84–91. [CrossRef] [PubMed]
34. Mantis, N.J.; Rol, N.; Corthésy, B. Secretory IgA's Complex Roles in Immunity and Mucosal Homeostasis in the Gut. *Mucosal. Immunol.* **2011**, *4*, 603–611. [CrossRef] [PubMed]
35. Pietrzak, B.; Tomela, K.; Olejnik-Schmidt, A.; Mackiewicz, A.; Schmidt, M. Secretory Iga in Intestinal Mucosal Secretions as an Adaptive Barrier against Microbial Cells. *Int. J. Mol. Sci.* **2020**, *21*, 9254. [CrossRef] [PubMed]
36. Guo, Y.; Wang, B.; Wang, T.; Gao, L.; Yang, Z.J.; Wang, F.F.; Shang, H.W.; Hua, R.; Xu, J.D. Biological Characteristics of Il-6 and Related Intestinal Diseases. *Int. J. Biol. Sci.* **2020**, *17*, 204–219. [CrossRef] [PubMed]
37. Feng, Q.; Chen, W.D.; Wang, Y.D. Gut Microbiota: An Integral Moderator in Health and Disease. *Front. Microbiol.* **2018**, *9*, 151. [CrossRef] [PubMed]
38. Davies, M.J.; Hawkins, C.L. The Role of Myeloperoxidase in Biomolecule Modification, Chronic Inflammation, and Disease. *Antioxid. Redox Signal* **2020**, *32*, 957–981. [CrossRef]
39. An, H.S.; Yoo, J.W.; Jeong, J.H.; Heo, M.; Hwang, S.H.; Jang, H.M.; Jeong, E.A.; Lee, J.; Shin, H.J.; Kim, K.E.; et al. Lipocalin-2 Promotes Acute Lung Inflammation and Oxidative Stress by Enhancing Macrophage Iron Accumulation. *Int. J. Biol. Sci.* **2023**, *19*, 1163–1177. [CrossRef] [PubMed]
40. Shin, H.J.; Jeong, E.A.; Lee, J.Y.; An, H.S.; Jang, H.M.; Ahn, Y.J.; Lee, J.; Kim, K.E.; Roh, G.S. Lipocalin-2 Deficiency Reduces Oxidative Stress and Neuroinflammation and Results in Attenuation of Kainic Acid-Induced Hippocampal Cell Death. *Antioxidants* **2021**, *10*, 100. [CrossRef]

41. Xiao, X.; Yeoh, B.S.; Vijay-Kumar, M. Lipocalin 2: An Emerging Player in Iron Homeostasis and Inflammation. *Annu. Rev. Nutr.* **2017**, *37*, 103–130. [CrossRef]
42. Li, Z.; Zhou, E.; Liu, C.; Wicks, H.; Yildiz, S.; Razack, F.; Ying, Z.; Kooijman, S.; Koonen, D.P.Y.; Heijink, M.; et al. Dietary Butyrate Ameliorates Metabolic Health Associated with Selective Proliferation of Gut Lachnospiraceae Bacterium 28-4. *JCI Insight* **2023**, *8*, e166655. [CrossRef] [PubMed]
43. Martens, E.C.; Chiang, H.C.; Gordon, J.I. Mucosal Glycan Foraging Enhances Fitness and Transmission of a Saccharolytic Human Gut Bacterial Symbiont. *Cell Host Microbe* **2008**, *4*, 447–457. [CrossRef] [PubMed]
44. Cox, L.M.; Blaser, M.J. Probiotic Compositions for Improving Metabolism and Immunity. US Patent 10,653,728 B2, 19 May 2020.
45. Bang, S.; Shin, Y.H.; Ma, X.; Park, S.M.; Graham, D.B.; Xavier, R.J.; Clardy, J. A Cardiolipin from Muribaculum Intestinale Induces Antigen-Specific Cytokine Responses. *J. Am. Chem. Soc.* **2023**, *145*, 23422–23426. [CrossRef] [PubMed]
46. Frankenberg, N.; Moser, J.; Jahn, D. Bacterial Heme Biosynthesis and Its Biotechnological Application. *Appl. Microbiol. Biotechnol.* **2003**, *63*, 115–127. [CrossRef] [PubMed]
47. Pascale, R.M.; Simile, M.M.; Calvisi, D.F.; Feo, C.F.; Feo, F. S-Adenosylmethionine: From the Discovery of Its Inhibition of Tumorigenesis to Its Use as a Therapeutic Agent. *Cells* **2022**, *11*, 409. [CrossRef] [PubMed]
48. Li, Q.; Cui, J.; Fang, C.; Liu, M.; Min, G.; Li, L. S-Adenosylmethionine Attenuates Oxidative Stress and Neuroinflammation Induced by Amyloid-β Through Modulation of Glutathione Metabolism. *J. Alzheimer's Dis.* **2017**, *58*, 549–558. [CrossRef]
49. Shandilya, S.; Kumar, S.; Kumar Jha, N.; Kumar Kesari, K.; Ruokolainen, J. Interplay of Gut Microbiota and Oxidative Stress: Perspective on Neurodegeneration and Neuroprotection. *J. Adv. Res.* **2022**, *38*, 223–244. [CrossRef] [PubMed]
50. Kunst, C.; Schmid, S.; Michalski, M.; Tümen, D.; Buttenschön, J.; Müller, M.; Gülow, K. The Influence of Gut Microbiota on Oxidative Stress and the Immune System. *Biomedicines* **2023**, *11*, 1388. [CrossRef] [PubMed]
51. Bastos, P.A.D.; Wheeler, R.; Boneca, I.G. Uptake, Recognition and Responses to Peptidoglycan in the Mammalian Host. *FEMS Microbiol. Rev.* **2021**, *45*, fuaa044. [CrossRef]
52. Wolf, A.J.; Underhill, D.M. Peptidoglycan Recognition by the Innate Immune System. *Nat. Rev. Immunol.* **2018**, *18*, 243–254. [CrossRef]
53. Tezuka, H.; Ohteki, T. Regulation of IgA Production by Intestinal Dendritic Cells and Related Cells. *Front. Immunol.* **2019**, *10*, 1891. [CrossRef] [PubMed]
54. Ni, S.; Yuan, X.; Cao, Q.; Chen, Y.; Peng, X.; Lin, J.; Li, Y.; Ma, W.; Gao, S.; Chen, D. Gut Microbiota Regulate Migration of Lymphocytes from Gut to Lung. *Microb. Pathog.* **2023**, *183*, 106311. [CrossRef] [PubMed]
55. Poole, L.G.; Beier, J.I.; Torres-Gonzales, E.; Schlueter, C.F.; Hudson, S.V.; Artis, A.; Warner, N.L.; Nguyen-Ho, C.T.; Dolin, C.E.; Ritzenthaler, J.D.; et al. Chronic + Binge Alcohol Exposure Promotes Inflammation and Alters Airway Mechanics in the Lung. *Alcohol* **2019**, *80*, 53–63. [CrossRef] [PubMed]
56. Hawkins, C.L.; Davies, M.J. Role of Myeloperoxidase and Oxidant Formation in the Extracellular Environment in Inflammation-Induced Tissue Damage. *Free Radic. Biol. Med.* **2021**, *172*, 633–651. [CrossRef] [PubMed]

antioxidants

MDPI

Review

The Impact of Oxidative Stress on the Epigenetics of Fetal Alcohol Spectrum Disorders

Sergio Terracina [1], Luigi Tarani [2], Mauro Ceccanti [3], Mario Vitali [4], Silvia Francati [1], Marco Lucarelli [1,5], Sabrina Venditti [6], Loredana Verdone [7], Giampiero Ferraguti [1,*,†] and Marco Fiore [8,*,†]

1 Department of Experimental Medicine, Sapienza University of Rome, 00185 Rome, Italy; marco.lucarelli@uniroma1.it (M.L.)
2 Department of Maternal Infantile and Urological Sciences, Sapienza University of Rome, 00185 Roma, Italy
3 SITAC, Società Italiana per il Trattamento dell'Alcolismo e le sue Complicanze, 00185 Rome, Italy; mauro.ceccanti@uniroma1.it
4 ASUR Marche, AV4, 60122 Ancona, Italy
5 Pasteur Institute Cenci Bolognetti Foundation, Sapienza University of Rome, 00185 Rome, Italy
6 Department of Biology and Biotechnologies Charles Darwin, Sapienza University, 00185 Rome, Italy
7 Institute of Molecular Biology and Pathology (IBPM-CNR), 00185 Rome, Italy
8 Institute of Biochemistry and Cell Biology (IBBC-CNR), Department of Sensory Organs, Sapienza University of Rome, 00185 Roma, Italy
* Correspondence: giampiero.ferraguti@uniroma1.it (G.F.); marco.fiore@cnr.it (M.F.)
† These authors contributed equally to this work.

Abstract: Fetal alcohol spectrum disorders (FASD) represent a continuum of lifelong impairments resulting from prenatal exposure to alcohol, with significant global impact. The "spectrum" of disorders includes a continuum of physical, cognitive, behavioral, and developmental impairments which can have profound and lasting effects on individuals throughout their lives, impacting their health, social interactions, psychological well-being, and every aspect of their lives. This narrative paper explores the intricate relationship between oxidative stress and epigenetics in FASD pathogenesis and its therapeutic implications. Oxidative stress, induced by alcohol metabolism, disrupts cellular components, particularly in the vulnerable fetal brain, leading to aberrant development. Furthermore, oxidative stress is implicated in epigenetic changes, including alterations in DNA methylation, histone modifications, and microRNA expression, which influence gene regulation in FASD patients. Moreover, mitochondrial dysfunction and neuroinflammation contribute to epigenetic changes associated with FASD. Understanding these mechanisms holds promise for targeted therapeutic interventions. This includes antioxidant supplementation and lifestyle modifications to mitigate FASD-related impairments. While preclinical studies show promise, further clinical trials are needed to validate these interventions' efficacy in improving clinical outcomes for individuals affected by FASD. This comprehensive understanding of the role of oxidative stress in epigenetics in FASD underscores the importance of multidisciplinary approaches for diagnosis, management, and prevention strategies. Continued research in this field is crucial for advancing our knowledge and developing effective interventions to address this significant public health concern.

Keywords: FASD; epigenetics; oxidative stress; antioxidants; alcohol

Citation: Terracina, S.; Tarani, L.; Ceccanti, M.; Vitali, M.; Francati, S.; Lucarelli, M.; Venditti, S.; Verdone, L.; Ferraguti, G.; Fiore, M. The Impact of Oxidative Stress on the Epigenetics of Fetal Alcohol Spectrum Disorders. *Antioxidants* **2024**, *13*, 410. https://doi.org/10.3390/antiox13040410

Academic Editor: Alessandra Napolitano

Received: 6 March 2024
Revised: 26 March 2024
Accepted: 27 March 2024
Published: 28 March 2024

1. Fetal Alcohol Spectrum Disorders (FASD)

Fetal alcohol spectrum disorders (FASD) represent a spectrum of lifelong, debilitating conditions that result from prenatal exposure to alcohol [1–4]. This diverse range of disorders encompasses a continuum of physical, cognitive, behavioral, and developmental impairments which can have profound and lasting effects on individuals throughout their lives [5–7]. The term "spectrum" reflects the wide variability in the impact of alcohol exposure on fetal development, with some individuals experiencing more severe manifestations than others. The critical period of vulnerability is during pregnancy, particularly during the first trimester, when organ systems are rapidly developing [8]. The main risk factors for FASD are increased fetal exposure to alcohol and sustained alcohol intake during any trimester of pregnancy, genetic predisposition, maternal lower socioeconomic statuses and smoking, and paternal chronic alcohol use [9–14].

Alcohol is able to freely cross the placenta during pregnancy and enter the growing fetus through the umbilical cord; the different quantities, defense efficiency, and excretion of maternal and fetal enzymes allow for alcohol to have a lengthy effect on the fetus [15]. Alcohol is a teratogen substance acting through various mechanisms, including direct damage of its metabolites, reactive oxygen species (ROS) generated as byproducts of cytochrome P450 family 2 subfamily E member 1 (CYP2E1), decreased endogenous antioxidant levels, mitochondrial damage, lipid peroxidation, disrupted neuronal cell–cell adhesion, placental vasoconstriction, inhibition of cofactors required for fetal growth and development, and epigenetic changes (Figure 1). Alcohol interferes with the development of cells and tissues in the fetus [16]. It disrupts the process of cell division and differentiation, leading to abnormal growth and development of various organs, especially the brain. Early observations supported alcohol, rather than acetaldehyde, being the more important teratogen, and specific genetic susceptibility differences to alcohol-related birth defects were found (e.g., alcohol dehydrogenase-2*3 allele protects against alcohol-related birth defects) [17]

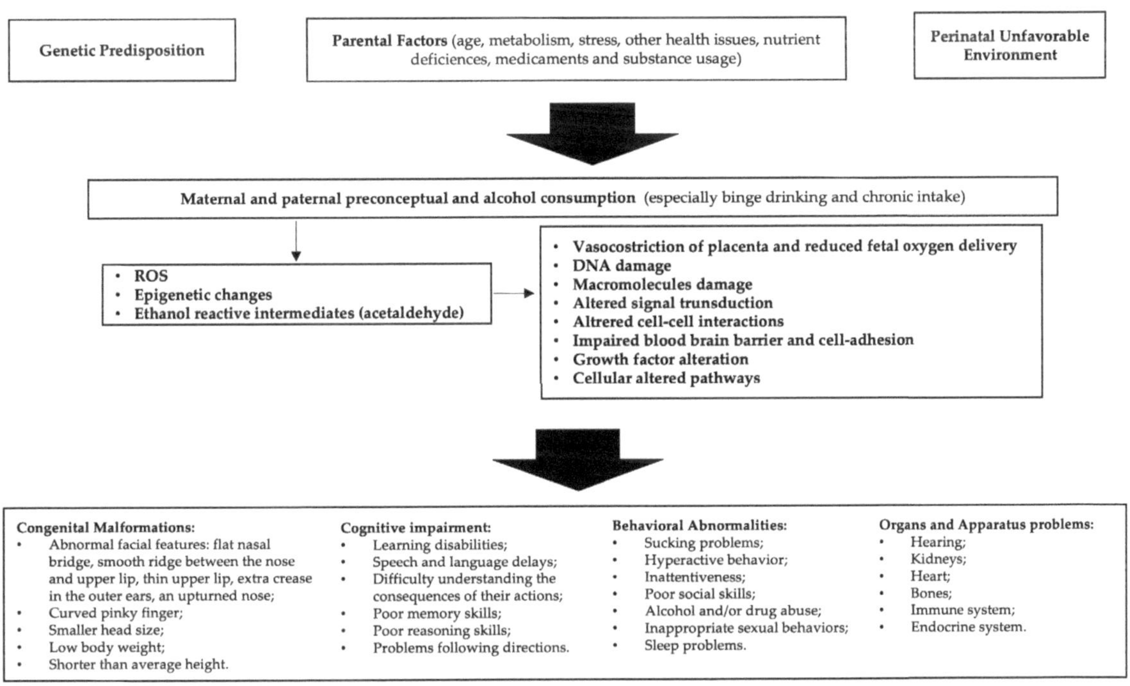

Figure 1. Etiopathogenesis of fetal alcohol spectrum disorders (FASD): In predisposed individuals, the risk of FASD is high in the case of alcohol abuse. Decreased endogenous antioxidant levels and mitochondrial damage may result in reduced compensation for the increased reactive oxygen species (ROS) generated by alcohol metabolism. Furthermore, epigenetic changes due to oxidative stress and acetaldehyde activity lead to cellular alterations that ultimately cause manifestations of FASD.

Thus, prenatal exposure to alcohol can interfere with the normal growth and development of the fetus, leading to a myriad of challenges that may manifest in infancy, childhood, adolescence, and adulthood. The severity of FASD can be influenced by factors such as the timing, amount, and pattern of alcohol consumption. Individual genetic and environmental factors also play a significant role [18]. The hallmark features of FASD include physical anomalies, cognitive deficits, and behavioral issues [19]. Physical characteristics may include facial abnormalities, growth deficiencies, and organ malformations [20]. Cognitive impairments often encompass difficulties in learning, memory, attention, and problem-solving skills [21]. Behavioral challenges can range from hyperactivity and impulsivity to social and emotional difficulties [22]. The intricate interplay of these components makes the diagnosis and management of FASD a complex and multidisciplinary task. Prevention is paramount, and education about the risks of alcohol consumption during pregnancy is crucial [23]. Unfortunately, FASD remains a significant public health concern globally, affecting individuals from all walks of life [19].

To address the complexities of FASD, a comprehensive approach is required. This involves collaboration among healthcare professionals, educators, policymakers, and community support systems. Early intervention and appropriate support services can enhance the quality of life for individuals with FASD, providing them with the tools they need to navigate the challenges associated with their unique conditions. As our understanding of FASD continues to evolve, ongoing research and advocacy efforts are essential to raise awareness, improve diagnostic methods, and develop effective interventions to mitigate the impact of prenatal alcohol exposure on individuals and their families.

2. FASD Epigenetics

Epigenetics, a captivating and rapidly advancing field within the realm of genetics, unveils the intricate dance between genes and the environment, fundamentally shaping the destiny of living organisms [24–26]. At its core, epigenetics explores the heritable changes in gene activity that occur without alterations to the underlying deoxyribonucleic acid (DNA) sequence. This field revolutionizes our understanding of how external factors, spanning from lifestyle choices to environmental exposures, can imprint molecular marks on the genome, influencing gene expression and, consequently, the phenotype.

The term 'epigenetics' itself underscores the pivotal role of these processes. It translates to 'above' or 'on top of' genetics [27–29]. Unlike the unalterable DNA code, epigenetic modifications act as dynamic regulators, orchestrating the symphony of gene expression in response to various internal and external cues. These modifications include DNA methylation, histone modification, and non-coding ribonucleic acid (RNA) molecules, collectively influencing the accessibility of genes to the cellular machinery responsible for transcription [30,31]. The impact of epigenetics extends far beyond the individual organism, as these marks can be passed down through generations, heralding the era of transgenerational inheritance [4,32,33]. This phenomenon challenges the conventional view that genetic information flows strictly through the DNA sequence, introducing a dynamic layer of complexity to our understanding of heredity.

Consequently, the study of epigenetics not only elucidates the molecular intricacies governing development and cellular function, but also sheds light on the potential intergenerational consequences of environmental exposure. In this expansive landscape, researchers delve into the epigenetic mechanisms underpinning health and disease. From the early stages of embryonic development to the intricate regulation of tissue-specific gene expression, epigenetic processes play a pivotal role in determining cellular identity and function [34–36]. Moreover, aberrations in epigenetic regulation have been implicated in a myriad of diseases, including cancer, neurodegenerative disorders, and metabolic conditions, providing a new avenue for therapeutic exploration. As scientists continue to unravel the epigenetic tapestry, they grapple with the ethical implications and societal ramifications of this knowledge.

The dynamic nature of epigenetic modifications prompts questions about the potential reversibility of epigenetic changes and the development of interventions to modulate these processes for therapeutic purposes [37–39]. The intersection of science, ethics, and medicine in the realm of epigenetics underscores the need for careful consideration and responsible stewardship as we navigate the uncharted territories of this revolutionary field.

The epigenetics of FASD represents a compelling area of research that delves into the molecular mechanisms underlying the long-term effects of prenatal alcohol exposure on gene regulation [40,41]. Furthermore, FASD, resulting from maternal alcohol consumption during pregnancy, encompass a range of developmental, cognitive, and behavioral abnormalities. Understanding how alcohol-induced epigenetic changes contribute to the varied and often severe phenotypic outcomes is crucial for developing targeted interventions and therapies [42–47]. Most of the studies on FASD epigenetics have been published in recent decades, and the majority have been conducted on animal models [48]. One of the key epigenetic modifications associated with FASD is DNA methylation [15,48–50]. Studies have revealed alterations in the methylation patterns of specific genes involved in neural development and function in individuals with FASD.

For instance, genes related to neuronal migration, synaptogenesis, and neurotransmitter regulation may undergo abnormal DNA methylation, leading to disruptions in neural circuitry and function [51,52]. The dynamic nature of DNA methylation makes it a potential biomarker for assessing the severity and persistence of FASD-related impairments. Histone modifications, another critical facet of epigenetics, play a role in orchestrating the three-dimensional structure of chromatin and regulating gene accessibility [53]. Prenatal alcohol exposure has been linked to changes in histone acetylation and methylation patterns, particularly in genes associated with neurodevelopment [54].

Altered histone modifications can influence the expression of genes involved in learning, memory, and behavioral regulation, contributing to the cognitive and behavioral deficits observed in individuals with FASD. Non-coding RNAs, such as microRNAs, also emerge as key players in the epigenetic landscape of FASD [55,56]. These small RNA molecules can post-transcriptionally regulate gene expression, and their dysregulation has been implicated in the pathogenesis of neurodevelopmental disorders [57]. Studies suggest that alcohol exposure during pregnancy can disrupt the expression of specific microRNAs, potentially contributing to the aberrant gene expression patterns associated with FASD [55].

The transgenerational aspect of epigenetics adds an additional layer of complexity to the study of FASD [11,31]. Emerging evidence suggests that prenatal alcohol exposure can induce epigenetic changes that persist across generations, influencing the susceptibility of offspring to FASD-related outcomes [4]. This transgenerational epigenetic inheritance underscores the importance of considering not only the immediate consequences of prenatal alcohol exposure, but also its potential impact on future generations.

Understanding the epigenetic landscape of FASD holds promise for the development of targeted interventions and therapeutic strategies. By unraveling the molecular mechanisms through which alcohol exposure induces lasting epigenetic changes, researchers aim to identify potential targets for intervention and prevention, ultimately improving the quality of life for individuals affected by FASD and potentially mitigating the risk of FASD in future generations.

3. Oxidative Stress and FASD

Alcohol causes FASD by interfering with molecular pathways associated with increased oxidative stress, altered organ development, and changes in epigenetic gene expression control during fetal development [58]. Oxidative stress, characterized by an imbalance between reactive oxygen species (ROS) and the body's ability to neutralize them, plays a significant role in the pathogenesis of FASD, leading to potential damage to key cellular components during the development phase of the fetus [59,60]. Actually, during pregnancy the hypoxic condition leads to an increased likelihood of free radical formation, triggering oxidative stress and inflammation associated not only with preterm delivery and gestational diabetes mellitus, but also with epigenetic alterations and placental disorders [61,62]. Thus, pregnancy is effectively characterized by an increased risk of complications associated with increased oxidative stress. When alcohol is metabolized in the liver, it generates

ROS as byproducts (including superoxide radicals and hydrogen peroxide), leading to elevated levels of ROS that can overwhelm the body's antioxidant defense systems and result in oxidative stress [63,64]. ROS can cause damage to cellular structures such as lipids, proteins, and DNA of developing fetal tissues, including the brain, which is particularly vulnerable to oxidative stress because of the rich lipid composition and the high metabolic rate. Furthermore, oxidative stress can impact mitochondrial function, trigger inflammatory responses, and disrupt normal cellular processes, including neuronal migration, synaptogenesis, and myelination [65,66].

In fact, the fetal body has defenses against ROS. Specifically, it can produce endocrine antioxidative enzymes, such as catalase, providing critical protection. It can also activate mechanisms to repair damaged cellular and genetic components, such as oxoguanine glycosylase 1 (OGG1), activated in the case of DNA. Additionally, the fetal body can reduce the risk of damage by producing products like the fetal nuclear factor erythroid 2-related factor 2 (Nrf2), an ROS-sensing protein that upregulates an array of proteins, including antioxidative enzymes and DNA repair proteins [64].

In particular, oxidative stress plays a major role in the epigenetic changes associated with FASD [64,67]. In fact, it has been associated with alterations in DNA methylation patterns and expression of miRNAs, as well as histone modifications, shifting gene accessibility and expression in patients affected by FASD and neurodevelopmental disorders. Furthermore, as stated before, oxidative stress can directly cause damage to DNA and its components, leading to mutations potentially affecting the expression of genes critical for brain development and function. Mitochondrial dysfunction also may contribute to epigenetic changes, as mitochondria play a key role in providing the intermediate metabolites necessary to generate and modify epigenetic marks in the nucleus, which in turn can regulate the expression of mitochondrial proteins [68]. In the context of FASD, neuroinflammation may contribute to epigenetic changes that modulate the expression of genes involved in neurodevelopment.

Studying the impact of oxidative stress on FASD epigenetics holds great potential for advancing our understanding of the disorder, identifying diagnostic markers, and improving the management of this incurable disease.

4. Epigenetics and Oxidative Stress

The FASD risk is likely increased in children who are genetically and environmentally predisposed, especially in the case of enhanced pathways for ROS formation and/or deficient pathways for ROS detoxification or DNA repair [69].

As stated before, alcohol has the potential to alter gene expression by impacting DNA methylation processes [70,71]. This occurs by enhancing the breakdown and reduction of methyl groups, leading to the disruption of subsequent S-adenosylmethionine (SAM)-dependent transmethylation reactions in the folate pathway, which are crucial for DNA methylation [8]. Additionally, alcohol influences nucleosomal remodeling by initiating histone modifications. It also impacts the expression of microRNA. Furthermore, both maternal and paternal preconceptual alcohol exposures induce mitochondrial dysfunction and a heightened response to oxidative stress in developing organs. This is achieved by metabolizing ethanol into acetaldehyde, facilitated by enzymes like alcohol dehydrogenase, cytochrome P450-CYP2E1, and catalase [59,64,72]. This process generates ROS and reactive nitrogen species (RNS), altering the cells' internal redox balance, leading to neuronal cell death and modified gene expression due to DNA oxidation. Mitochondrial dysfunction and mitochondrial DNA (mtDNA) damage, which are also hallmarks of aging, are key events in FASD [73,74].

Indeed, alcohol can induce mtDNA damage, resulting in increased oxidative stress and alterations in the mtDNA repair protein 8-oxoguanine DNA glycosylase-1 (OGG1) [75]. Therefore, pregnancy inherently heightens susceptibility to oxidative stress, and this risk is further increased by alcohol consumption, leading to various adverse outcomes. These include impaired development, abnormal placental function, and several complications such as pre-eclampsia, recurrent pregnancy loss, fetal anomalies, intrauterine growth restriction, and, in severe cases, fetal demise [76]. In response to the uncontrolled rise in RNS/ROS levels, the body relies on trace elements involved in both non-enzymatic and enzymatic defense mechanisms.

These elements, namely, copper (Cu), zinc (Zn), manganese (Mn), and selenium (Se), play crucial roles. Assessing ROS may benefit from the use of marker proteins like malondialdehyde (MDA), superoxide dismutase (SOD), glutathione peroxidase (GPx), glutathione reductase (GR), catalase (CAT), and glutathione (GSH) [77]. These markers serve as indirect indicators of the intensity of oxidative stress and can provide insights into potential pregnancy complications. Prenatal alcohol exposure can alter the Mammalian Target of Rapamycin (mTOR) signaling pathway, resulting in increased oxidative stress [78]. mTOR plays a major role in modulating protein synthesis and autophagy, which are necessary for proper fetal development. In fact, mTOR alterations have recently been implicated in FASD etiology, as long-lasting effects following alcohol exposure include impaired hippocampal and synapse formation, and reduced brain size, as well as cognitive, behavioral, and memory impairments [79].

The brain is particularly susceptible to generating ROS, including superoxide anions, hydrogen peroxide, and hydroxyl radicals [80]. This susceptibility arises due to the brain's elevated metabolic rate for oxygen consumption. Its cells utilize about 20% of the oxygen consumed by the entire organism. Additionally, brain tissues contain high levels of unsaturated fatty acids, which serve as substrates for the production of ROS. Moreover, certain brain regions contain elevated levels of iron, and various neurotransmitters, such as dopamine, levodopa, serotonin, and norepinephrine, have a tendency to react spontaneously with oxygen [81].

It is important to note that antioxidant enzyme activity, including superoxide dismutase, catalase, and glutathione peroxidase, is generally lower in the brain than in organs like the liver or kidney [82,83]. Furthermore, even though oxidative stress plays a role in normal fetal development, its imbalance caused by alcohol consumption and the higher susceptibility of fetal cells leads to neurotoxic effects. Hence, antioxidants such as vitamin E, vitamin C, and glutathione play crucial roles in FASD treatment. Their ability to counteract the harmful effects of oxidative stress has the potential to mitigate or prevent some of the neurological and developmental issues caused by prenatal alcohol exposure (Figure 2).

Ethanol-induced oxidative stress can also cause damage to DNA, resulting in genetic mutations within individual cells [69]. This damage can lead to the immortalization and multiplication of cells, potentially resulting in cancer development after birth. Alternatively, ethanol-induced oxidative stress can lead to direct or indirect alterations in the epigenetic makeup of DNA, histones, or RNA across multiple cells.

These modifications can influence the expression of genes and contribute to teratogenesis, leading to birth defects and abnormalities in neurodevelopment after birth. Moreover, paternal consumption of alcohol before conception triggers epigenetic alterations in male sperm. This is facilitated by ROS generation and accelerated breakdown of substances, leading to the loss of methyl groups. These changes disrupt SAM-dependent transmethylation reactions in the folate pathway, crucial for DNA methylation. Additionally, there is restructuring of nucleosomes via modifications to histones and abnormal expression of microRNAs [4].

Research has specifically focused on several neurotransmitters, insulin resistance, alterations of the hypothalamic–pituitary–adrenal (HPA) axis, abnormal glycosylation of several proteins, oxidative stress, nutritional antioxidants, and various epigenetic factors [84]. Prenatal alcohol consumption is also associated with a widespread increase in the neuroendocrine stress response, regulated by the HPA axis [85,86]. This response influences drinking behavior and is linked to epigenetic changes in neurotrophins and POMC genes, impacting pathways that regulate mood, emotion, and serotonergic function. Recent studies have found a correlation between mtDNA damage and phenotypical abnormalities associated with FASD. This suggests that the amount of damaged mtDNA in fetal brain-derived exosomes may serve as a marker to predict FASD risk in fetuses [75]. Moreover, IGF-1 might reduce alcohol-caused mtDNA damage and neuronal apoptosis.

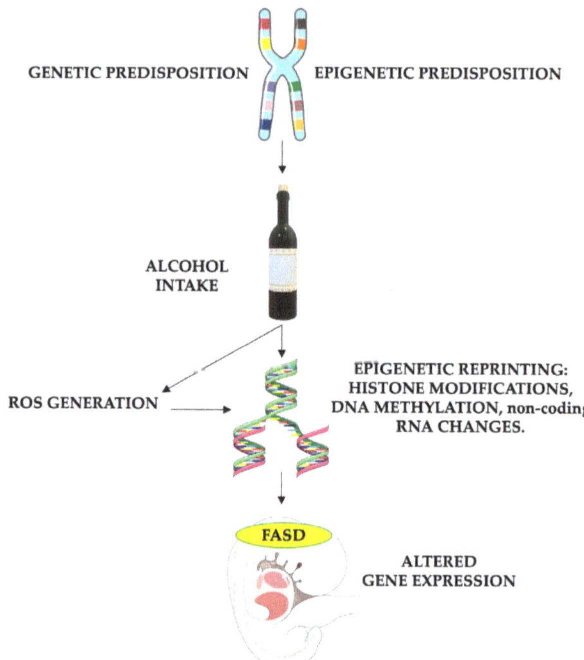

Figure 2. The role of oxidative stress in causing epigenetic modifications in FASD patients. The risk of FASD is increased in children who are genetically and environmentally predisposed. Alcohol intake in these patients leads to particularly high damage, enhancing reactive oxygen species (ROS) formation and altering DNA repair. Furthermore, both alcohol and oxidative stress have the potential to alter gene expression by impacting histone methylation and acetylation, DNA methylation processes (through the reduction of methyl groups and disruption of SAM-dependent transmethylation reactions in the folate pathway), and non-coding RNA expression. These epigenetic changes cause altered gene expression, leading to fetal abnormalities associated with FASD. FASD stands for fetal alcohol spectrum disorders, and ROS stands for reactive oxygen species. Parts of the figure were drawn using pictures from Servier Medical Art and Microsoft PowerPoint 365, Version 2112 (https://www.microsoft.com/microsoft-365, accessed on 21 March 2024). Servier Medical Art by Servier is licensed under a Creative Commons Attribution 3.0 Unported License (https://creativecommons.org/licenses/by/3.0/, accessed on 21 March 2024).

5. FASD Current Treatment

The management and treatment of FASD require a comprehensive approach due to its multifaceted nature and varied impacts on individuals and families. Randomized controlled trials emphasize the effectiveness of a combined approach involving interventions at the parental, child, and school levels [87,88]. Given the complexity of FASD and its diverse clinical manifestations, a multimodal treatment strategy is strongly advocated. This approach engages individuals with FASD, their families, and their educational institutions, integrating pharmacological, cognitive–behavioral, and psychoeducational interventions [89]. Primarily, interventions aim to enhance developmental outcomes and mitigate secondary conditions, recognizing that alcohol's harmful effects extensively affect the central nervous systems of individuals with FASD [90]. Thus, interventions must initially target primary disabilities, such as executive functioning and memory impairments.

To achieve this, it is crucial to tailor interventions to each patient's specific neurocognitive symptom profile through multidisciplinary assessment [91,92].

Pharmacological treatments are often necessary, especially for comorbid emotional and behavioral disorders. However, the effectiveness of medications such as stimulants for ADHD symptoms in individuals with FASD may vary due to the unique sensitivity of their alcohol-damaged brains [93,94]. Recent studies have identified behavioral symptom clusters in FASD, each requiring specific pharmacological approaches for management [95]. Behavioral and educational interventions play a pivotal role in addressing adaptive behaviors, learning difficulties, and emotional challenges. These evidence-based approaches encompass behavioral therapy (BT), educational therapy (ET), and interventions targeting neurocognitive functioning [96].

For instance, interventions like computerized progressive attention training (CPAT) and cognitive control therapy (CCT) aim to improve attention, memory, and self-regulation [97,98]. Additionally, mentorship programs like Wellness, Resilience, and Partnership (WRaP) assist individuals with FASD in vocational, educational, and social skill development. Social skills training, academic support, and parenting training are essential components of FASD management. Programs like Children's Friendship Training (CFT) enhance social skills, while language and literacy training (LLT) and math intervention programs address academic deficits [99–102]. Parenting training programs provide vital support for caregivers, helping them understand and cope with the challenges of raising a child with FASD, ultimately improving family dynamics and reducing caregiver stress [103,104].

6. Therapeutic Implications

Early diagnosis and intervention can help manage the symptoms and improve the quality of life for individuals affected by FASD, but a cure is not available for this disease [105,106]. Antioxidants are commonly employed to protect the fetus against ethanol teratogenicity [107,108]. Indeed, while the optimal therapeutic strategy is complete abstinence from alcohol during pregnancy, various substances have been shown to reduce the production of ROS in these patients and to lessen the frequency of severe FASD manifestations [72,109–113]. On the other hand, considering that epigenetic changes are potentially reversible through pharmaceutical interventions, there is an opportunity to develop drugs targeting specific epigenetic mechanisms involved in regulating gene expression. This could have significant clinical relevance [114].

Mitigating oxidative stress through strategies like antioxidant supplementation or lifestyle modifications may potentially modulate FASD-associated epigenetic modifications, improving clinical outcomes. In a recent study, glutathione supplementation was shown to inhibit the effects of prenatal alcohol exposure. This led to improved survival, reduced incidence of morphological defects (especially congenital heart abnormalities), and prevention of global hypomethylation of DNA in heart tissues [115]. Moreover, targeting the effects of oxidative stress on epigenetics, along with the ROS-generating pathways, may offer new avenues for therapeutic interventions in FASD [116].

However, currently, the best therapeutic approach for patients affected by FASD remains unclear. It often involves prenatal administration of antioxidants, food supplements, folic acid, choline, neuroactive peptides, and neurotrophic growth factors. Studies have shown that avoiding comorbidities and addressing the family system can significantly improve the quality of life for individuals with FASD [15,117]. Moreover, many other products with antioxidant activity have been effectively tested. Particularly, those that act on the methionine metabolic cycle have taken the spotlight in recent years [118,119] (see Table 1).

Table 1. Role of antioxidants and epigenetics in FASD treatment. FAS, fetal alcohol syndrome; FASD, fetal alcohol spectrum disorder; NA, not applicable; TLR4, toll-like receptor 4.

Population	Intervention	Outcome	Reference
FAS patients and animal models	Prenatal antioxidant administration food supplements, folic acid, choline, neuroactive peptides, neurotrophic growth factors, and lifestyle interventions.	The treatment options for FAS have recently started to be explored, although none are currently approved clinically. Furthermore, avoiding comorbidities and addressing the family system can significantly improve the quality of life.	[15]
Animal models	Responsiveness to various stimuli including perinatal care, diet, and physical activity.	Environments richer in prenatal care, or richer in stimuli, or giving the possibility of practice a specific skill (e.g., motor abilities), or providing a diet richer in antioxidants, tend to minimize the noxious effects of alcohol exposure, suggesting the plasticity of the central nervous system when there are favorable contextual factors and timely therapeutic interventions.	[72]
Cell cultures, animal models	Vitamin E, β-carotene, flavonoids, and folic acid	Antioxidants have neuroprotective effects and prevent ethanol teratogenicity.	[106]
Animal models and FASD patients	Polyphenols, carotenoids, thioredoxins, vitamin E	Although further studies are needed to better understand the relationship between oxidative stress and pediatric diseases, evidence encourages future therapeutic strategies.	[108]
Animal models	Resveratrol	Potential use as a dietary supplement to prevent damage due to oxidative stress associated with chronic alcohol abuse.	[109]
Cell cultures, animal models	Astaxanthin, ascorbic acid (Vitamin C), Vitamin E, β-carotene, (-)-epigallocatechin-3-gallate (EGCG), omega-3 fatty acids, folic acid, neurotrophic factor-9	We have many interventions effective against oxidative stress associated with FASD, but most evidence comes from animal models; more clinical trials are needed to show whether or not antioxidants may act against FASD damage.	[110]
Animal models	Astaxanthin	Protective effect on FASD, acting against oxidative stress- and TLR4 signaling-associated inflammatory reaction.	[111]
Animal models	Polyphenols	Polyphenol supplementation partially counteracts the pro-oxidant effects of alcohol.	[113]
Animal embryo models	Glutathione	Glutathione supplementation protects from heart defects and global DNA hypomethylation induced by prenatal alcohol exposure.	[115]
Cell cultures, animal models	NA	Targeting the alcohol-mediated epigenetics effects may offer new avenues for therapeutic interventions in FASD.	[116]
Animal models	Folic acid, selenium	These two antioxidants may play a major role in FASD management by acting on the methionine metabolic cycle.	[118]
Animal embryo models	Methyl donor betaine	Supplementation with the methyl donor betaine prevents congenital defects induced by prenatal alcohol exposure.	[119]

Therapies targeting specific epigenetic pathways affected by prenatal alcohol exposure may also help alleviate FASD-related impairments. Unfortunately, most evidence supporting the beneficial effects of therapeutic approaches acting on both ROS and epigenetic pathways comes from murine models, with human clinical trials still being notably scarce. Additional clinical trials are needed to determine the extent to which antioxidants contribute to mitigating FASD damage and to assess the actual impact of their epigenetic modulatory effects on the management and efficacy of treating these patients [110].

The role of oxidative stress on epigenetics in FASD underscores the complex interplay between environmental exposures, genetic predisposition, molecular mechanisms, and clinical outcomes. Further research in this area is necessary to fully comprehend the implications for the diagnosis, prognosis, and treatment of FASD.

7. Discussion

The primary objective of the study discussed in this paper was to explore the intricate relationship between oxidative stress and epigenetics in the pathogenesis of FASD and its therapeutic implications. FASD represents a spectrum of lifelong impairments resulting from prenatal exposure to alcohol, presenting significant challenges due to their diverse manifestations, ranging from physical abnormalities to cognitive and behavioral deficits [120]. Prenatal exposure to alcohol disrupts normal fetal development, leading to a myriad of health problems, including facial abnormalities, growth deficiencies, and organ malformations [121].

Additionally, cognitive impairments, such as difficulties in learning, memory, attention, and problem-solving skills, are common among individuals with FASD [20,122]. Behavioral challenges may include hyperactivity, impulsivity, and social/emotional difficulties. These issues not only affect the individuals with FASD, but also have broader implications for their families and communities, highlighting the urgent need for effective prevention and intervention strategies. The projected lifespan for individuals with FAS is approximately 34 years (with a 95% confidence range of 31 to 37 years), with external causes contributing significantly (44%) to mortality. These external causes encompass suicide (15%), accidents (14%), and substance-related fatalities involving illegal drugs or alcohol poisoning (7%), among other factors [123].

As a safe dose of alcohol use during pregnancy has not been established, it is recommended that pregnant women abstain completely from alcohol to prevent FASD. Unfortunately, identifying women at risk remains challenging, and the diagnosis tends to be overlooked or delayed, lacking adequate public acknowledgment [60,124]. This oversight in diagnosing has substantial social and economic repercussions, escalating challenges in education, employment, and social interactions and leading to increased dependency on social services and healthcare systems [125].

Fetal cellular alterations of epigenetic patterns and susceptibility to reactive oxygen species (ROS) appear to play a major role in causing fetal changes. The molecular bases of FASD involve oxidative stress, characterized by an imbalance between ROS and antioxidant defense systems induced by alcohol metabolism. This oxidative stress leads to cellular damage, particularly in the vulnerable fetal brain, resulting in disruptions in development. Moreover, oxidative stress is implicated in epigenetic changes, including alterations in DNA methylation, histone modifications, and microRNA expression, influencing gene regulation in individuals with FASD [64,69].

These epigenetic changes can influence gene regulation, contributing to the varied phenotypic outcomes observed in individuals with FASD. It has been suggested that the risk of FASD is increased in genetically predisposed progeny, particularly in cases of heightened oxidative stress [69].

Prevention should be the primary focus to reduce this preventable disease. Unfortunately, deterrence and educational campaigns appear to have failed in definitively reducing alcohol use during pregnancy [126]. The role of oxidative stress in epigenetics in FASD has significant implications for prevention and treatment.

Early diagnosis and prompt treatment significantly enhance the quality of life for FASD patients [15,127]. Current treatment options for FASD involve supportive approaches such as motivational interviewing and the community-reinforcement approach. There is potential for proactive maternal nutritional intervention, including prenatal administration of antioxidant supplements, folic acid, choline, neuroactive peptides, and neurotrophic growth factors [20,128,129]. Recent suggestions indicate that targeting specific epigenetic mechanisms involved in regulating gene expression could hold significant clinical relevance for individuals with FASD [114]. Additionally, emerging epigenetic tools might be utilized as preventive, diagnostic, and therapeutic markers.

Understanding these mechanisms presents opportunities for targeted therapeutic interventions, such as antioxidant supplementation and lifestyle modifications, to alleviate the detrimental impact of alcohol on fetal development and mitigate FASD-related impairments [115]. Further clinical trials are essential to validate the efficacy of these interventions in humans and to assess their impact on epigenetic modifications associated with FASD.

8. Conclusions

In conclusion, the study of oxidative stress and epigenetics in FASD provides valuable insights into the intricate interplay between environmental exposures, genetic predisposition, molecular mechanisms, and clinical outcomes. By unraveling these mechanisms, researchers aim to develop targeted interventions and therapeutic strategies to mitigate the impact of prenatal alcohol exposure on individuals and their families. Continued research in this field is essential for advancing our understanding of FASD and for developing effective prevention and treatment approaches to address this global health challenge.

Future approaches to FASD prevention and treatment may involve multidisciplinary strategies targeting both oxidative stress and epigenetic pathways. Therapeutic interventions aimed at modulating epigenetic changes associated with prenatal alcohol exposure hold promise for improving clinical outcomes and enhancing the quality of life for individuals affected by FASD. Additionally, efforts to raise awareness, improve diagnostic methods, and develop effective interventions are essential for addressing this significant public health concern on a global scale.

Author Contributions: Conceptualization, S.T., G.F. and M.F.; methodology, S.T., M.V., L.V., M.L., S.F., G.F. and M.F.; validation, S.T., G.F. and M.F.; formal analysis, S.T., G.F., M.V., L.V., M.L., S.F., L.T., M.C. and M.F.; investigation, M.F. and S.V.; resources, M.F. and S.V.; data curation, S.T., G.F., L.V., M.V., M.L., S.F., L.T., M.C. and M.F.; writing—original draft preparation, S.T., G.F. and M.F.; writing—review and editing, S.T., G.F., L.T., M.C., S.V. and M.F.; supervision, S.T., G.F., L.T., M.C. and M.F. All authors have read and agreed to the published version of the manuscript.

Funding: This research received no external funding.

Institutional Review Board Statement: Not applicable.

Informed Consent Statement: Not applicable.

Data Availability Statement: Not applicable.

Acknowledgments: We thank SITAC, Società Italiana per il Trattamento dell'Alcolismo e le sue Complicanze, Rome, Italy, Sapienza University of Rome, Italy and IBBC-CNR, Rome, Italy.

Conflicts of Interest: The authors declare no conflicts of interest.

References

1. Mattson, S.N.; Crocker, N.; Nguyen, T.T. Fetal alcohol spectrum disorders: Neuropsychological and behavioral features. *Neuropsychol. Rev.* **2011**, *21*, 81–101. [CrossRef] [PubMed]
2. Greenmyer, J.R.; Popova, S.; Klug, M.G.; Burd, L. Fetal alcohol spectrum disorder: A systematic review of the cost of and savings from prevention in the United States and Canada. *Addiction* **2020**, *115*, 409–417. [CrossRef] [PubMed]
3. Vorgias, D.; Bernstein, B. *Fetal Alcohol Syndrome*; StatPearls Publishing: Treasure Island, FL, USA, 2021.
4. Terracina, S.; Ferraguti, G.; Tarani, L.; Messina, M.P.; Lucarelli, M.; Vitali, M.; De Persis, S.; Greco, A.; Minni, A.; Polimeni, A.; et al. Transgenerational Abnormalities Induced by Paternal Preconceptual Alcohol Drinking. Findings from Humans and Animal Models. *Curr. Neuropharmacol.* **2021**, *19*, 1158–1173. [CrossRef] [PubMed]
5. Kalberg, W.O.; Buckley, D. FASD: What types of intervention and rehabilitation are useful? *Neurosci. Biobehav. Rev.* **2007**, *31*, 278–285. [CrossRef]
6. Center for Disease Control and Prevention. Data & Statistics: Prevalence of FASDs. 2023. Available online: https://www.cdc.gov/ncbddd/fasd/data.html (accessed on 21 March 2024).
7. Haycock, P.C. Fetal alcohol spectrum disorders: The epigenetic perspective. *Biol. Reprod.* **2009**, *81*, 607–617. [CrossRef]
8. Ciafrè, S.; Ferraguti, G.; Greco, A.; Polimeni, A.; Ralli, M.; Ceci, F.M.; Ceccanti, M.; Fiore, M. Alcohol as an early life stressor: Epigenetics, metabolic, neuroendocrine and neurobehavioral implications. *Neurosci. Biobehav. Rev.* **2020**, *118*, 654–668. [CrossRef] [PubMed]
9. Popova, S.; Lange, S.; Shield, K.; Mihic, A.; Chudley, A.E.; Mukherjee, R.A.S.; Bekmuradov, D.; Rehm, J. Comorbidity of fetal alcohol spectrum disorder: A systematic review and meta-analysis. *Lancet* **2016**, *387*, 978–987. [CrossRef] [PubMed]
10. Rasmussen, C.; Andrew, G.; Zwaigenbaum, L.; Tough, S. Neurobehavioural outcomes of children with fetal alcohol spectrum disorders: A Canadian perspective. *Paediatr. Child. Health* **2008**, *13*, 185–191. [PubMed]
11. Mead, E.A.; Sarkar, D.K. Fetal alcohol spectrum disorders and their transmission through genetic and epigenetic mechanisms. *Front. Genet.* **2014**, *5*, 154. [CrossRef]
12. Burd, L.; Blair, J.; Dropps, K. Prenatal alcohol exposure, blood alcohol concentrations and alcohol elimination rates for the mother, fetus and newborn. *J. Perinatol.* **2012**, *32*, 652–659. [CrossRef]
13. Keen, C.L.; Uriu-Adams, J.Y.; Skalny, A.; Grabeklis, A.; Grabeklis, S.; Green, K.; Yevtushok, L.; Wertelecki, W.W.; Chambers, C.D. The plausibility of maternal nutritional status being a contributing factor to the risk for fetal alcohol spectrum disorders: The potential influence of zinc status as an example. *BioFactors* **2010**, *36*, 125–135. [CrossRef]

14. King, J.C.; Fabro, S. Alcohol consumption and cigarette smoking: Effect on pregnancy. *Clin. Obstet. Gynecol.* **1983**, *26*, 437–448. [CrossRef]
15. Gupta, K.K.; Gupta, V.K.; Shirasaka, T. An Update on Fetal Alcohol Syndrome—Pathogenesis, Risks, and Treatment. *Alcohol. Clin. Exp. Res.* **2016**, *40*, 1594–1602. [CrossRef]
16. Miller, L.; Shapiro, A.M.; Wells, P.G. Embryonic catalase protects against ethanol-initiated DNA oxidation and teratogenesis in acatalasemic and transgenic human catalase-expressing mice. *Toxicol. Sci.* **2013**, *134*, 400–411. [CrossRef]
17. McCarver, D.G.; Thomasson, H.R.; Martier, S.S.; Sokol, R.J.; Li, T. Alcohol dehydrogenase-2*3 allele protects against alcohol-related birth defects among African Americans. *J. Pharmacol. Exp. Ther.* **1997**, *283*, 1095–1101.
18. Tunc-Ozcan, E.; Sittig, L.J.; Harper, K.M.; Graf, E.N.; Redei, E.E. Hypothesis: Genetic and epigenetic risk factors interact to modulate vulnerability and resilience to FASD. *Front. Genet.* **2014**, *5*, 261. [CrossRef]
19. Popova, S.; Charness, M.E.; Burd, L.; Crawford, A.; Hoyme, H.E.; Mukherjee, R.A.S.; Riley, E.P.; Elliott, E.J. Fetal alcohol spectrum disorders. *Nat. Rev. Dis. Prim.* **2023**, *9*, 11. [CrossRef]
20. Breton-Larrivée, M.; Elder, E.; Legault, L.; Langford-Avelar, A.; MacFarlane, A.J.; McGraw, S. Mitigating the detrimental developmental impact of early fetal alcohol exposure using a maternal methyl donor-enriched diet. *FASEB J.* **2023**, *37*, e22829. [CrossRef]
21. Ceccanti, M.; Hamilton, D.; Coriale, G.; Carito, V.; Aloe, L.; Chaldakov, G.; Romeo, M.; Ceccanti, M.; Iannitelli, A.; Fiore, M. Spatial learning in men undergoing alcohol detoxification. *Physiol. Behav.* **2015**, *149*, 324–330. [CrossRef]
22. Novick Brown, N.; Greenspan, S. Diminished culpability in fetal alcohol spectrum disorders (FASD). *Behav. Sci. Law.* **2022**, *40*, 1–13. [CrossRef]
23. Ismail, S.; Buckley, S.; Budacki, R.; Jabbar, A.; Gallicano, G.I. Screening, Diagnosing and Prevention of Fetal Alcohol Syndrome: Is This Syndrome Treatable. *Dev. Neurosci.* **2010**, *32*, 91–100. [CrossRef]
24. Fuso, A.; Lucarelli, M. CpG and non-CpG methylation in the diet-epigenetics-neurodegeneration connection. *Curr. Nutr. Rep.* **2019**, *8*, 74–82. [CrossRef]
25. Nicolia, V.; Lucarelli, M.; Fuso, A. Environment, epigenetics and neurodegeneration: Focus on nutrition in Alzheimer's disease. *Exp. Gerontol.* **2015**, *68*, 8–12. [CrossRef]
26. Crouch, J.; Shvedova, M.; Thanapaul, R.J.R.S.; Botchkarev, V.; Roh, D. Epigenetic Regulation of Cellular Senescence. *Cells* **2022**, *11*, 672. [CrossRef]
27. Li, Y. Modern epigenetics methods in biological research. *Methods* **2021**, *187*, 104–113. [CrossRef]
28. Fuso, A.; Ferraguti, G.; Grandoni, F.; Ruggeri, R.; Scarpa, S.; Strom, R.; Lucarelli, M. Early demethylation of non-CpG, CpC-rich, elements in the myogenin 5′-flanking region: A priming effect on the spreading of active demethylation? *Cell Cycle* **2010**, *9*, 3965–3976. [CrossRef]
29. Mitro, N.; Verdeguer, F.; Perissi, V. Editorial: Epigenetics and metabolism. *Front. Endocrinol. (Lausanne)* **2024**, *15*, 1373368. [CrossRef]
30. Smith, I.M.; Mydlarz, W.K.; Mithani, S.K.; Califano, J.A. DNA global hypomethylation in squamous cell head and neck cancer associated with smoking, alcohol consumption and stage. *Int. J. Cancer* **2007**, *121*, 1724–1728. [CrossRef]
31. Finegersh, A.; Rompala, G.R.; Martin, D.I.; Homanics, G.E. Drinking beyond a lifetime: New and emerging insights into paternal alcohol exposure on subsequent generations. *Alcohol* **2015**, *49*, 461–470. [CrossRef]
32. Raia, T.; Armeli, F.; Cavallaro, R.A.; Ferraguti, G.; Businaro, R.; Lucarelli, M.; Fuso, A. Perinatal S-Adenosylmethionine Supplementation Represses PSEN1 Expression by the Cellular Epigenetic Memory of CpG and Non-CpG Methylation in Adult TgCRD8 Mice. *Int. J. Mol. Sci.* **2023**, *24*, 11675. [CrossRef]
33. Sarkar, D.K. Male germline transmits fetal alcohol epigenetic marks for multiple generations: A review. *Addict. Biol.* **2016**, *21*, 23–34. [CrossRef]
34. Monti, N.; Cavallaro, R.A.; Stoccoro, A.; Nicolia, V.; Scarpa, S.; Kovacs, G.G.; Fiorenza, M.T.; Lucarelli, M.; Aronica, E.; Ferrer, I.; et al. CpG and non-CpG Presenilin1 methylation pattern in course of neurodevelopment and neurodegeneration is associated with gene expression in human and murine brain. *Epigenetics* **2020**, *15*, 781–799. [CrossRef] [PubMed]
35. Pierandrei, S.; Truglio, G.; Ceci, F.; Del Porto, P.; Bruno, S.M.; Castellani, S.; Conese, M.; Ascenzioni, F.; Lucarelli, M. Dna methylation patterns correlate with the expression of scnn1a, scnn1b, and scnn1g (Epithelial sodium channel, enac) genes. *Int. J. Mol. Sci.* **2021**, *22*, 3754. [CrossRef]
36. Ho, S.M.; Johnson, A.; Tarapore, P.; Janakiram, V.; Zhang, X.; Leung, Y.K. Environmental epigenetics and its implication on disease risk and health outcomes. *ILAR J.* **2012**, *53*, 289–305. [CrossRef]
37. Blaconà, G.; Raso, R.; Castellani, S.; Pierandrei, S.; Del Porto, P.; Ferraguti, G.; Ascenzioni, F.; Conese, M.; Lucarelli, M. Downregulation of epithelial sodium channel (ENaC) activity in cystic fibrosis cells by epigenetic targeting. *Cell. Mol. Life Sci.* **2022**, *79*, 257. [CrossRef] [PubMed]
38. Lomberk, G.A.; Iovanna, J.; Urrutia, R. The promise of epigenomic therapeutics in pancreatic cancer. *Epigenomics* **2016**, *8*, 831–842. [CrossRef]
39. Santaló, J.; Berdasco, M. Ethical implications of epigenetics in the era of personalized medicine. *Clin. Epigenetics* **2022**, *14*, 44. [CrossRef] [PubMed]
40. Ramsay, M. Genetic and epigenetic insights into fetal alcohol spectrum disorders. *Genome Med.* **2010**, *2*, 27. [CrossRef] [PubMed]
41. Basavarajappa, B.S. Epigenetics in fetal alcohol spectrum disorder. *Prog. Mol. Biol. Transl. Sci.* **2023**, *197*, 211–239. [CrossRef]

42. Legault, L.M.; Bertrand-Lehouillier, V.; McGraw, S. Pre-implantation alcohol exposure and developmental programming of FASD: An epigenetic perspective. *Biochem. Cell Biol.* **2018**, *96*, 117–130. [CrossRef]
43. Basavarajappa, B.S.; Subbanna, S. Epigenetic mechanisms in developmental alcohol-induced neurobehavioral deficits. *Brain Sci.* **2016**, *6*, 12. [CrossRef]
44. Kaminen-Ahola, N. Fetal alcohol spectrum disorders: Genetic and epigenetic mechanisms. *Prenat. Diagn.* **2020**, *40*, 1185–1192. [CrossRef]
45. Mahnke, A.H.; Miranda, R.C.; Homanics, G.E. Epigenetic mediators and consequences of excessive alcohol consumption. *Alcohol* **2017**, *60*, 1–6. [CrossRef]
46. Okazaki, S.; Otsuka, I.; Shinko, Y.; Horai, T.; Hirata, T.; Yamaki, N.; Sora, I.; Hishimoto, A. Epigenetic Clock Analysis in Children with Fetal Alcohol Spectrum Disorder. *Alcohol. Clin. Exp. Res.* **2021**, *45*, 329–337. [CrossRef]
47. Liyanage, V.; Curtis, K.; Zachariah, R.; Chudley, A.; Rastegar, M. Overview of the Genetic Basis and Epigenetic Mechanisms that Contribute to FASD Pathobiology. *Curr. Top. Med. Chem.* **2016**, *17*, 808–828. [CrossRef]
48. Ungerer, M.; Knezovich, J.; Ramsay, M. In utero alcohol exposure, epigenetic changes, and their consequences. *Alcohol. Res. Curr. Rev.* **2013**, *35*, 37–46.
49. Oei, J.L. Alcohol use in pregnancy and its impact on the mother and child. *Addiction* **2020**, *115*, 2148–2163. [CrossRef]
50. Ouko, L.A.; Shantikumar, K.; Knezovich, J.; Haycock, P.; Schnugh, D.J.; Ramsay, M. Effect of alcohol consumption on CpG methylation in the differentially methylated regions of H19 and IG-DMR in male gametes—Implications for fetal alcohol spectrum disorders. *Alcohol. Clin. Exp. Res.* **2009**, *33*, 1615–1627. [CrossRef]
51. Chang, R.C.; Skiles, W.M.; Chronister, S.S.; Wang, H.; Sutton, G.I.; Bedi, Y.S.; Snyder, M.; Long, C.R.; Golding, M.C. DNA methylation-independent growth restriction and altered developmental programming in a mouse model of preconception male alcohol exposure. *Epigenetics* **2017**, *12*, 841–853. [CrossRef]
52. Haycock, P.C.; Ramsay, M. Exposure of mouse embryos to ethanol during preimplantation development: Effect on DNA methylation in the h19 imprinting control region. *Biol. Reprod.* **2009**, *81*, 618–627. [CrossRef]
53. Jasiulionis, M.G. Abnormal epigenetic regulation of immune system during aging. *Front. Immunol.* **2018**, *9*, 197. [CrossRef] [PubMed]
54. Lussier, A.A.; Bodnar, T.S.; Weinberg, J. Intersection of Epigenetic and Immune Alterations: Implications for Fetal Alcohol Spectrum Disorder and Mental Health. *Front. Neurosci.* **2021**, *15*, 788630. [CrossRef] [PubMed]
55. Rompala, G.R.; Mounier, A.; Wolfe, C.M.; Lin, Q.; Lefterov, I.; Homanics, G.E. Heavy chronic intermittent ethanol exposure alters small noncoding RNAs in mouse sperm and epididymosomes. *Front. Genet.* **2018**, *9*, 32. [CrossRef] [PubMed]
56. Bedi, Y.; Chang, R.C.; Gibbs, R.; Clement, T.M.; Golding, M.C. Alterations in sperm-inherited noncoding RNAs associate with late-term fetal growth restriction induced by preconception paternal alcohol use. *Reprod. Toxicol.* **2019**, *87*, 11–20. [CrossRef] [PubMed]
57. Fuso, A.; Raia, T.; Orticello, M.; Lucarelli, M. The complex interplay between DNA methylation and miRNAs in gene expression regulation. *Biochimie* **2020**, *173*, 12–16. [CrossRef] [PubMed]
58. Roozen, S.; Ehrhart, F. Fetal alcohol spectrum disorders and the risk of crime. *Handb. Clin. Neurol.* **2023**, *197*, 197–204.
59. Wentzel, P.; Rydberg, U.; Eriksson, U.J. Antioxidative treatment diminishes ethanol-induced congenital malformations in the rat. *Alcohol. Clin. Exp. Res.* **2006**, *30*, 1752–1760. [CrossRef] [PubMed]
60. Derme, M.; Piccioni, M.G.; Brunelli, R.; Crognale, A.; Denotti, M.; Ciolli, P.; Scomparin, D.; Tarani, L.; Paparella, R.; Terrin, G.; et al. Oxidative Stress in a Mother Consuming Alcohol during Pregnancy and in Her Newborn: A Case Report. *Antioxidants* **2023**, *12*, 1216. [CrossRef] [PubMed]
61. Tossetta, G.; Fantone, S.; Piani, F.; Crescimanno, C.; Ciavattini, A.; Giannubilo, S.R.; Marzioni, D. Modulation of NRF2/KEAP1 Signaling in Preeclampsia. *Cells* **2023**, *12*, 1545. [CrossRef]
62. Auvinen, P.; Vehviläinen, J.; Marjonen, H.; Modhukur, V.; Sokka, J.; Wallén, E.; Rämö, K.; Ahola, L.; Salumets, A.; Otonkoski, T.; et al. Chromatin modifier developmental pluripotency associated factor 4 (DPPA4) is a candidate gene for alcohol-induced developmental disorders. *BMC Med.* **2022**, *20*, 495. [CrossRef]
63. Giorgi, C.; Marchi, S.; Simoes, I.C.M.; Ren, Z.; Morciano, G.; Perrone, M.; Patalas-Krawczyk, P.; Borchard, S.; Jędrak, P.; Pierzynowska, K.; et al. Mitochondria and Reactive Oxygen Species in Aging and Age-Related Diseases. *Int. Rev. Cell Mol. Biol.* **2018**, *340*, 209–344. [CrossRef] [PubMed]
64. Wells, P.G.; Bhatia, S.; Drake, D.M.; Miller-Pinsler, L. Fetal oxidative stress mechanisms of neurodevelopmental deficits and exacerbation by ethanol and methamphetamine. *Birth Defects Res. Part. C—Embryo Today Rev.* **2016**, *108*, 108–130. [CrossRef] [PubMed]
65. Wallace, D.C. A mitochondrial paradigm of metabolic and degenerative diseases, aging, and cancer: A dawn for evolutionary medicine. *Annu. Rev. Genet.* **2005**, *39*, 359–407. [CrossRef] [PubMed]
66. Weinberg, F.; Ramnath, N.; Nagrath, D. Reactive oxygen species in the tumor microenvironment: An overview. *Cancers* **2019**, *11*, 1191. [CrossRef] [PubMed]
67. Seitz, H.K.; Mueller, S. Alcohol and cancer: An overview with special emphasis on the role of acetaldehyde and cytochrome P450 2E1. *Adv. Exp. Med. Biol.* **2015**, *815*, 59–70. [CrossRef] [PubMed]
68. Matilainen, O.; Quirós, P.M.; Auwerx, J. Mitochondria and Epigenetics—Crosstalk in Homeostasis and Stress. *Trends Cell Biol.* **2017**, *27*, 453–463. [CrossRef] [PubMed]

69. Bhatia, S.; Drake, D.M.; Miller, L.; Wells, P.G. Oxidative stress and DNA damage in the mechanism of fetal alcohol spectrum disorders. *Birth Defects Res.* **2019**, *111*, 714–748. [CrossRef] [PubMed]
70. Starkman, B.G.; Sakharkar, A.J.; Pandey, S.C. Epigenetics-beyond the genome in alcoholism. *Alcohol. Res.* **2012**, *34*, 293–305.
71. Sarkar, D.K.; Gangisetty, O.; Wozniak, J.R.; Eckerle, J.K.; Georgieff, M.K.; Foroud, T.M.; Wetherill, L.; Wertelecki, W.; Chambers, C.D.; Riley, E.; et al. Persistent Changes in Stress-Regulatory Genes in Pregnant Women or Children Exposed Prenatally to Alcohol. *Alcohol. Clin. Exp. Res.* **2019**, *43*, 1887–1897. [CrossRef]
72. Romero-Márquez, J.M.; Navarro-Hortal, M.D.; Jiménez-Trigo, V.; Muñoz-Ollero, P.; Forbes-Hernández, T.Y.; Esteban-Muñoz, A.; Giampieri, F.; Noya, I.D.; Bullón, P.; Vera-Ramírez, L.; et al. An Olive-Derived Extract 20% Rich in Hydroxytyrosol Prevents β-Amyloid Aggregation and Oxidative Stress, Two Features of Alzheimer Disease, via SKN-1/NRF2 and HSP-16.2 in Caenorhabditis elegans. *Antioxidants* **2022**, *11*, 629. [CrossRef]
73. Chu, J.; Tong, M.; Monte, S.M. Chronic ethanol exposure causes mitochondrial dysfunction and oxidative stress in immature central nervous system neurons. *Acta Neuropathol.* **2007**, *113*, 659–673. [CrossRef] [PubMed]
74. Higuchi, S.; Matsushita, S.; Murayama, M.; Takagi, S.; Hayashida, M. Alcohol and aldehyde dehydrogenase polymorphisms and the risk for alcoholism. *Am. J. Psychiatry* **1995**, *152*, 1219–1221. [CrossRef] [PubMed]
75. Darbinian, N.; Darbinyan, A.; Merabova, N.; Kassem, M.; Tatevosian, G.; Amini, S.; Goetzl, L.; Selzer, M.E. In utero ethanol exposure induces mitochondrial DNA damage and inhibits mtDNA repair in developing brain. *Front. Neurosci.* **2023**, *17*, 958. [CrossRef] [PubMed]
76. Grzeszczak, K.; Łanocha-Arendarczyk, N.; Malinowski, W.; Ziętek, P.; Kosik-Bogacka, D. Oxidative Stress in Pregnancy. *Biomolecules* **2023**, *13*, 1768. [CrossRef] [PubMed]
77. Singh, S.; Verma, S.K.; Kumar, S.; Ahmad, M.K.; Nischal, A.; Singh, S.K.; Dixit, R.K. Evaluation of Oxidative Stress and Antioxidant Status in Chronic Obstructive Pulmonary Disease. *Scand. J. Immunol.* **2017**, *85*, 130–137. [CrossRef] [PubMed]
78. Gutala, R.; Wang, J.; Kadapakkam, S.; Hwang, Y.; Ticku, M.; Li, M.D. Microarray analysis of ethanol-treated cortical neurons reveals disruption of genes related to the ubiquitin-proteasome pathway and protein synthesis. *Alcohol. Clin. Exp. Res.* **2004**, *28*, 1779–1788. [CrossRef] [PubMed]
79. Carabulea, A.L.; Janeski, J.D.; Naik, V.D.; Chen, K.; Mor, G.; Ramadoss, J. A multi-organ analysis of the role of mTOR in fetal alcohol spectrum disorders. *FASEB J.* **2023**, *37*, e22897. [CrossRef] [PubMed]
80. Sokoloff, L. Energetics of functional activation in neural tissues. *Neurochem. Res.* **1999**, *24*, 321–329. [CrossRef]
81. Gerlach, M.; Ben-Shachar, D.; Riederer, P.; Youdim, M.B.H. Altered brain metabolism of iron as a cause of neurodegenerative diseases? *J. Neurochem.* **1994**, *63*, 793–807. [CrossRef]
82. Floyd, R.A.; Carney, J.M. Free radical damage to protein and DNA: Mechanisms involved and relevant observations on brain undergoing oxidative stress. *Ann. Neurol.* **1992**, *32*, S22–S27. [CrossRef]
83. Bergamini, C.; Gambetti, S.; Dondi, A.; Cervellati, C. Oxygen, Reactive Oxygen Species and Tissue Damage. *Curr. Pharm. Des.* **2005**, *10*, 1611–1626. [CrossRef] [PubMed]
84. Bakoyiannis, I.; Gkioka, E.; Pergialiotis, V.; Mastroleon, I.; Prodromidou, A.; Vlachos, G.D.; Perrea, D. Fetal alcohol spectrum disorders and cognitive functions of young children. *Rev. Neurosci.* **2014**, *25*, 631–639. [CrossRef] [PubMed]
85. Rompala, G.R.; Finegersh, A.; Homanics, G.E. Paternal preconception ethanol exposure blunts hypothalamic-pituitary-adrenal axis responsivity and stress-induced excessive fluid intake in male mice. *Alcohol* **2016**, *53*, 19–25. [CrossRef] [PubMed]
86. Pearson, J.; Tarabulsy, G.M.; Bussières, E.-L. Foetal programming and cortisol secretion in early childhood: A meta-analysis of different programming variables. *Infant. Behav. Dev.* **2015**, *40*, 204–215. [CrossRef] [PubMed]
87. Racine, E.; Bell, E.; Di Pietro, N.C.; Wade, L.; Illes, J. Evidence-Based Neuroethics for Neurodevelopmental Disorders. *Semin. Pediatr. Neurol.* **2011**, *18*, 21–25. [CrossRef] [PubMed]
88. Olson, H.C.; Oti, R.; Gelo, J.; Beck, S. "Family matters:" Fetal alcohol spectrum disorders and the family. *Dev. Disabil. Res. Rev.* **2009**, *15*, 235–249. [CrossRef] [PubMed]
89. Roozen, S.; Black, D.; Peters, G.J.Y.; Kok, G.; Townend, D.; Nijhuis, J.G.; Koek, G.H.; Curfs, L.M.G. Fetal Alcohol Spectrum Disorders (FASD): An Approach to Effective Prevention. *Curr. Dev. Disord. Rep.* **2016**, *3*, 229–234. [CrossRef] [PubMed]
90. Pei, J.; Flannigan, K. Interventions for Fetal Alcohol Spectrum Disorder: Meeting Needs Across the Lifespan. *Int. J. Neurorehabilitation* **2016**, *3*, 1–9. [CrossRef]
91. Ordenewitz, L.K.; Weinmann, T.; Schlüter, J.A.; Moder, J.E.; Jung, J.; Kerber, K.; Greif-Kohistani, N.; Heinen, F.; Landgraf, M.N. Evidence-based interventions for children and adolescents with fetal alcohol spectrum disorders—A systematic review. *Eur. J. Paediatr. Neurol.* **2021**, *33*, 50–60. [CrossRef]
92. Christine Christensen Løhaugen, G.; Cecilie Tveiten, A.; Skranes, J. Interventions for children and adolescents with Fetal Alcohol Spectrum Disorders (FASD). In *Handbook of Substance Misuse and Addictions*; Patel, V.B., Preedy, V.R., Eds.; Springer International Publishing: Cham, Switzerland, 2022; pp. 1–28, ISBN 978-3-030-67928-6.
93. Nazarova, V.A.; Sokolov, A.V.; Chubarev, V.N.; Tarasov, V.V.; Schiöth, H.B. Treatment of ADHD: Drugs, psychological therapies, devices, complementary and alternative methods as well as the trends in clinical trials. *Front. Pharmacol.* **2022**, *13*, 1066988. [CrossRef]
94. Ritfeld, G.J.; Kable, J.A.; Holton, J.E.; Coles, C.D. Psychopharmacological Treatments in Children with Fetal Alcohol Spectrum Disorders: A Review. *Child. Psychiatry Hum. Dev.* **2022**, *53*, 268–277. [CrossRef] [PubMed]

95. Mela, M.; Okpalauwaekwe, U.; Anderson, T.; Eng, J.; Nomani, S.; Ahmed, A.; Barr, A.M. The utility of psychotropic drugs on patients with Fetal Alcohol Spectrum Disorder (FASD): A systematic review. *Psychiatry Clin. Psychopharmacol.* **2018**, *28*, 436–445. [CrossRef]
96. Paley, B.; O'Connor, M.J. Behavioral interventions for children and adolescents with fetal alcohol spectrum disorders. *Alcohol. Res. Health* **2011**, *34*, 64–75.
97. Riley, E.P.; McGee, C.L. Fetal alcohol spectrum disorders: An overview with emphasis on changes in brain and behavior. *Exp. Biol. Med.* **2005**, *230*, 357–365. [CrossRef] [PubMed]
98. Kerns, K.A.; MacSween, J.; Vander Wekken, S.; Gruppuso, V. Investigating the efficacy of an attention training programme in children with foetal alcohol spectrum disorder. *Dev. Neurorehabil.* **2010**, *13*, 413–422. [CrossRef] [PubMed]
99. Kable, J.A.; Taddeo, E.; Strickland, D.; Coles, C.D. Community translation of the Math Interactive Learning Experience Program for children with FASD. *Res. Dev. Disabil.* **2015**, *39*, 1–11. [CrossRef] [PubMed]
100. Adnams, C.M.; Sorour, P.; Kalberg, W.O.; Kodituwakku, P.; Perold, M.D.; Kotze, A.; September, S.; Castle, B.; Gossage, J.; May, P.A. Language and literacy outcomes from a pilot intervention study for children with fetal alcohol spectrum disorders in South Africa. *Alcohol* **2007**, *41*, 403–414. [CrossRef]
101. Frankel, F.; Myatt, R. *Children's Friendship Training*; Taylor & Francis Ltd.: London, UK, 2013; pp. 1–198. [CrossRef]
102. Laugeson, E.A.; Paley, B.; Schonfeld, A.M.; Carpenter, E.M.; Frankel, F.; O'Connor, M.J. Adaptation of the children's friendship training program for children with Fetal Alcohol Spectrum Disorders. *Child. Fam. Behav. Ther.* **2007**, *29*, 57–69. [CrossRef]
103. Leenaars, L.S.; Denys, K.; Henneveld, D.; Rasmussen, C. The impact of fetal alcohol spectrum disorders on families: Evaluation of a family intervention program. *Community Ment. Health J.* **2012**, *48*, 431–435. [CrossRef]
104. Kable, J.A.; Coles, C.D.; Taddeo, E. Socio-cognitive habilitation using the math interactive learning experience program for alcohol-affected children. *Alcohol. Clin. Exp. Res.* **2007**, *31*, 1425–1434. [CrossRef]
105. Sambo, D.; Goldman, D. Genetic influences on Fetal Alcohol Spectrum Disorder. *Genes* **2023**, *14*, 195. [CrossRef] [PubMed]
106. Freeman, J.; Condon, C.; Hamilton, S.; Mutch, R.C.; Bower, C.; Watkins, R.E. Challenges in Accurately Assessing Prenatal Alcohol Exposure in a Study of Fetal Alcohol Spectrum Disorder in a Youth Detention Center. *Alcohol. Clin. Exp. Res.* **2019**, *43*, 309–316. [CrossRef] [PubMed]
107. Cohen-Kerem, R.; Koren, G. Antioxidants and fetal protection against ethanol teratogenicity. I. Review of the experimental data and implications to humans. *Neurotoxicol. Teratol.* **2003**, *25*, 1–9. [CrossRef] [PubMed]
108. Micangeli, G.; Menghi, M.; Profeta, G.; Tarani, F.; Mariani, A.; Petrella, C.; Barbato, C.; Ferraguti, G.; Ceccanti, M.; Tarani, L.; et al. The Impact of Oxidative Stress on Pediatrics Syndromes. *Antioxidants* **2022**, *11*, 1983. [CrossRef]
109. Petrella, C.; Carito, V.; Carere, C.; Ferraguti, G.; Ciafrè, S.; Natella, F.; Bello, C.; Greco, A.; Ralli, M.; Mancinelli, R.; et al. Oxidative stress inhibition by resveratrol in alcohol-dependent mice. *Nutrition* **2020**, *79–80*, 110783. [CrossRef] [PubMed]
110. Zhang, Y.; Wang, H.; Li, Y.; Peng, Y. A review of interventions against fetal alcohol spectrum disorder targeting oxidative stress. *Int. J. Dev. Neurosci.* **2018**, *71*, 140–145. [CrossRef] [PubMed]
111. Zheng, D.; Li, Y.; He, L.; Tang, Y.; Li, X.; Shen, Q.; Yin, D.; Peng, Y. The protective effect of astaxanthin on fetal alcohol spectrum disorder in mice. *Neuropharmacology* **2014**, *84*, 13–18. [CrossRef] [PubMed]
112. Davis, K.; Desrocher, M.; Moore, T. Fetal Alcohol Spectrum Disorder: A Review of Neurodevelopmental Findings and Interventions. *J. Dev. Phys. Disabil.* **2011**, *23*, 143–167. [CrossRef]
113. Carito, V.; Ceccanti, M.; Cestari, V.; Natella, F.; Bello, C.; Coccurello, R.; Mancinelli, R.; Fiore, M. Olive polyphenol effects in a mouse model of chronic ethanol addiction. *Nutrition* **2017**, *33*, 65–69. [CrossRef]
114. Ferraguti, G.; Terracina, S.; Micangeli, G.; Lucarelli, M.; Tarani, L.; Ceccanti, M.; Spaziani, M.; D'Orazi, V.; Petrella, C.; Fiore, M. NGF and BDNF in pediatrics syndromes. *Neurosci. Biobehav. Rev.* **2023**, *145*, 105015. [CrossRef]
115. Jawaid, S.; Strainic, J.P.; Kim, J.; Ford, M.R.; Thrane, L.; Karunamuni, G.H.; Sheehan, M.M.; Chowdhury, A.; Gillespie, C.A.; Rollins, A.M.; et al. Glutathione Protects the Developing Heart from Defects and Global DNA Hypomethylation Induced by Prenatal Alcohol Exposure. *Alcohol. Clin. Exp. Res.* **2021**, *45*, 69–78. [CrossRef] [PubMed]
116. Ciafrè, S.; Carito, V.; Ferraguti, G.; Greco, A.; Chaldakov, G.N.; Fiore, M.; Ceccanti, M. How alcohol drinking affects our genes: An epigenetic point of view. *Biochem. Cell Biol.* **2019**, *97*, 345–356. [CrossRef] [PubMed]
117. Denny, L.A.; Coles, S.; Blitz, R. Fetal Alcohol Syndrome and Fetal Alcohol Spectrum Disorders. *Am. Fam. Physician* **2017**, *96*, 515–522. [CrossRef] [PubMed]
118. Ojeda, L.; Nogales, F.; Murillo, L.; Carreras, O. The role of folic acid and selenium against oxidative damage from ethanol in early life programming: A review. *Biochem. Cell Biol.* **2018**, *96*, 178–188. [CrossRef] [PubMed]
119. Karunamuni, G.; Sheehan, M.M.; Doughman, Y.Q.; Gu, S.; Sun, J.; Li, Y.; Strainic, J.P.; Rollins, A.M.; Jenkins, M.W.; Watanabe, M. Supplementation with the Methyl Donor Betaine Prevents Congenital Defects Induced by Prenatal Alcohol Exposure. *Alcohol. Clin. Exp. Res.* **2017**, *41*, 1917–1927. [CrossRef] [PubMed]
120. Hoyme, H.E.; Kalberg, W.O.; Elliott, A.J.; Blankenship, J.; Buckley, D.; Marais, A.S.; Manning, M.A.; Robinson, L.K.; Adam, M.P.; Abdul-Rahman, O.; et al. Updated clinical guidelines for diagnosing fetal alcohol spectrum disorders. *Pediatrics* **2016**, *138*, e20154256. [CrossRef]
121. Fainsod, A.; Bendelac-Kapon, L.; Shabtai, Y. Fetal Alcohol Spectrum Disorder: Embryogenesis Under Reduced Retinoic Acid Signaling Conditions. *Subcell. Biochem.* **2020**, *95*, 197–225. [CrossRef] [PubMed]

122. Ceci, F.M.; Ferraguti, G.; Petrella, C.; Greco, A.; Ralli, M.; Iannitelli, A.; Carito, V.; Tirassa, P.; Chaldakov, G.N.; Messina, M.P.; et al. Nerve Growth Factor in Alcohol Use Disorders. *Curr. Neuropharmacol.* **2020**, *19*, 45–60. [CrossRef] [PubMed]
123. Thanh, N.X.; Jonsson, E. Life expectancy of people with fetal alcohol syndrome. *J. Popul. Ther. Clin. Pharmacol.* **2016**, *23*, e53–e59.
124. May, P.A.; Gossage, J.P. Maternal risk factors for fetal alcohol spectrum disorders: Not as simple as it might seem. *Alcohol Res. Health* **2011**, *34*, 15–26.
125. Thanh, N.X.; Jonsson, E.; Salmon, A.; Sebastianski, M. Incidence and prevalence of fetal alcohol spectrum disorder by sex and age group in Alberta, Canada. *J. Popul. Ther. Clin. Pharmacol.* **2014**, *21*, e395–e404. [PubMed]
126. Popova, S.; Lange, S.; Probst, C.; Gmel, G.; Rehm, J. Estimation of national, regional, and global prevalence of alcohol use during pregnancy and fetal alcohol syndrome: A systematic review and meta-analysis. *Lancet Glob. Health* **2017**, *5*, e290–e299. [CrossRef] [PubMed]
127. Ferraguti, G.; Ciolli, P.; Carito, V.; Battagliese, G.; Mancinelli, R.; Ciafrè, S.; Tirassa, P.; Ciccarelli, R.; Cipriani, A.; Messina, M.P.; et al. Ethylglucuronide in the urine as a marker of alcohol consumption during pregnancy: Comparison with four alcohol screening questionnaires. *Toxicol. Lett.* **2017**, *275*, 49–56. [CrossRef] [PubMed]
128. Steane, S.E.; Cuffe, J.S.M.; Moritz, K.M. The role of maternal choline, folate and one-carbon metabolism in mediating the impact of prenatal alcohol exposure on placental and fetal development. *J. Physiol.* **2023**, *601*, 1061–1075. [CrossRef] [PubMed]
129. May, P.A.; Marais, A.-S.; Kalberg, W.O.; de Vries, M.M.; Buckley, D.; Hasken, J.M.; Snell, C.L.; Barnard Röhrs, R.; Hedrick, D.M.; Bezuidenhout, H.; et al. Multifaceted case management during pregnancy is associated with better child outcomes and less fetal alcohol syndrome. *Ann. Med.* **2023**, *55*, 926–945. [CrossRef]

antioxidants

MDPI

Article

Nucleoredoxin Redox Interactions Are Sensitized by Aging and Potentiated by Chronic Alcohol Consumption in the Mouse Liver

Osiris Germán Idelfonso-García [1,2,*], Brisa Rodope Alarcón-Sánchez [1,3], Dafne Guerrero-Escalera [1], Norma Arely López-Hernández [1], José Luis Pérez-Hernández [4], Ruth Pacheco-Rivera [5], Jesús Serrano-Luna [3], Osbaldo Resendis-Antonio [6,7], Erick Andrés Muciño-Olmos [8], Diana Ivette Aparicio-Bautista [9], Gustavo Basurto-Islas [10], Rafael Baltiérrez-Hoyos [11,12], Verónica Rocío Vásquez-Garzón [11,12], Saúl Villa-Treviño [3], Pablo Muriel [13], Héctor Serrano [2], Julio Isael Pérez-Carreón [1] and Jaime Arellanes-Robledo [1,12,*]

1 Laboratory of Liver Diseases, National Institute of Genomic Medicine—INMEGEN, Mexico City 14610, Mexico; brisa.alarcon@cinvestav.mx (B.R.A.-S.); dguerrero@inmegen.gob.mx (D.G.-E.); nalopez@inmegen.gob.mx (N.A.L.-H.); jiperez@inmegen.gob.mx (J.I.P.-C.)
2 Department of Health Sciences, Div CBS, Metropolitan Autonomous University-Iztapalapa Campus, Mexico City 09340, Mexico; hser@xanum.uam.mx
3 Department of Cell Biology, Center for Research and Advanced Studies of the National Polytechnic Institute—CINVESTAV-IPN, Mexico City 07360, Mexico; jesus.serrano@cinvestav.mx (J.S.-L.); svilla@cell.cinvestav.mx (S.V.-T.)
4 Department of Gastroenterology and Hepatology, Hospital General de México "Dr. Eduardo Liceaga", Mexico City 06720, Mexico; jose.luis.perezhe@pemex.com
5 Laboratory of Molecular Diagnostics, Department of Biochemistry, National School of Biological Sciences of the National Polytechnic Institute, Mexico City 07738, Mexico; rpachecor@ipn.mx
6 Laboratory of Human Systems Biology, National Institute of Genomic Medicine—INMEGEN, Mexico City 14610, Mexico; oresendis@inmegen.gob.mx
7 Coordination of Scientific Research—CIC, Research Support Network—RAI, Center for Complexity Sciences—C3, National Autonomous University of Mexico—UNAM, Mexico City 04510, Mexico
8 Division of Translational Cancer Research, Department of Laboratory Medicine, Lund University, 22381 Lund, Sweden; erick.mucino_olmos@med.lu.se
9 Laboratory of Genomics of Bone Metabolism, National Institute of Genomic Medicine—INMEGEN, Mexico City 14610, Mexico; daparicio@inmegen.gob.mx
10 Department of Science and Engineering, University of Guanajuato, Leon 37670, Guanajuato, Mexico; gustavo.basurto@ugto.mx
11 Laboratory of Fibrosis and Cancer, Faculty of Medicine and Surgery, 'Benito Juárez' Autonomous University of Oaxaca—UABJO, Oaxaca 68120, Mexico; rbaltierrezho@conacyt.mx (R.B.-H.); vrvasquezga@conacyt.mx (V.R.V.-G.)
12 Deputy Directorate of Humanistic and Scientific Research, National Council of Humanities, Sciences and Technologies—CONAHCYT, Mexico City 03940, Mexico
13 Laboratory of Experimental Hepatology, Department of Pharmacology, Center for Research and Advanced Studies of the National Polytechnic Institute—CINVESTAV-IPN, Mexico City 07360, Mexico; pmuriel@cinvestav.mx
* Correspondence: oidelfonso@inmegen.edu.mx (O.G.I.-G.); jarellanes@inmegen.gob.mx (J.A.-R.); Tel.: +52-55-5350-1900 (ext. 1218) (J.A.-R.)

check for updates

Citation: Idelfonso-García, O.G.; Alarcón-Sánchez, B.R.; Guerrero-Escalera, D.; López-Hernández, N.A.; Pérez-Hernández, J.L.; Pacheco-Rivera, R.; Serrano-Luna, J.; Resendis-Antonio, O.; Muciño-Olmos, E.A.; Aparicio-Bautista, D.I.; et al. Nucleoredoxin Redox Interactions Are Sensitized by Aging and Potentiated by Chronic Alcohol Consumption in the Mouse Liver. *Antioxidants* **2024**, *13*, 257. https://doi.org/10.3390/antiox13030257

Academic Editor: Marco Fiore

Received: 29 January 2024
Revised: 10 February 2024
Accepted: 14 February 2024
Published: 20 February 2024

Abstract: Aging is characterized by increased reactive species, leading to redox imbalance, oxidative damage, and senescence. The adverse effects of alcohol consumption potentiate aging-associated alterations, promoting several diseases, including liver diseases. Nucleoredoxin (NXN) is a redox-sensitive enzyme that targets reactive oxygen species and regulates key cellular processes through redox protein–protein interactions. Here, we determine the effect of chronic alcohol consumption on NXN-dependent redox interactions in the liver of aged mice. We found that chronic alcohol consumption preferentially promotes the localization of NXN either into or alongside senescent cells, declines its interacting capability, and worsens the altered interaction ratio of NXN with FLII, MYD88, CAMK2A, and PFK1 proteins induced by aging. In addition, carbonylated protein and cell proliferation increased, and the ratios of collagen I and collagen III were inverted. Thus, we demonstrate an emerging phenomenon associated with altered redox homeostasis during aging, as shown by the declining capability of NXN to interact with partner proteins, which is enhanced by chronic alcohol consumption in the mouse liver. This evidence opens an attractive window to elucidate the consequences of both aging and chronic alcohol consumption on the downstream signaling pathways regulated by NXN-dependent redox-sensitive interactions.

Keywords: alcoholic liver disease; cellular senescence; oxidative stress; protein carbonylation

1. Introduction

Aging is a natural process that causes progressive loss of tissue and organ function over time and is closely associated with systemic inflammation, among many other diseases [1]. Alcohol abuse is rapidly increasing among adults over 65 years old, a population more sensitive to the adverse effects of alcohol because oxidative stress and DNA damage resulting from alcohol misuse exacerbate age-associated diseases, such as liver diseases, and increase the risk of cancer [2].

Several well-known altered molecular mechanisms that trigger human aging, such as increased reactive oxygen species (ROS) production, inflammation, immune system decrease, and DNA damage response, among others, have been identified [3]. As a result, various tissues may enter into a permanent cell cycle arrest where both endogenous and exogenous stress response mechanisms contribute to this end throughout life [4]. Most cells cease to proliferate but remain metabolically active for a long period, called senescent cells [5]. It has been well established that the increased number of senescent cells in the liver strongly impacts its functioning and regeneration [6]. An increased rate of senescent cells decreases liver function during alcohol-related liver disease progression [4]. In addition, senescent cells secrete a large number of proinflammatory cytokines, chemokines, growth factors and proteases that may affect neighboring cells, a phenomenon known as the senescence-associated secretory phenotype (SASP) [7]. This phenomenon potentiates ROS production and oxidative DNA damage as ROS might attack nitrogenous bases, which are associated with premature aging when they are not completely repaired, and eventually promote malignancy [8].

The liver is a complex metabolic organ essential for maintaining whole-body homeostasis via regulation of energy metabolism, xenobiotic and endobiotic clearance, and molecular biosynthesis [9]. Thus, chronic liver insults might severely influence organ functioning, which increases with age [10]. Alcohol-related liver disease (ALD), a major cause of morbidity and mortality worldwide, encompasses a spectrum of altered hepatic manifestations ranging from simple steatosis to steatohepatitis, cirrhosis, and eventually hepatocellular carcinoma (HCC) [11]. One of the key players in ALD progression is oxidative stress since chronic alcohol consumption increases ROS, which induces lipid peroxidation and protein carbonylation, decreases the hepatic antioxidant defense, leading to hepatocellular damage [12], and promotes biomolecule modification, such as protein carbonylation [13]. Interestingly, although the activity of cytochrome P450-2E1, an enzyme involved in ethanol metabolism, decreases with age [14], it is implicated in promoting aging-related hepatic disease by increasing protein carbonylation and oxidative DNA damage, among other oxidation-associated damages [15].

Nucleoredoxin (NXN), a redox-sensitive and multifunctional enzyme, acts as an oxidoreductase by targeting ROS [16] and is necessary for the integrity of antioxidant systems [17]. It regulates different key cellular processes through redox-sensitive protein–protein interactions, including innate immunity, inflammation, proliferation, glycolysis, and neuronal plasticity, among others [18], which largely depend on the oxidative stress status of the cells [17]. Notably, some NXN interactions and their downstream signaling pathways are disrupted by ROS producers, such as H_2O_2 [19], acetaldehyde [20], and ethanol [21,22]. Moreover, NXN regulates signaling pathways that have been linked to the establishment of cellular senescence, such as the Wingless (WNT)/β-catenin and Toll-like receptor 4 (TLR4)/myeloid differentiation primary response 88 (MYD88) pathways [23,24].

Of interest, the impairment of redox homeostasis in aging may cause covalent modifications to deacetylases, which modulate signaling pathways such as WNT/β-catenin [25]. Therefore, an unexplored area is whether NXN-dependent redox-sensitive interactions, which might be targeted by oxidative stress induced by both aging and chronic alcohol consumption, are sensitized by aging and enhanced by chronic alcohol consumption in the mouse liver. Thus, using an ALD in vivo model [26], in this investigation we determine the effect of chronic alcohol consumption on NXN-dependent redox interactions in the liver of aged mice.

2. Materials and Methods

2.1. Experimental Design

Female C57BL/6J mice were obtained from the Animal Production and Experimentation Unit of the Center for Research and Advanced Studies of the National Polytechnic Institute (UPEAL-CINVESTAV-IPN; Mexico City, Mexico) and were fed ad libitum and housed in a controlled environment with a 12 h light/dark cycle. Animal experimentation was following the Institutional Animal Use and Care Committee of CINVESTAV-IPN with the approved protocol No. 0114-14. To study ALD, animals were subjected to a protocol of chronic ethanol consumption, as previously reported [26]. Three groups (n = 5 per group) of animals aged 7 weeks, and 12 and 18 months, were first subjected to a six-day ramp-up period divided into two days per ethanol dosage, i.e., 5% (v/v) was increased per dose until reaching 20% (v/v) ethanol in 20% (w/v) sucrose. Then, animals were kept at 20% ethanol ad libitum for 8 weeks as the only source of drinking fluid (Figure 1). In the last week of ethanol consumption, animals received a single dose (1 mg/kg, i.p.) of lipopolysaccharide (LPS). The LPS dose was selected and used based on previous studies in which the acute effects of LPS in ALD models were evaluated [21,22]. Simultaneously, three groups of matched control animals received tap water and a standard diet ad libitum. Finally, animals were euthanized by exsanguination under isoflurane anesthesia, and immediately, liver pieces were divided for protein isolation, and frozen in liquid nitrogen for cryopreservation, and then stored at −75 °C for further analysis. Other pieces were fixed in 4% formalin and embedded in paraffin for histological or immunostaining analyses. The animal body weights were recorded at the beginning and at the end of alcohol consumption, while the percentage of relative liver weight was recorded immediately after euthanasia (Table 1).

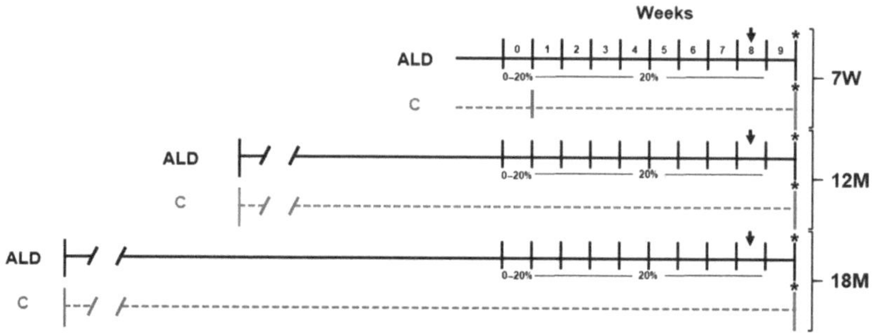

Figure 1. Schematic representation of ALD model in young and aged mice. Animal groups represented by continuous black lines received 20% (v/v) ethanol for 8 weeks, after which they received a single dose of LPS (1 mg/kg, i.p.) in the last week, which is indicated by a black arrow. Animal groups represented by dashed gray lines did not receive either ethanol or LPS. All animals were euthanized one week after the end of alcohol consumption, indicated by an asterisk (*). n = 5 animals/group. ALD, alcohol liver disease model; C, control; 7W, 7-week-old mice; 12M, 12-month-old mice; 18M, 18-month-old mice.

2.2. Histopathology and Immunodetection Analyses

Liver tissues fixed in 4% formalin were embedded in paraffin, cut into 3-μm-thick slices, deparaffinized and gradually rehydrated. For histological examination and fibrosis detection, sections were processed by standard hematoxylin and eosin (H&E) and Masson's trichrome (M'sT) staining, respectively.

For immunohistochemistry (IHC) and immunofluorescence (IF) analyses; after tissue rehydration, antigens were unmasked and processed according to the Mouse on Mouse HRP-Polymer Bundle kit (MM510; BioCare Medical. Concord, CA, USA). Next, for IHC analysis, primary antibodies against collagen 3 alpha 1 (COL3A1; 1:200; NB600-594SS, Novus Biologicals; Centennial, CO, USA), interleukin 6 (IL-6; 1:100; 21865-1-AP; Proteintech; Rosemont, IL, USA), KI67 and collagen 1 alpha 1 (KI67; 1:200 and COL1A1; 1:200; GTX16667 and GTX112731, respectively; GeneTex; Irvine, CA, USA), were incubated overnight at 4 °C in the blocking solution provided in the kit (MM510; BioCare Medical). After a standard staining protocol using the HRP/DAB Mouse/Rabbit ImmunoDetector Detection System (BSB 0005, BioSB; Santa Barbara, CA, USA), sections were lightly counterstained with hematoxylin, dehydrated, and mounted. The positive label in liver tissue was quantified in 20 randomly selected 20× fields per individual sample, totaling 100 images per treatment group in a total of 600 images using ImageJ 1.5 software. For IF analyses, anti-proliferating cell nuclear antigen (PCNA; 10205-2-AP; Proteintech; Rosemont, IL, USA) and anti-poly (ADP-ribose) (PAR; 1:200; ALX804220R100, Enzo. Farmingdale, NY, USA), were incubated overnight at 4 °C. Tissues were then incubated for 1 h at room temperature (RT) in the dark with Alexa Fluor 488 anti-rabbit and anti-mouse antibodies (1:300; ab150077 and ab150117, respectively; Abcam; Cambridge, MA, USA). Nuclei were stained with DAPI (1:1000; 62247, Thermo Scientific; Rockford, IL, USA) for 5 min at RT. Finally, tissues were rinsed, drained, and fixed using Fluoroshield Mounting Medium (ab104135; Abcam; Cambridge, MA, USA). For quantifying the area and positive nuclei, 20 randomly selected 20× fields per individual sample were captured, totaling 100 images per treatment group, and 600 total images. The label was quantified using ImageJ 1.5 software. Images for IHC and IF analysis were captured using a ZEISS Axio-A1 microscope (Oberkochen, Germany).

Double-labeling for detecting NXN and phospho-H2A.X (p-H2A.X) was performed using the IF protocol on frozen tissue. Frozen 7-μm-thick liver sections were rinsed twice in ice-cold PBS, fixed with 3.7% formaldehyde in PBS (pH 7.4) for 10 min at RT, washed, and incubated for 50 min in washing solution [0.25% Triton X-100 in PBS (PBST)]. Sections were blocked with 1% bovine serum albumin (BSA) in PBST for 30 min at RT. Then, anti-p-H2A.X (Ser139) (1:50, 2577S; Cell Signaling; Danvers, MA, USA) antibody was incubated overnight at 4 °C in blocking solution. After washing, sections were incubated with Alexa Fluor 488 goat anti-rabbit (1:400, ab150077; Abcam; Cambridge, MA, USA) diluted in blocking solution for 1 h at RT in the dark. Then, anti-NXN (1:25, GTX107039; GeneTex; Irvine, CA, USA) antibody was incubated for 4 h at 4 °C in a blocking solution. After washing, sections were incubated with Alexa Fluor 568 goat anti-rabbit (1:1000, A-11079; Thermo Scientific; Rockford, IL, USA) diluted in blocking solution for 1 h at RT in the dark. Nuclear DNA was stained with DAPI (1:1000, 62247; Thermo Scientific; Rockford, IL, USA) for 5 min at RT. Finally, tissues were rinsed, drained, and fixed using Fluoroshield Mounting Medium (ab104137; Abcam; Cambridge, MA, USA). Pictures were captured using a Leica TCS SP8 AOBS (Acousto-optical Beam Splitter) DMI6000 confocal microscope (Wetzlar, Germany), provided by the Confocal Microscopy Unit at the Cell Biology Department of CINVESTAV-IPN (CONAHCYT–Mexico, Mexico City).

2.3. Senescence-Associated β-Galactosidase Activity In Situ (SA-β-gal)

SA-β-gal detection was performed, as previously described [27]. Briefly, frozen 10-μm thick liver sections were mounted on glass slides, fixed with fixing solution (1 mM $MgCl_2$, 0.1 M EGTA, and 0.2% glutaraldehyde) for 5 min, washed with deionized water and incubated at 37 °C for 24 h with β-gal staining solution (0.4 M citric acid/sodium phosphate buffer, pH 6), 5 mM potassium ferrocyanide, 5 mM potassium ferricyanide, 1 M $MgCl_2$, and 1 mg/mL X-gal). Sections were then counterstained with hematoxylin and dehydrated, and SA-β-gal-positive cells were captured accounting for 20 randomly selected either 40 or 100× fields per individual sample, i.e., 100 images per treatment group, and 600 total images. Images were captured using a ZEISS Axio-A1 microscope (Oberkochen, Germany).

To detect SA-β-gal activity alongside NXN protein in the same tissue section, frozen sections were treated for β-gal activity detection as previously described [27]. Briefly, after incubation with β-gal staining solution, sections were blocked with 1% BSA in PBS for 2 h and incubated for 1 h at 4 °C with anti-NXN (1:150; GTX107039, GeneTex; Irvine, CA, USA) antibody. Endogenous peroxidase activity was blocked for 12 min at RT; then, ImmunoDetector Biotinylated Link was added for 10 min, and then secondary InmunoDetector HRP-label conjugated antibodies and the DAB-Plus Substrate Kit were used for signal detection (BSB 0005, BioSB; Santa Barbara, CA, USA). Sections were then counterstained with hematoxylin and dehydrated, and SA-β-gal-positive cells were captured from 10 randomly selected 20, 40 and 100× fields per individual sample, totaling 50 images per treatment group, and 300 total images. Images were captured using a ZEISS Axio-A1 microscope (Oberkochen, Germany).

2.4. Western Blot (WB) Analysis

Total protein extracts were prepared in a lysis buffer as previously described [21]. Lysates were incubated for 30 min under shaking and centrifuged at 16,000× g for 10 min at 4 °C, and supernatants were stored at −70 °C for further analyses. After protein quantification, equivalent amounts of proteins were analyzed in Laemmli buffer by SDS–PAGE and transferred to a PVDF membrane. The antibodies used were anti-pH2A.X (2577S; Cell Signaling; Danvers, MA, USA), anti-NXN (GTX107039, GeneTex; Irvine, CA, USA), and anti-GAPDH (60004-1-Ig; Proteintech; Rosemont, IL, USA). Protein loading was confirmed by analyzing the blots with anti-GAPDH. Densitometric analyses were performed using ImageJ 1.5.29 software.

2.5. Determination of Carbonylated Proteins

The carbonyl content of proteins was detected using the OxyBlot protein oxidation detection kit (S7150; Sigma-Aldrich, Purchase, NY, USA) according to the manufacturer's instructions, with minor modifications. To derivatize carbonyl groups for WB analysis, equal volumes of tissue lysate (10 mg/mL) were mixed in 12% SDS, followed by 2 volumes of 20 mM 2,4-dinitrophenylhydrazine in 10% trifluoroacetic acid. The mixture was incubated in the dark at RT for 20 min and the OxyBlot neutralization solution was added to stop the reaction; a nonderivative control per group was included. The primary antibody was an anti-rabbit polyclonal against dinitrophenylhydrazone, and the HRP-conjugated secondary antibody was used at 1:300 dilution. Bands revealed in the full lanes were quantified by densitometry analysis using the ImageJ 1.5.29 software.

2.6. Immunoprecipitation (IP) Assay

Total proteins from liver tissues were processed in a lysis buffer containing 20 mmol/L Tris-Cl, pH 7.8, 137 mmol/L NaCl, 1% (*v*/*v*) Nonidet P40, 10% (*v*/*v*) glycerol, and 2 mmol/L EDTA (all from Sigma; St. Louis, MO, USA), and supplied with Complete and PhoStop (Roche; Basel, Switzerland). Proteins of interest were immunoprecipitated with a mixture of protein A/G Plus-Agarose complex beads (sc2003, Santa Cruz Biotechnology; Santa Cruz, CA, USA) and rabbit anti-NXN (GTX107039; GeneTex; Irvine, CA, USA). Precipitated and coprecipitated proteins were analyzed by WB and normalized by reanalyzing the blots for the constant regions (heavy chains) of the immunoglobulins used for IP assays.

To detect precipitated and coprecipitated proteins, anti-NXN (AF-5718, Novus Biologicals; Centennial, CO, USA), anti-flightless (FLII) and anti-phosphofructokinase 1 (PFK1) (sc-21716 and sc-166722, respectively, Santa Cruz Biotechnology; Santa Cruz, CA, USA), anti-MYD88 (4283S, Cell Signaling; Danvers, MA, USA), and anti-calcium/calmodulin-dependent protein kinase II type alpha (CAMK2A) (SAB4503250, Merck; Darmstadt, Germany) antibodies were used. To avoid cross-reactivity with the denatured heavy or light chains of the antibodies used for IP assays, goat anti-mouse and mouse anti-rabbit light chain-specific secondary antibodies (115-035-174 and 211-032-171, respectively; Jackson ImmunoResearch, Baltimore, PA, USA) were used.

2.7. Statistical Analysis

Statistical analyses were performed using GraphPad Prism 8.0 software (GraphPad, La Jolla, CA, USA). Five animal per group were included and data calculation was performed using a one-way ANOVA test for multiple comparisons. Data are expressed as the mean ± SE. Differences were considered significant when $p < 0.05$.

3. Results

3.1. Chronic Alcohol Consumption Increases Mice Body Weight and the Relative Liver Weight during Aging

Although aged mice subjected to the ALD model recorded higher body weights (both $p < 0.0001$) at the end of ethanol consumption, i.e., 12- and 18-month-old mice, as compared with the onset of treatment, that of control groups were also significantly increased (both $p < 0.0001$). Body weight of young animals was not different at the end of the experiment. Interestingly, the percentage of the relative liver weight was only increased in 18-month old mice subjected to the ALD model ($p < 0.0001$) as compared with its respective control (Table 1).

Table 1. Effect of chronic ethanol consumption on mice body weight and relative liver weight.

Group	Age	Body Weight (g) Initial	Body Weight (g) Final	Percentage of Relative Liver Weight
C	7W	15.28 ± 0.16	15.64 ± 0.14	5.53 ± 0.12
	12M	24.93 ± 2.60	29.92 ± 1.39 *	5.65 ± 0.66
	18M	32.00 ± 2.08	36.00 ± 1.05 *	4.38 ± 0.28
ALD	7W	17.66 ± 0.37	18.44 ± 0.30	4.97 ± 0.16
	12M	26.40 ± 0.58	30.34 ± 0.81 *	6.14 ± 0.31
	18M	26.92 ± 0.68	30.30 ± 0.47 *	6.25 ± 0.64 $

Values represent the average ± SE. * Statistically different from the initial body weight; $ Statistically different from the respective control group; $p < 0.0001$.

3.2. Chronic Alcohol Consumption Modifies Extracellular Matrix Components during Aging

Histopathological analysis by H&E staining showed a mild inflammatory infiltrate alongside disarrangement of the hepatocyte cords in 18-month-old mice of both control and ALD groups (Figure 2A). As shown in Figure 2B, M'sT staining revealed a slight signal of collagen fibers in the liver tissues of 18-month-old mice subjected to the ALD model. Then, by IHC analysis, we analyzed COL1A1 and COL3A1 levels, two key components of the extracellular matrix (ECM) altered during the progression of liver fibrosis [28]. As shown in Figure 2C, the level of COL1A1 was increased ($p = 0.0158$) in 12-month-old control mice as compared with that in young control mice, i.e., 7-week-old mice; notably, in the ALD model, COL1A1 was increased in 12- and 18-month-old mice (both $p < 0.0001$). Contrastingly, the level of COL3A1 in 12- and 18-month-old control mice was significantly (both $p < 0.0001$) increased, but in the ALD model, COL3A1 was decreased ($p < 0.0001$) in both 12- and 18-month-old mice as compared with young mice subjected to the ALD model.

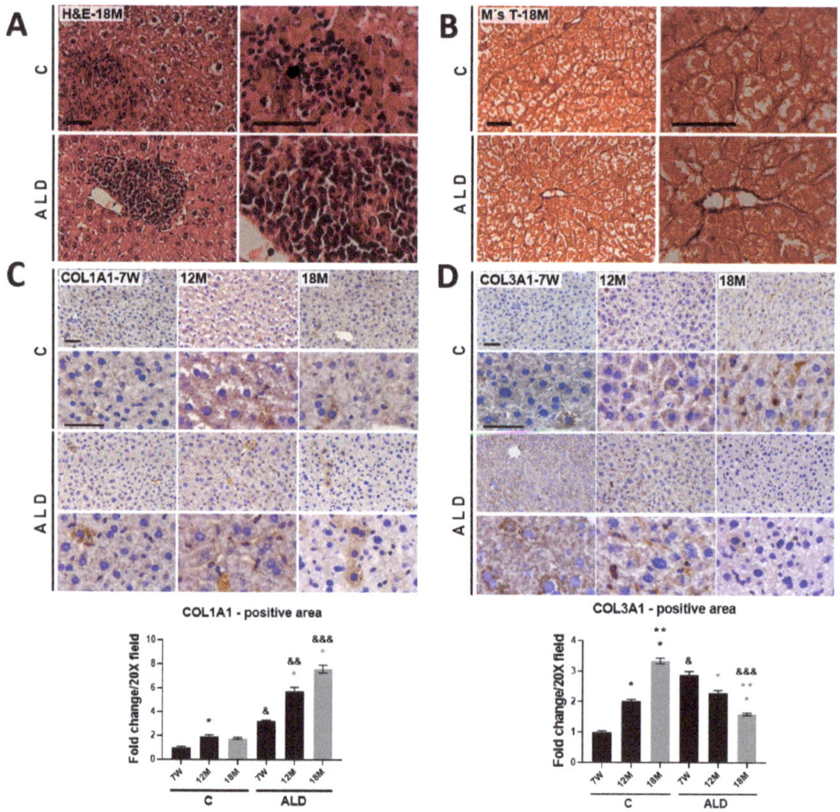

Figure 2. Extracellular matrix components determination in the liver of aged mice. (**A**) H&E staining. (**B**) M'sT staining. Magnification is 40 and 100× and the scale bar indicates 50 μm. (**C**,**D**) IHC analysis and signal quantification of COL1A1 and COL3A1 proteins. Magnification is 40 and 100× and the scale bar indicates 50 μm. Bars represent mean ± SE. * Significantly different from 7W, ** from 12M compared to the C group, * from 7W, ** from 12M compared to the ALD group, & from 7W, && from 12M, &&& from 18M compared to the respective control group; $p < 0.05$. $n = 5$ animals/group. ALD, alcohol liver disease model; C, control; 7W, 7-week-old mice; 12M, 12-month-old mice; 18M, 18-month-old mice.

3.3. *Cell Proliferation Is Increased by Chronic Alcohol Consumption during Aging*

The evidence has shown that the regenerative capability of the liver decreases with age [29]. Thus, to determine the effect of chronic alcohol consumption on aged livers, we measured the proliferation markers PCNA and KI67 levels. IF analysis revealed that PCNA-positive cells were decreased in both 12- and 18-month-old control mice (both $p < 0.0001$) as compared to young control animals; however, the ALD model increased PCNA-positive cells in 18-month-old mice ($p = 0.0472$), but they were decreased in 12-month-old mice ($p = 0.0023$) as compared with young animals. PCNA-positive cells were decreased in 7-week-old mice subjected to the ALD model ($p < 0.0001$) as compared with the respective young controls (Figure 3A). IHC analysis revealed that the number of KI67-positive cells in 12- and 18-month-old control mice decreased (both $p < 0.0001$) as compared with that in young control animals, while in the ALD model, this number increased ($p < 0.0001$) in 18-month-old mice as compared with that in young animals. Interestingly, young and 12-month-old mice subjected to the ALD model showed a reduced number of KI67-positive cells ($p < 0.0001$ and $p = 0.0039$, respectively) as compared with their respective untreated controls (Figure 3B).

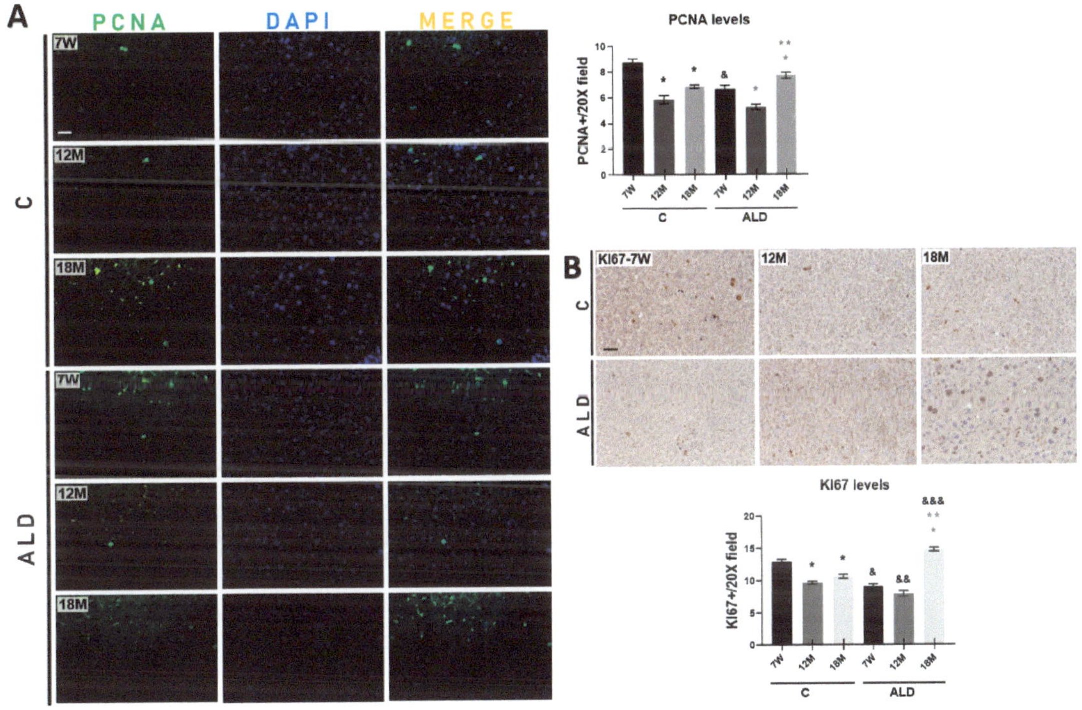

Figure 3. Effect of chronic alcohol consumption on cell proliferation in liver tissue of young and aged mice. (**A**) IF analysis and signal quantification of PCNA protein. Magnification is 20× and the scale bar indicates 100 μm. (**B**) IHC analysis and signal quantification of KI67 protein. Magnification is 20× and the scale bar indicates 100 μm. Bars represent the mean ± SE. * Significantly different from 7W, * from 7W, ** from 12M compared to the ALD group, & from 7W, && from 12M, &&& from 18M compared to the respective control group; $p < 0.05$. $n = 5$ animals/group. ALD, alcohol liver disease model; C, control; 7W, 7-week-old mice; 12M, 12-month-old mice; 18M, 18-month-old mice.

3.4. Chronic Alcohol Consumption Increases Cellular Senescence Markers in the Liver of Aged Mice

An early event in the double-strand DNA break (DSB) response is the rapid recruitment and activation of PARP1, resulting in PAR polymerization, an early marker of DNA damage, followed by histone recruitment to DNA damage sites to stimulate chromatin remodeling and DNA repair [30]. To determine the status of DNA damage associated with aging and/or promoted by chronic alcohol consumption, we quantified PAR-positive cells by IF analysis. Results showed that PAR increased in both 12- and 18-month-old control mice livers (both $p < 0.0001$) as compared with young animals. Of note, the ALD model induced PAR levels in the livers of both 12- and 18-month-old mice (both $p < 0.0001$) as compared with young animals, but the ALD model also strongly induced PAR levels ($p < 0.0001$) in young mice as compared with young control animals, as well as in both 12- and 18-month-old mice livers (both $p < 0.0001$) as compared with their respective untreated controls (Figure 4A).

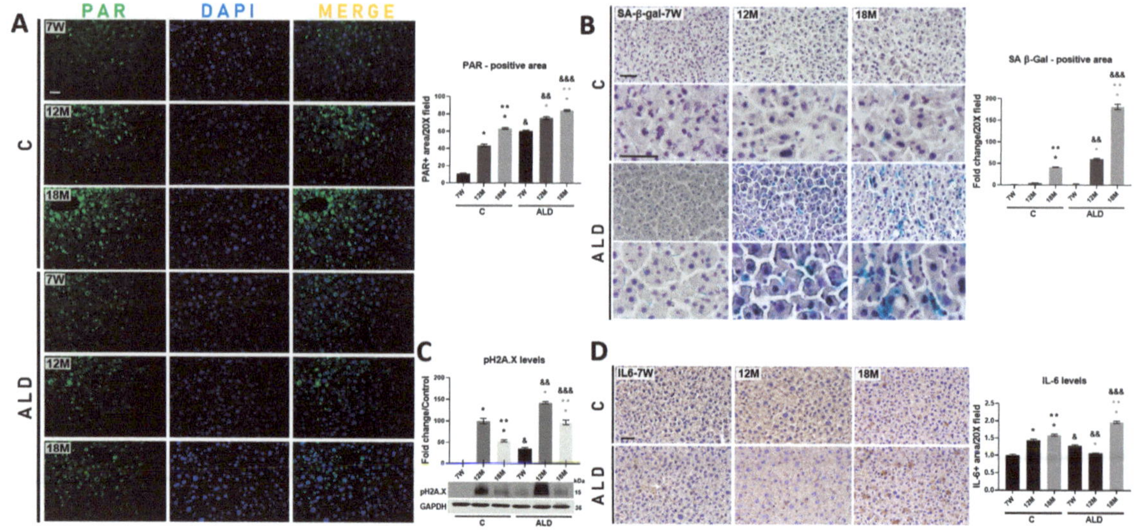

Figure 4. Determination of DNA damage and cellular senescence markers induced by chronic alcohol consumption in the liver of young and aged mice. (**A**) IF analysis and signal quantification of PAR protein. Magnification is 20× and the scale bar indicates 100 µm. (**B**) Histochemical analysis and signal quantification of SA-β-gal activity. Magnification is 20 and 40×, and the scale bar indicates 50 µm. (**C**) WB analysis and intensity quantification of pH2A.X protein. pH2A.X protein levels were normalized with those of GAPDH protein, which was used as a loading control. (**D**) IHC analysis and signal quantification of IL-6 protein. Magnification is 20× and the scale bar indicates 50 µm. Bars represent the mean ± SE. * Significantly different from 7W, ** from 12M compared to the C group, * from 7W, ** from 12M compared to the ALD group, & from 7W, && from 12M, &&& from 18M compared to the respective control group; $p < 0.05$. $n = 5$ animals/group. ALD, alcohol liver disease model; C, control; 7W, 7-week-old mice; 12M, 12-month-old mice; 18M, 18-month-old mice.

Since there is no single unequivocal marker of cellular senescence, we determined the presence of several well-known markers associated with this phenomenon, such as SA-β-gal activity, pH2A.X, and IL-6 [7]. As expected, histochemical analysis revealed that the activity of the SA-β-gal enzyme increased in the livers of 18 month-aged control mice as compared with 12-month-old and young animals; interestingly, SA-β-gal activity was strongly induced in the livers of both 12- and 18-month-old mice subjected to the ALD model as compared with young animals and with their respective untreated controls (Figure 4B). WB analysis revealed that pH2A.X levels increased in both 12- and 18-month-old control mice (both $p < 0.0001$); however, as shown in Figure 4C, chronic alcohol consumption induced the level of this protein in all animal groups subjected to this scheme, namely 7-week-old, 12- and 18-month-old mice (all $p < 0.0001$). In addition, IHC analysis demonstrated that the IL-6-positive area increased in the livers of 12- and 18-month-old control mice (both $p < 0.0001$) compared with those of young mice. The ALD model increased its level only in 18-month-old mice ($p < 0.0001$) as compared with 7-week-old and 12 month-aged mice, as well as with its respective untreated control (Figure 4D).

3.5. Chronic Alcohol Consumption Promotes the Preferential Localization of NXN Either into or Alongside Senescent Cells in the Liver of Aged Mice

Based on the increased activity of the SA-β-gal enzyme in the liver of 18-month-old mice (Figure 4B), we then determined whether there was a possible relationship between senescence markers and NXN in the liver tissue of aged mice. Double staining of NXN with pH2A.X protein and SA-β-gal activity by IHC and IF analyses, respectively, in the liver of 18-month-old mice revealed that NXN was preferentially localized in the nucleus and cytoplasm of either senescent or adjacent cells in animals subjected to the ALD model (Figure 5A,B).

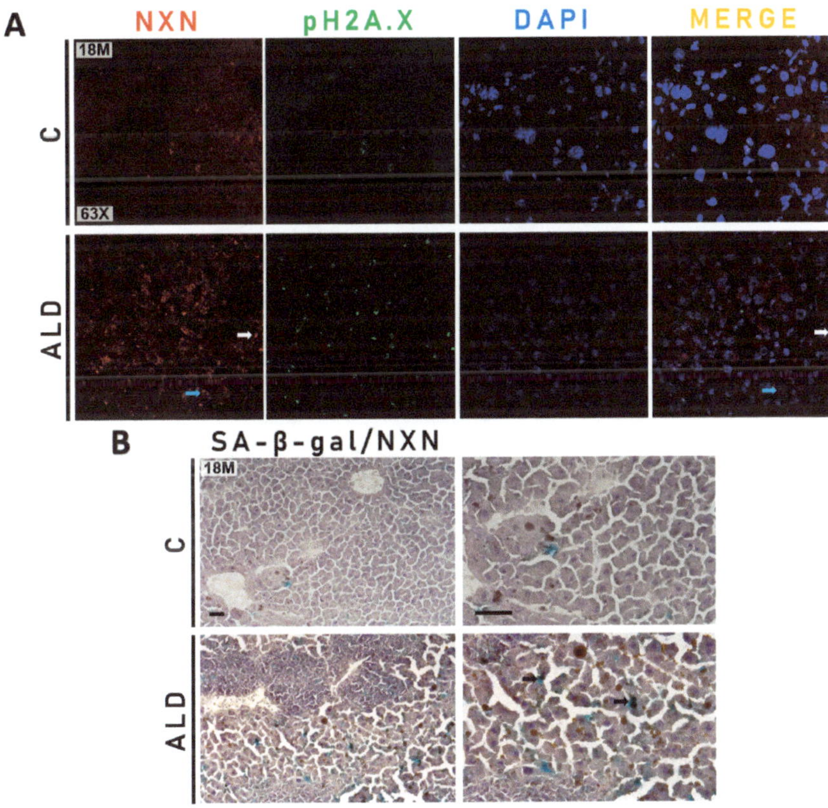

Figure 5. Co-labeling of NXN with pH2A.X and SA-β-gal activity in the liver of aged mice (**A**) Double staining of NXN and pH2A.X proteins by IF analysis. White and blue arrows point out nuclear and cytoplasmic localization of NXN, respectively. Magnification is 63×. (**B**) Double staining of SA-β-gal activity and NXN protein by IHC and histochemical analyses, respectively. Black arrows indicate double staining of SA-β-gal activity and NXN. Magnification is 20 and 40× and the scale bar indicates 50 µm. Pictures show tissues from 18-month-old mice. ALD, alcohol liver disease model; C, control.

3.6. Carbonylated Proteins Are Increased by Chronic Alcohol Consumption in the Liver of Aged Mice

Among a wide range of ROS-derived modifications, carbonylation of biomolecules is a major hallmark of oxidative stress and determines the oxidation degree associated with cellular damage, aging, and several age-related disorders [31]. Carbonylated proteins analysis showed that the baseline observed in young control animals was increased ($p = 0.0197$) in 18-month-old control mice; interestingly, chronic alcohol consumption promoted the level of carbonylated proteins in the liver of all animal groups of the ALD model, namely 7-week-old, 12- and 18-month-old mice ($p < 0.0001$, $p < 0.0001$ and $p = 0.0006$, respectively) as compared with their respective untreated controls (Figure 6A). The measurement of NXN protein level revealed that while no changes were observed among control groups, chronic alcohol consumption stimulated its level in 7-week-old mice subjected to the ALD model ($p = 0.0047$); this phenomenon was not observed in both 12- and 18-month-old mice (Figure 6B).

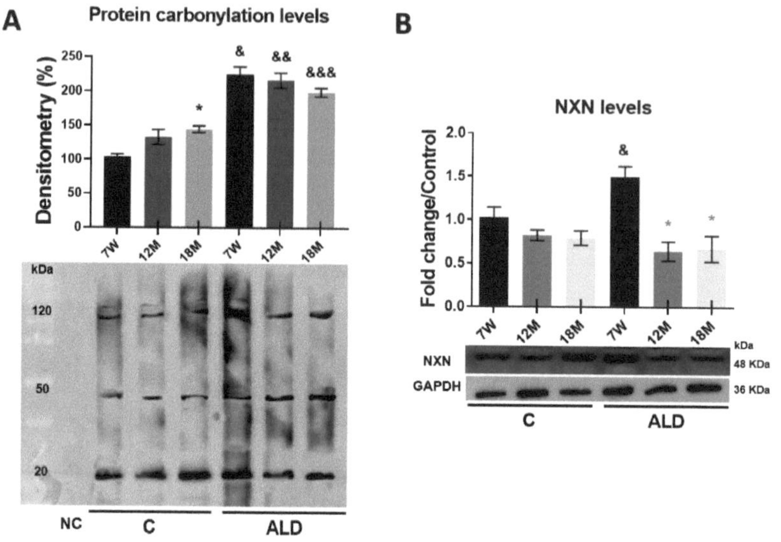

Figure 6. Effect of chronic alcohol consumption on protein carbonylation and NXN protein levels in the liver of young and aged mice. Two pools of three individual samples each were made by mixing equal amounts of proteins. (**A**) Determination of carbonylated proteins by OxyBlot analysis. (**B**) WB analysis of NXN protein. NXN protein level was normalized with that of GAPDH, which was used as a loading control. Bars represent the mean ± SE. * Significantly different from 7W, * from 7W, & from 7W, && from 12M, &&& from 18M compared to the respective control group; $p < 0.05$. n = 5 animals/group. ALD, alcohol liver disease model; C, control; 7W, 7-week-old mice; 12M, 12-month-old mice; 18M, 18-month-old mice.

3.7. *Chronic Alcohol Consumption Worsens Aging-Promoted Alteration of NXN-Dependent Interaction Ratios in the Mouse Liver*

It has been well established that NXN interacts, in a redox-dependent manner, with several proteins such as FLII, MYD88, CAMK2A, and PFK1, among others, to regulate the downstream activity of different signaling pathways and, as a consequence, contributes to cellular redox homeostasis [18]; furthermore, some of the NXN redox-sensitive interactions are affected by chronic alcohol consumption [21,32]. Thus, to determine whether aging modifies the ratio of NXN/FLII, NXN/MYD88, NXN/CAMK2A, and NXN/PFK1 redox-sensitive interactions and whether chronic alcohol consumption worsens the aging-induced alterations in the liver, we performed IP analyses from total protein extracts. Results demonstrated that NXN was precipitated (Figure 7A,B), and FLII, MYD88, CAMK2A, and PFK1 were efficiently coprecipitated by an anti-NXN antibody (Figure 7A). Of note, in 18-month-old control mice, the NXN/FLII interaction ratio was strongly ($p < 0.0001$) increased, but the strongest increment of this interaction ratio was observed in the 7-week-old mice subjected to the ALD model, which was gradually decreased in both 12- and 18-month-old mice (both $p < 0.0001$). Notably, in 18-month-old mice subjected to chronic alcohol consumption, this interaction ratio was also decreased ($p < 0.0001$) as compared with its respective untreated control (Figure 7A,C). The NXN/MYD88 interaction ratio was found to be strongly decreased ($p < 0.0001$) in 18-month-old control mice as compared with both young and 12-month-old control mice; however, it was similarly reduced in young and aged mice subjected to the ALD model, even below the interaction ratio observed in their respective controls, especially in 7-week-old and 12-month-old mice ($p = 0.0001$ and $p < 0.0001$, respectively) as compared with their respective untreated control mice (Figure 7A,D). On the other hand, the interaction ratio of the NXN/CAMK2A complex was induced in 7-week-old mice ($p = 0.0018$) of the ALD model as compared with its untreated control, but it gradually decreased, reaching significance in 18-month-old mice ($p = 0.0426$) as compared with 7-week-old mice. Untreated controls showed a non-significant increasing trend during aging (Figure 7A,E). Finally, the interaction ratio of the NXN/PFK1 complex in untreated animals showed a gradual and significant ($p = 0.0121$) decrease in 18-month-old mice as compared with the young animals; contrarily, this interaction complex was induced by chronic alcohol consumption in 18-month-old mice ($p < 0.0001$) as compared with its respective untreated control ($p < 0.0001$), as well as compared with 7-week-old and 12-month-old mice ($p = 0.0001$ and $p = 0.0011$, respectively) subjected to the ALD model (Figure 7A,F).

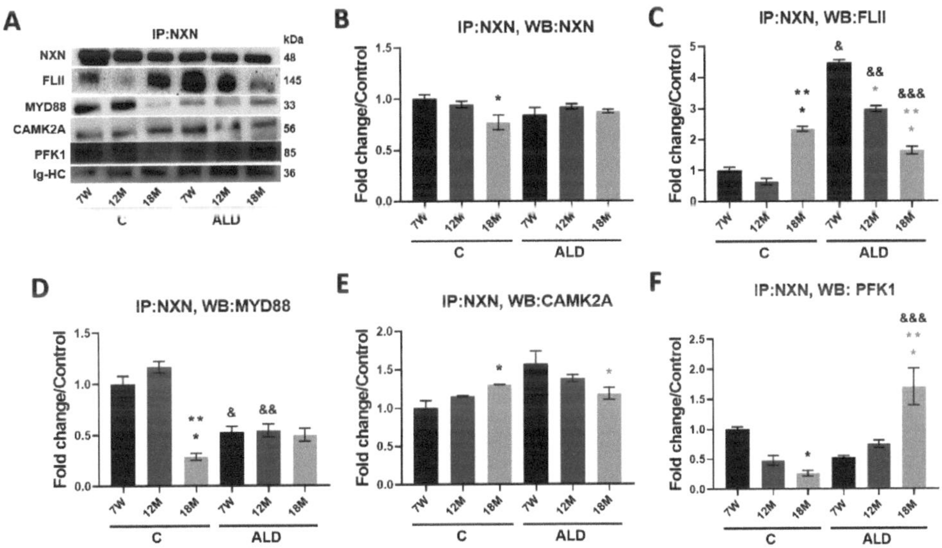

Figure 7. Effect of chronic alcohol consumption on NXN/FLII, NXN/MYD88, NXN/CAMK2A and NXN/PFK1 interaction in the liver of young and aged mice. Total protein extracted from mice was incubated to precipitate NXN in complex with its interacting proteins by using an anti-NXN antibody as indicated in materials and methods section. Two pools of three individual samples each were made by mixing equal amounts of proteins. (**A**) WB analyses were used to detect immunoprecipitated NXN and co-precipitated FLII, MYD88, CAMK2A and PFK1 proteins. Densitometry quantification of (**B**) NXN; (**C**) FLII; (**D**) MYD88; (**E**) CAMK2A and (**F**) PFK1 protein intensity. Immunoglobulin heavy chain (Ig-HC) bands, used for IP assay (anti-NXN), were detected to normalize protein levels. Bars represent the mean ± SE. * Significantly different from 7W, ** from 12M compared to the C group, * from 7W, ** from 12M compared to the ALD group, [&] from 7W, [&&] from 12M, [&&&] from 18M compared to the respective control group; $p < 0.05$. $n = 5$ animals/group. ALD, alcohol liver disease model; C, control; 7W, 7-week-old mice; 12M, 12-month-old mice; 18M, 18-month-old mice.

4. Discussion

Deterioration of liver functioning due to aging contributes to the progression of age-related liver disease [33]. A well-described mechanism is that during aging, hepatic stellate cells adopt aging-related changes and secrete various inflammatory and tumor-promoting factors, including IL-1β, IL-6 and CXCL7, to induce neutrophil infiltration. Then, neutrophil-derived excessive oxidative stress induces DSB and restricts the proliferation of liver progenitor cells, leading to the impairment of liver regeneration [34]. Accumulating evidence indicates that an aged liver is more susceptible to injury due to alcohol abuse [35], which exacerbates oxidative stress and unbalances the redox status, playing a critical role in the alcohol-mediated cellular fate [36]. Since NXN is a redox-sensitive enzyme that has been proposed as a master regulator of cellular redox homeostasis [18], here we determine the effect of chronic alcohol consumption on NXN-dependent redox interactions in the liver of aged mice and identify that some NXN redox interactions are sensitized by aging and worsened by chronic alcohol consumption, which might potentiate the liver malfunctioning as discussed below.

Following hepatic injury, hepatic stellate cells undergo a complex activation process and become the leading source for the increased and irregular deposition of ECM components, leading to fibrosis [37]. During a normal fibrillogenesis process, while the level of COL1A1 is decreased, that of COL3A1 is increased; however, in liver fibrotic tissues, this homeostatic collagen I/III ratio is inverted [28]. This altered ratio can be found in the liver of aged individuals and might be associated with a chronic low-grade proinflammatory state [38]. Although M'sT staining did not clearly detect deposition of collagen fibers, IHC analyses revealed that chronic alcohol consumption increased COL1A1 levels and decreased that COL3A1 levels in the livers of aged mice. This phenomenon was accompanied by a mild inflammatory infiltrate and disarrangement of the hepatocyte cords (Figure 2). Although collagen fibers were not detected by M'sT staining, a less sensitive procedure than immunodetection, this evidence indicates that both the reversal of the collagen I/III ratio and the slight inflammatory infiltrate were favored by aging in the liver of animals subjected to the ALD model, as some of the initial alterations might progress to a worse condition if the liver continues to be exposed to chronic insult by either a single or multiple promoting factors for a more extended time.

Cell proliferation is an essential process for the growth and maintenance of tissue homeostasis. Some widely used proliferation markers, such as PCNA and KI67 proteins, have been helpful for identifying alterations in the cell proliferation process [39]. Although the liver is an organ with high proliferative potential that promotes its regeneration after partial tissue removal or cellular damage, a phenomenon known as compensatory proliferation, its regenerative capability declines with aging [40]. It has been reported that chronic alcohol consumption for 24 months, increases liver cell proliferation from the onset up to 7 months of alcohol consumption; interestingly, at 12 months, it is decreased but at 18 months, the proliferation ratio is similar to controls [41]. As expected, our results show that KI67 and PCNA levels decreased in aged-untreated animals, i.e., 12- and 18-month-old mice; however, chronic alcohol consumption increased their protein levels in 18-month-old mice (Figure 3). Of note, this result is in concordance with the increased relative liver weight observed in the same animal group (Table 1). A plausible explanation for this contrasting result when compared with the previously reported data [41], is that the compensatory proliferation induced by the hepatotoxic effects of chronic alcohol consumption, is boosted in the initial times of alcohol exposure and afterward, the cells adapt to alcohol cytotoxicity, as occurs with exposure to high levels of LPS promoted by ethanol-induced intestinal barrier dysfunction [42]. This is because at the time of evaluation, in our ALD model, the animals had been subjected to alcohol consumption only for the first 8 weeks at 18-month-old, but in the study by Mendenhall and coworkers [41], the animals had been fed alcohol for 18 months at the same age. Thus, the induction of compensatory proliferation by alcohol might be associated with the onset of exposure of liver cells to chronic alcohol consumption.

Ethanol-induced oxidative stress might lead to cellular senescence by increasing ROS production and reducing the activity of the cellular antioxidant system, which increase DNA damage [12]. In response to DNA damage, DNA-dependent PARP, especially PARP-1, quickly recognizes both single- and DSB ends. Then, PAR is synthesized, a DNA damage signaling molecule that allows a rapid and efficient cellular evaluation of DNA damage and concentrates key factors of the single-strand break repair pathway at the damage site [30]. Interestingly, it has been reported that the inactivation of *Parp-1* gene in mice leads to acceleration of aging [43]. Our IF analysis revealed that PAR level was significantly increased in untreated aged mice but was exacerbated by chronic alcohol consumption in both young and aged mice (Figure 4A). In vitro assays have shown that senescence, a stress response that universally occurs within tissues under pathophysiological conditions, is stimulated in hepatocytes by ethanol, suggesting that it plays a key role in ALD [44];

moreover, in hepatocyte cultures exposed to ethanol the SA-β-gal activity, a widely used biomarker of cellular senescence, is increased [45]. Interestingly, our results showed that the activity of SA-β-gal was increased in untreated 18-month-old mice, but this increment was observed from 12 months on and potentiated in 18-month-old mice subjected to the ALD model (Figure 4B), indicating that chronic alcohol consumption exacerbated the appearance of senescent cells in the liver of aged mice.

Telomere shortening is an important event in senescence, which correlates with senescence-associated heterochromatin foci, a facultative heterochromatin domain associated with irreversible cell cycle arrest [46]. pH2A.X, a replacement histone, is the second most common marker of cellular senescence after SA-β-gal activity and is the most sensitive marker of DSB and telomere shortening [47]. Our results demonstrated that pH2A.X level was induced in the livers of 12- and 18-month-old mice of both untreated and treated with ethanol, but it was higher in those that consumed ethanol (Figure 4C). Moreover, most senescent cells may exhibit a SASP and overproduce massive proinflammatory cytokines and growth factors. IL-6 is one of the basic proinflammatory cytokines in SASP components [48]. It has been reported that in the livers of aged rats, *Il-6* mRNA is increased mainly in Kupffer cells, suggesting that these cells might be the main source of this cytokine in aged livers [49]. In addition, when acute injury occurs, the liver loses hepatocellular functions and induces compensatory regeneration, and IL-6, along with growth factors, promotes hepatocyte proliferation by increasing both DNA synthesis and genomic instability [50]. Our results showed that chronic alcohol consumption increased the IL-6 level in aged mice (Figure 4D). Altogether, these data indicate that the DNA repair mechanism is activated by aging itself in an attempt to reverse aging-associated cell damage, but it is potentiated by the cytotoxic effect of chronic alcohol consumption as an early damage event in both aged and young liver cells; thus, this damage leads to the establishment of cellular senescence, which might contribute to accelerating ALD progression during aging.

Oxidation causes structural damage to macromolecules that accumulates with age and contributes to disruption in signaling pathways, inadvertently resulting in cellular aging. It has been shown that redox-regulated signaling networks control SASP, which plays an important role in driving age-related diseases [51]. NXN is a redox-sensitive enzyme that maintains redox homeostasis, regulates several signaling pathways, and targets ROS [18]. Thus, we determined the co-labeling of NXN with pH2A.X and SA-β-gal by IF and IHC analyses, respectively, to identify a possible relationship between cellular senescence and NXN, a redox-sensitive enzyme that has not been previously associated with this cellular process. Our results showed that NXN either co-localized with or was preferentially detected alongside senescent cells (Figure 5). Since NXN targets ROS [18], this evidence strongly suggests that NXN is localized within or near senescent cells to protect them from oxidative stress produced by the senescent cells themselves and/or by ethanol metabolism.

Protein carbonylation is an irreversible and irreparable oxidative post-translational modification that yields a reactive carbonyl moiety. It has been shown that protein carbonylation is associated with several human diseases, increases with age, and is linked to the age-dependent depletion of specific enzymes [52]. Notably, protein carbonylation is increased in the liver of young mice exposed to chronic alcohol consumption for four months [53]. We observed that protein carbonylation was increased in untreated 18-month-old mice, but alcohol consumption administered for only eight weeks increased the level of carbonylated proteins in the liver of both young and aged mice, indicating that protein carbonylation is an early oxidative modification induced by ethanol, and aging contributed to potentiating this phenomenon (Figure 6A). A plausible mechanism in aged livers is that increased ROS production by ethanol consumption impairs mitochondria structure and function and increases the content of carbonylated proteins in mitochondrial but not those in the cell cytosol, as previously reported [12]. On the other hand, although NXN has not been implicated in cellular senescence, it may contribute based on its role in ROS and cellular redox homeostasis regulation [18]. Our results showed that NXN protein levels were strongly induced in the livers of young mice that consumed alcohol but not in aged mice (Figure 6B). A plausible explanation for this phenomenon is that in young animals, NXN still has the full capability to fight the increased oxidative stress and reverse the oxidative damage induced by alcohol. However, the enzymatic activity of NXN had already declined in aged mice, as occurs with alcohol-metabolizing liver enzymes in elderly people [15]. This decline might result from the oxidative modification of the NXN enzyme promoted by the carbonylation induced by chronic alcohol consumption, an intriguing hypothesis that needs to be addressed.

Nucleoredoxin interacts in a redox-dependent manner with seven proteins so far, regulates several cellular processes, and has been associated with different pathologies [18]. As a first approach to elucidating whether its protein–protein interaction ratios are modified with aging, we evaluated the interaction status of NXN with FLII, MYD88, CAMK2A and PFK1 proteins, which regulate innate immunity, inflammation, neuronal plasticity and glycolysis, respectively. While MYD88 is considered a hub of the inflammatory signaling pathways downstream of TLRs and other receptors [54], FLII, which also interacts with MYD88, is a multifunctional protein and has recently been identified as an emerging regulator of inflammation [55]. During a proinflammatory stimulus, MYD88 is recruited to TLR4, leading to the activation of NF-κB to transcribe genes involved in innate immunity and inflammation. To avoid unnecessary hyperactivation of the TLR4/MYD88 pathway, FLII hijacks MYD88 from TLR4 through NXN, forming the FLII/NXN/MYD88 complex [56]. We found that FLII increased its interaction ratio with NXN in the liver of young mice, but this interaction was strongly decreased in those aged mice subjected to the ALD model, while the MYD88 interaction ratio was unchanged in these animal groups. Of note, in untreated 18-month-old mice, the FLII/NXN interaction ratio was increased (Figure 7C,D). This evidence strongly suggests that chronic alcohol consumption decreases NXN's efficiency to interact with FLII, likely due to the increased protein carbonylation induced by chronic alcohol consumption, which might be affecting redox-sensitive proteins such as NXN, a phenomenon that might be sensitized by aging but exacerbated by chronic alcohol consumption.

Recently, it was reported that NXN interacts with and oxidizes CAMK2A to maintain its oxidative status and activity in neurons [57]. CAMK2A, a serine/threonine kinase, is highly abundant in the brain and plays several roles mediated by the regulation of intracellular Ca^{2+}, such as neurotransmitter synthesis, ion channel regulation, cell division, muscle contractility, and gene transcription [18,57]. To our knowledge, its role in the liver has not been investigated. We observed that the NXN/CAMK2A interaction ratio decreased in aged mice subjected to the ALD model (Figure 7E), which might be a consequence of the impaired oxidase capability of NXN initiated by aging and increased by chronic alcohol consumption, affecting CAMK2A-dependent roles in the liver that remain to be investigated.

Phosphofructokinase-1 is a crucial regulator enzyme of glycolysis associated with the promotion of cancer cell proliferation since it is elevated in cancer tissues [58]. It has been shown that NXN interacts with PFK1 and stabilizes its oligomerization to regulate its catalytic activity. Because NXN deficiency increases NADPH and reduces glutathione levels, two major cellular antioxidants generated through the pentose phosphate pathway, it has been proposed that PFK1 makes cells more resistant to oxidative stress. Although it is still unknown how NXN might stabilize PFK1 oligomerization, it has been proposed that there might be a joint effect among NXN activity, glycosylation, phosphorylation and acylation since PFK1 is also subject to these post-translational modifications [59]. We found that the NXN/PFK1 interaction ratio decreased in untreated aged mice but increased in those subjected to chronic alcohol consumption (Figure 7F). Thus, this evidence suggests that a mechanism by which chronic alcohol consumption during aging increases cell proliferation (Figure 3) is through PFK1 stabilization by the contribution of NXN as well as some post-translational modifications. Further analysis will confirm this proposal.

5. Conclusions

We provide evidence showing that chronic alcohol consumption worsens the alteration of NXN-dependent redox-sensitive interactions promoted by aging in the mouse liver, such as that of NXN with FLII, MYD88, CAMK2A and PFK1 proteins. Although NXN was preferentially localized either into or alongside senescent cells, which could be an attempt to protect the liver tissue from the adverse effects of ethanol, its interacting capability had already declined with aging. This was accompanied by an increase in SA-β-gal activity, PAR, pH2A.X, and IL-6 levels, protein carbonylation and cell proliferation, as well as the modification of the collagen I and collagen III ratios (Figure 8). Moreover, given that this evidence, to our knowledge, is the first to show the involvement of NXN in aging, our finding opens a window to escalate research addressed in elucidating how deeply the NXN interactions and their downstream signaling pathways are involved in aging and in the challenges that the liver faces when exposed to toxic agents, such as ethanol. Particularly, as it is well-known in the general population, ethanol intake is an independent predictor of other liver diseases, such as cirrhosis and HCC in patients bearing chronic hepatitis C virus infection, and of death in patients bearing either hepatitis C or B virus infection [60]. Thus, it is plausible that the disruption of NXN interactions can be potentiated by the concomitant effects of ethanol chronic intake and hepatitis virus infection. Although this proposal was not addressed and might be a limitation of our research, it is also an encouraging hypothesis arising from the current findings. Thus, the involvement of NXN in aging represents an emerging and attractive phenomenon that should continue to be investigated.

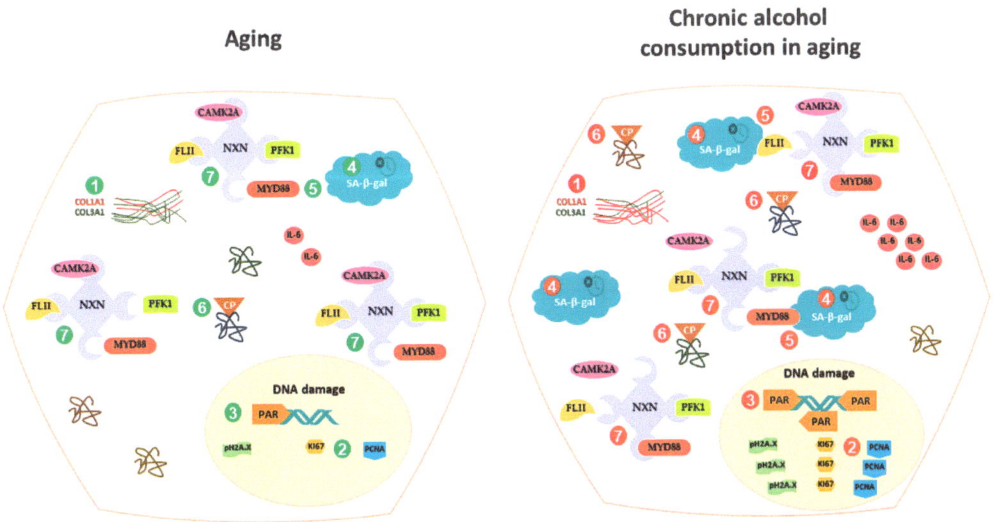

Figure 8. Schematic summary of chronic alcohol consumption effects on NXN-dependent redox-sensitive interactions during aging. Chronic alcohol consumption worsens the adverse effects of aging on (1) COL1A1/COL3A1 ratio, (2) cell proliferation, (3) DNA damage, (4) SA-β-gal activity, (5) NXN protein level and localization, (6) Protein carbonylation, and (7) the interaction of NXN with FLII, MYD88, CAMK2A and PFK1 proteins in the mouse liver. CAMK2A, calcium/calmodulin-dependent protein kinase II type alpha; COL1A1, collagen 1 alpha 1; COL3A1, collagen 3 alpha 1; CP, carbonylated protein; FLII, flightless; IL-6, interleukin 6; MYD88, myeloid differentiation primary response 88; NXN, nucleoredoxin; PAR, poly (ADP-ribose); PCNA, proliferating cell nuclear antigen; PFK1, phosphofructokinase 1; pH2AX, phospho-H2A.X (Ser139); SA-β-gal, senescence-associated β-galactosidase activity.

Author Contributions: Conceptualization, O.G.I.-G., B.R.A.-S. and J.A.-R.; investigation, O.G.I.-G., B.R.A.-S. and J.A.-R.; writing—original draft preparation, O.G.I.-G., B.R.A.-S. and J.A.-R.; writing—review and editing, O.G.I.-G., B.R.A.-S., V.R.V.-G., R.B.-H., P.M., S.V.-T., H.S., J.I.P.-C. and J.A.-R.; visualization, O.G.I.-G., B.R.A.-S. and J.A.-R.; supervision, D.G.-E., N.A.L.-H., J.L.P.-H., R.P.-R., J.S.-L., O.R.-A., E.A.M.-O., D.I.A.-B., G.B.-I., V.R.V.-G., R.B.-H., P.M., S.V.-T., H.S., J.I.P.-C. and J.A.-R.; Funding, V.R.V.-G., P.M., S.V.-T. and J.A.-R. All authors performed writing—review and editing and have read and approved the final manuscript. All authors have read and agreed to the published version of the manuscript.

Funding: This work was funded by CONAHCYT, grant CF2019-53358 to V.R.V.-G., S.V.T-., P.-M., and J.A.-R.; by INMEGEN, grant 06/2017/I to JAR; and by UPEAL-CINVESTAV-IPN, grant 0114-14 to S.V.-T.

Institutional Review Board Statement: Animal experimentation was following the Institutional Animal Use and Care Committee of CINVESTAV-IPN as well as the approved protocol No. 0114-14.

Informed Consent Statement: Not applicable.

Data Availability Statement: Data are contained within the article.

Acknowledgments: O.G.I.-G. thanks the PhD Program in Experimental Biology of the Department of Health Sciences, Div CBS—UAM-I, as well as the CONAHCYT-Mexico for awarding the doctoral fellowship No. 431419. B.R.A.-S. thanks to CONAHCYT-Mexico for awarding the master and doctoral fellowships No. 484737 and 752715, respectively. V.R.V.-G., R.B.-H., and J.A.-R. express their sincere thanks to the IxM-CONAHCYT program. All authors express their gratitude to S. Hernández-García, E. Romo-Medina, A. Cruz-Hernández, C. Hernández-Chávez, and JL Cruz-Colín for their support in administrative and laboratory procedures; to UPEAL-CINVESTAV-IPN staff, namely J. Fernández-Hernández, R. Leyva-Muñoz, R. Gaxiola-Centeno, BE Chávez-Álvarez, A. Rojo-García, MA López-López, I. Zavala-Mejia and ME Zúñiga-Alcántara, for the assistance in animal breeding, care and handling; to María A. Cabañas-Cortés for her technical support in capturing confocal microscopy images; to INMEGEN Histology Unit staff, namely NB Gabiño-López, SA Román-González, and R. Nava-Monroy, for the assistance in paraffin-embedding and tissue sectioning; and to the confocal microscopy unit at the Cell Biology Department of CINVESTAV-IPN, Zacatenco (CONAHCYT-Mexico, grant: 300062).

Conflicts of Interest: The authors declare no conflict of interest.

References

1. Flatt, T. A New Definition of Aging? *Front. Genet.* **2012**, *3*, 148. [CrossRef]
2. White, A.M.; Orosz, A.; Powell, P.A.; Koob, G.F. Alcohol and Aging—An Area of Increasing Concern. *Alcohol* **2023**, *107*, 19–27. [CrossRef] [PubMed]
3. Shadyab, A.H.; LaCroix, A.Z. Genetic Factors Associated with Longevity: A Review of Recent Findings. *Ageing Res. Rev.* **2015**, *19*, 1–7. [CrossRef]
4. Aravinthan, A.D.; Alexander, G.J.M. Senescence in Chronic Liver Disease: Is the Future in Aging? *J. Hepatol.* **2016**, *65*, 825–834. [CrossRef]
5. He, S.; Sharpless, N.E. Senescence in Health and Disease. *Cell* **2017**, *169*, 1000–1011. [CrossRef] [PubMed]
6. Ferreira-Gonzalez, S.; Lu, W.Y.; Raven, A.; Dwyer, B.; Man, T.Y.; O'Duibhir, E.; Lewis, P.J.S.; Campana, L.; Kendall, T.J.; Bird, T.G.; et al. Paracrine Cellular Senescence Exacerbates Biliary Injury and Impairs Regeneration. *Nat. Commun.* **2018**, *9*, 1020. [CrossRef] [PubMed]
7. Kudlova, N.; De Sanctis, J.B.; Hajduch, M. Cellular Senescence: Molecular Targets, Biomarkers, and Senolytic Drugs. *Int. J. Mol. Sci.* **2022**, *23*, 4168. [CrossRef]
8. Maldonado, E.; Morales-Pison, S.; Urbina, F.; Solari, A. Aging Hallmarks and the Role of Oxidative Stress. *Antioxidants* **2023**, *12*, 651. [CrossRef] [PubMed]
9. Rui, L. Energy Metabolism in the Liver. *Compr. Physiol.* **2014**, *4*, 177–197.
10. Sheedfar, F.; Di Biase, S.; Koonen, D.; Vinciguerra, M. Liver Diseases and Aging: Friends or Foes? *Aging Cell* **2013**, *12*, 950–954. [CrossRef]
11. Asrani, S.K.; Devarbhavi, H.; Eaton, J.; Kamath, P.S. Burden of Liver Diseases in the World. *J. Hepatol.* **2019**, *70*, 151–171. [CrossRef]
12. Dey, A.; Cederbaum, A.I. Alcohol and Oxidative Liver Injury. *Hepatology* **2006**, *43* (Suppl. S1), S63–S74. [CrossRef]
13. Shearn, C.T.; Orlicky, D.J.; Saba, L.M.; Shearn, A.H.; Petersen, D.R. Increased Hepatocellular Protein Carbonylation in Human End-Stage Alcoholic Cirrhosis. *Free Radic. Biol. Med.* **2015**, *89*, 1144–1153. [CrossRef]

14. Abdelmegeed, M.A.; Choi, Y.; Ha, S.K.; Song, B.J. Cytochrome P450-2e1 Promotes Aging-Related Hepatic Steatosis, Apoptosis and Fibrosis through Increased Nitroxidative Stress. *Free Radic. Biol. Med.* **2016**, *91*, 188–202. [CrossRef]
15. Meier, P.; Seitz, H.K. Age, Alcohol Metabolism and Liver Disease. *Curr. Opin. Clin. Nutr. Metab. Care* **2008**, *11*, 21–26. [CrossRef] [PubMed]
16. Kurooka, H.; Kato, K.; Minoguchi, S.; Takahashi, Y.; Ikeda, J.; Habu, S.; Osawa, N.; Buchberg, A.M.; Moriwaki, K.; Shisa, H.; et al. Cloning and Characterization of the Nucleoredoxin Gene That Encodes a Novel Nuclear Protein Related to Thioredoxin. *Genomics* **1997**, *39*, 331–339. [CrossRef] [PubMed]
17. Kneeshaw, S.; Keyani, R.; Delorme-Hinoux, V.; Imrie, L.; Loake, G.J.; Le Bihan, T.; Reichheld, J.P.; Spoel, S.H. Nucleoredoxin Guards against Oxidative Stress by Protecting Antioxidant Enzymes. *Proc. Natl. Acad. Sci. USA* **2017**, *114*, 8414–8419. [CrossRef] [PubMed]
18. Idelfonso-García, O.G.; Alarcón-Sánchez, B.R.; Vásquez-Garzón, V.R.; Baltiérrez-Hoyos, R.; Villa-Treviño, S.; Muriel, P.; Serrano, H.; Pérez-Carreón, J.I.; Arellanes-Robledo, J. Is Nucleoredoxin a Master Regulator of Cellular Redox Homeostasis? Its Implication in Different Pathologies. *Antioxidants* **2022**, *11*, 670. [CrossRef] [PubMed]
19. Funato, Y.; Michiue, T.; Asashima, M.; Miki, H. The Thioredoxin-Related Redox-Regulating Protein Nucleoredoxin Inhibits Wnt-Beta-Catenin Signalling through Dishevelled. *Nat. Cell Biol.* **2006**, *8*, 501–508. [CrossRef] [PubMed]
20. Arellanes-Robledo, J.; Reyes-Gordillo, K.; Shah, R.; Dominguez-Rosales, J.A.; Hernandez-Nazara, Z.H.; Ramirez, F.; Rojkind, M.; Lakshman, M.R. Fibrogenic Actions of Acetaldehyde Are Beta-Catenin Dependent but Wingless Independent: A Critical Role of Nucleoredoxin and Reactive Oxygen Species in Human Hepatic Stellate Cells. *Free Radic. Biol. Med.* **2013**, *65*, 1487–1496. [CrossRef] [PubMed]
21. Arellanes-Robledo, J.; Reyes-Gordillo, K.; Ibrahim, J.; Leckey, L.; Shah, R.; Lakshman, M.R. Ethanol Targets Nucleoredoxin/Dishevelled Interactions and Stimulates Phosphatidylinositol 4-Phosphate Production In Vivo and In Vitro. *Biochem. Pharmacol.* **2018**, *156*, 135–146. [CrossRef]
22. Alarcon-Sanchez, B.R.; Guerrero-Escalera, D.; Rosas-Madrigal, S.; Aparicio-Bautista, D.I.; Reyes-Gordillo, K.; Lakshman, M.R.; Ortiz-Fernandez, A.; Quezada, H.; Medina-Contreras, O.; Villa-Trevino, S.; et al. Nucleoredoxin Interaction with Flightless-I/Actin Complex Is Differentially Altered in Alcoholic Liver Disease. *Basic Clin. Pharmacol. Toxicol.* **2020**, *127*, 389–404. [CrossRef]
23. Gu, Z.; Tan, W.; Feng, G.; Meng, Y.; Shen, B.; Liu, H.; Cheng, C. Wnt/Beta-Catenin Signaling Mediates the Senescence of Bone Marrow-Mesenchymal Stem Cells from Systemic Lupus Erythematosus Patients through the P53/P21 Pathway. *Mol. Cell. Biochem.* **2014**, *387*, 27–37. [CrossRef]
24. Feng, G.; Zheng, K.; Cao, T.; Zhang, J.; Lian, M.; Huang, D.; Wei, C.; Gu, Z.; Feng, X. Repeated Stimulation by Lps Promotes the Senescence of Dpscs via Tlr4/Myd88-Nf-Kappab-P53/P21 Signaling. *Cytotechnology* **2018**, *70*, 1023–1035. [CrossRef] [PubMed]
25. Bellanti, F.; Vendemiale, G. The Aging Liver: Redox Biology and Liver Regeneration. *Antioxid. Redox Signal.* **2021**, *35*, 832–847. [CrossRef] [PubMed]
26. Koch, O.R.; Fusco, S.; Ranieri, S.C.; Maulucci, G.; Palozza, P.; Larocca, L.M.; Cravero, A.A.; Farre, S.M.; De Spirito, M.; Galeotti, T.; et al. Role of the Life Span Determinant P66(Shca) in Ethanol-Induced Liver Damage. *Lab. Investig.* **2008**, *88*, 750–760. [CrossRef] [PubMed]
27. Dimri, G.P.; Lee, X.; Basile, G.; Acosta, M.; Scott, G.; Roskelley, C.; Medrano, E.E.; Linskens, M.; Rubelj, I.; Pereira-Smith, O.; et al. A Biomarker That Identifies Senescent Human Cells in Culture and in Aging Skin In Vivo. *Proc. Natl. Acad. Sci. USA* **1995**, *92*, 9363–9367. [CrossRef] [PubMed]
28. Karsdal, M.A.; Daniels, S.J.; Nielsen, S.H.; Bager, C.; Rasmussen, D.G.K.; Loomba, R.; Surabattula, R.; Villesen, I.F.; Luo, Y.; Shevell, D.; et al. Collagen Biology and Non-Invasive Biomarkers of Liver Fibrosis. *Liver Int.* **2020**, *40*, 736–750. [CrossRef] [PubMed]
29. Tang, C.; Chen, H.; Jiang, L.; Liu, L. Liver Regeneration: Changes in Oxidative Stress, Immune System, Cytokines, and Epigenetic Modifications Associated with Aging. *Oxidative Med. Cell. Longev.* **2022**, *2022*, 9018811. [CrossRef]
30. Schreiber, V.; Dantzer, F.; Ame, J.C.; de Murcia, G. Poly(Adp-Ribose): Novel Functions for an Old Molecule. *Nat. Rev. Mol. Cell Biol.* **2006**, *7*, 517–528. [CrossRef]
31. Fedorova, M.; Bollineni, R.C.; Hoffmann, R. Protein Carbonylation as a Major Hallmark of Oxidative Damage: Update of Analytical Strategies. *Mass Spectrom. Rev.* **2014**, *33*, 79–97. [CrossRef]
32. Arellanes-Robledo, J.; Ibrahim, J.; Reyes-Gordillo, K.; Shah, R.; Leckey, L.; Lakshman, M.R. Flightless-I Is a Potential Biomarker for the Early Detection of Alcoholic Liver Disease. *Biochem. Pharmacol.* **2021**, *183*, 114323. [CrossRef] [PubMed]
33. Kim, I.H.; Kisseleva, T.; Brenner, D.A. Aging and Liver Disease. *Curr. Opin. Gastroenterol.* **2015**, *31*, 184–191. [CrossRef] [PubMed]
34. Cheng, Y.; Wang, X.; Wang, B.; Zhou, H.; Dang, S.; Shi, Y.; Hao, L.; Luo, Q.; Jin, M.; Zhou, Q.; et al. Aging-Associated Oxidative Stress Inhibits Liver Progenitor Cell Activation in Mice. *Aging* **2017**, *9*, 1359–1374. [CrossRef] [PubMed]
35. Seitz, H.K.; Stickel, F. Alcoholic Liver Disease in the Elderly. *Clin. Geriatr. Med.* **2007**, *23*, 905–921, viii. [CrossRef] [PubMed]
36. Fujino, G.; Noguchi, T.; Takeda, K.; Ichijo, H. Thioredoxin and Protein Kinases in Redox Signaling. *Semin. Cancer Biol.* **2006**, *16*, 427–435. [CrossRef] [PubMed]
37. Osna, N.A.; Donohue, T.M., Jr.; Kharbanda, K.K. Alcoholic Liver Disease: Pathogenesis and Current Management. *Alcohol Res.* **2017**, *38*, 147–161.
38. Adrover, J.M.; Nicolas-Avila, J.A.; Hidalgo, A. Aging: A Temporal Dimension for Neutrophils. *Trends Immunol.* **2016**, *37*, 334–345. [CrossRef] [PubMed]

39. Bologna-Molina, R.; Mosqueda-Taylor, A.; Molina-Frechero, N.; Mori-Estevez, A.D.; Sanchez-Acuna, G. Comparison of the Value of Pcna and Ki-67 as Markers of Cell Proliferation in Ameloblastic Tumors. *Med. Oral Patol. Oral Cir. Bucal* **2013**, *18*, e174–e179. [CrossRef]
40. Pibiri, M. Liver Regeneration in Aged Mice: New Insights. *Aging* **2018**, *10*, 1801–1824. [CrossRef]
41. Mendenhall, C.L.; Rouster, S.D.; Roselle, G.A.; Grossman, C.J.; Ghosn, S.; Gartside, P. Impact of Chronic Alcoholism on the Aging Rat: Changes in Nutrition, Liver Composition, and Mortality. *Alcohol Clin. Exp. Res.* **1993**, *17*, 847–853. [CrossRef]
42. Rao, R. Endotoxemia and Gut Barrier Dysfunction in Alcoholic Liver Disease. *Hepatology* **2009**, *50*, 638–644. [CrossRef]
43. Piskunova, T.S.; Yurova, M.N.; Ovsyannikov, A.I.; Semenchenko, A.V.; Zabezhinski, M.A.; Popovich, I.G.; Wang, Z.Q.; Anisimov, V.N. Deficiency in Poly(Adp-Ribose) Polymerase-1 (Parp-1) Accelerates Aging and Spontaneous Carcinogenesis in Mice. *Curr. Gerontol. Geriatr. Res.* **2008**, *2008*, 754190. [CrossRef]
44. Dong, R.; Wang, X.; Wang, L.; Wang, C.; Huang, K.; Fu, T.; Liu, K.; Wu, J.; Sun, H.; Meng, Q. Yangonin Inhibits Ethanol-Induced Hepatocyte Senescence Via Mir-194/Fxr Axis. *Eur. J. Pharmacol.* **2021**, *890*, 173653. [CrossRef] [PubMed]
45. Lu, C.; Ge, T.; Shao, Y.; Cui, W.; Li, Z.; Xu, W.; Bao, X. Znf281 Drives Hepatocyte Senescence in Alcoholic Liver Disease by Reducing Hk2-Stabilized Pink1/Parkin-Mediated Mitophagy. *Cell Prolif.* **2023**, *56*, e13378. [CrossRef] [PubMed]
46. Narita, M.; Narita, M.; Krizhanovsky, V.; Nunez, S.; Chicas, A.; Hearn, S.A.; Myers, M.P.; Lowe, S.W. A Novel Role for High-Mobility Group a Proteins in Cellular Senescence and Heterochromatin Formation. *Cell* **2006**, *126*, 503–514. [CrossRef] [PubMed]
47. Sedelnikova, O.A.; Horikawa, I.; Zimonjic, D.B.; Popescu, N.C.; Bonner, W.M.; Barrett, J.C. Senescing Human Cells and Ageing Mice Accumulate DNA Lesions with Unrepairable Double-Strand Breaks. *Nat. Cell Biol.* **2004**, *6*, 168–170. [CrossRef]
48. Calder, P.C.; Bosco, N.; Bourdet-Sicard, R.; Capuron, L.; Delzenne, N.; Dore, J.; Franceschi, C.; Lehtinen, M.J.; Recker, T.; Salvioli, S.; et al. Health Relevance of the Modification of Low Grade Inflammation in Ageing (Inflammageing) and the Role of Nutrition. *Ageing Res. Rev.* **2017**, *40*, 95–119. [CrossRef] [PubMed]
49. Maeso-Diaz, R.; Ortega-Ribera, M.; Fernandez-Iglesias, A.; Hide, D.; Munoz, L.; Hessheimer, A.J.; Vila, S.; Frances, R.; Fondevila, C.; Albillos, A.; et al. Effects of Aging on Liver Microcirculatory Function and Sinusoidal Phenotype. *Aging Cell* **2018**, *17*, e12829. [CrossRef] [PubMed]
50. Nechemia-Arbely, Y.; Shriki, A.; Denz, U.; Drucker, C.; Scheller, J.; Raub, J.; Pappo, O.; Rose-John, S.; Galun, E.; Axelrod, J.H. Early Hepatocyte DNA Synthetic Response Posthepatectomy Is Modulated by Il-6 Trans-Signaling and Pi3k/Akt Activation. *J. Hepatol.* **2011**, *54*, 922–929. [CrossRef] [PubMed]
51. Chandrasekaran, A.; Idelchik, M.; Melendez, J.A. Redox Control of Senescence and Age-Related Disease. *Redox Biol.* **2017**, *11*, 91–102. [CrossRef]
52. Dalle-Donne, I.; Giustarini, D.; Colombo, R.; Rossi, R.; Milzani, A. Protein Carbonylation in Human Diseases. *Trends Mol. Med.* **2003**, *9*, 169–176. [CrossRef] [PubMed]
53. Shearn, C.T.; Pulliam, C.F.; Pedersen, K.; Meredith, K.; Mercer, K.E.; Saba, L.M.; Orlicky, D.J.; Ronis, M.J.; Petersen, D.R. Knockout of the Gsta4 Gene in Male Mice Leads to an Altered Pattern of Hepatic Protein Carbonylation and Enhanced Inflammation Following Chronic Consumption of an Ethanol Diet. *Alcohol Clin. Exp. Res.* **2018**, *42*, 1192–1205. [CrossRef] [PubMed]
54. Deguine, J.; Barton, G.M. Myd88: A Central Player in Innate Immune Signaling. *F1000Prime Rep.* **2014**, *6*, 97. [CrossRef] [PubMed]
55. Ruzehaji, N.; Mills, S.J.; Melville, E.; Arkell, R.; Fitridge, R.; Cowin, A.J. The Influence of Flightless I on Toll-Like-Receptor-Mediated Inflammation in a Murine Model of Diabetic Wound Healing. *BioMed Res. Int.* **2013**, *2013*, 389792. [CrossRef]
56. Hayashi, T.; Funato, Y.; Terabayashi, T.; Morinaka, A.; Sakamoto, R.; Ichise, H.; Fukuda, H.; Yoshida, N.; Miki, H. Nucleoredoxin Negatively Regulates Toll-Like Receptor 4 Signaling via Recruitment of Flightless-I to Myeloid Differentiation Primary Response Gene (88). *J. Biol. Chem.* **2010**, *285*, 18586–18593. [CrossRef]
57. Tran, B.N.; Valek, L.; Wilken-Schmitz, A.; Fuhrmann, D.C.; Namgaladze, D.; Wittig, I.; Tegeder, I. Reduced Exploratory Behavior in Neuronal Nucleoredoxin Knockout Mice. *Redox Biol.* **2021**, *45*, 102054. [CrossRef]
58. Yalcin, A.; Telang, S.; Clem, B.; Chesney, J. Regulation of Glucose Metabolism by 6-Phosphofructo-2-Kinase/Fructose-2,6-Bisphosphatases in Cancer. *Exp. Mol. Pathol.* **2009**, *86*, 174–179. [CrossRef]
59. Funato, Y.; Hayashi, T.; Irino, Y.; Takenawa, T.; Miki, H. Nucleoredoxin Regulates Glucose Metabolism via Phosphofructokinase 1. *Biochem. Biophys. Res. Commun.* **2013**, *440*, 737–742. [CrossRef]
60. Bellentani, S.; Scaglioni, F.; Ciccia, S.; Bedogni, G.; Tiribelli, C. Hcv, Hbv and Alcohol—The Dionysos Study. *Dig. Dis.* **2010**, *28*, 799–801. [CrossRef] [PubMed]

antioxidants

Article

Liposomal Glutathione Augments Immune Defenses against Respiratory Syncytial Virus in Neonatal Mice Exposed in Utero to Ethanol

Theresa W. Gauthier *, Xiao-Du Ping, Frank L. Harris and Lou Ann S. Brown

Department of Pediatrics, Division of Neonatal-Perinatal Medicine, Emory University, Atlanta, GA 30322, USA; xping@emory.edu (X.-D.P.); fharris@emory.edu (F.L.H.); lbrow03@emory.edu (L.A.S.B.)
* Correspondence: tgauthi@emory.edu; Fax: +1-404-727-3236

Abstract: We previously reported that maternal alcohol use increased the risk of sepsis in premature and term newborns. In the neonatal mouse, fetal ethanol (ETOH) exposure depleted the antioxidant glutathione (GSH), which promoted alveolar macrophage (AM) immunosuppression and respiratory syncytial virus (RSV) infections. In this study, we explored if oral liposomal GSH (LGSH) would attenuate oxidant stress and RSV infections in the ETOH-exposed mouse pups. C57BL/6 female mice were pair-fed a liquid diet with 25% of calories from ethanol or maltose–dextrin. Postnatal day 10 pups were randomized to intranasal saline, LGSH, and RSV. After 48 h, we assessed oxidant stress, AM immunosuppression, pulmonary RSV burden, and acute lung injury. Fetal ETOH exposure increased oxidant stress threefold, lung RSV burden twofold and acute lung injury threefold. AMs were immunosuppressed with decreased RSV clearance. However, LGSH treatments of the ETOH group normalized oxidant stress, AM immune phenotype, the RSV burden, and acute lung injury. These studies suggest that the oxidant stress caused by fetal ETOH exposure impaired AM clearance of infectious agents, thereby increasing the viral infection and acute lung injury. LGSH treatments reversed the oxidative stress and restored AM immune functions, which decreased the RSV infection and subsequent acute lung injury.

Keywords: fetal ethanol exposure; alveolar macrophage; respiratory syncytial virus; oxidant stress; liposomal glutathione

check for updates

Citation: Gauthier, T.W.; Ping, X.-D.; Harris, F.L.; Brown, L.A.S. Liposomal Glutathione Augments Immune Defenses against Respiratory Syncytial Virus in Neonatal Mice Exposed in Utero to Ethanol. *Antioxidants* **2024**, *13*, 137. https://doi.org/10.3390/antiox13020137

Academic Editor: Marco Fiore

Received: 29 December 2023
Revised: 18 January 2024
Accepted: 20 January 2024
Published: 23 January 2024

1. Introduction

In clinical studies, we previously reported that maternal alcohol use during pregnancy was associated with an increased risk of sepsis in both premature and term newborns [1,2]. Furthermore, in premature newborns weighing $\leq$ 1500 g at birth, maternal admittance of alcohol ingestion during pregnancy occurred in one third of pregnancies. In utero alcohol exposure in these low-birthweight premature newborns was associated with significantly increased odds of developing neonatal late-onset sepsis and bronchopulmonary dysplasia [1,2]. In addition to fetal alcohol spectrum disorder, other studies have demonstrated that alcohol consumption during pregnancy is associated with a range of adverse outcomes for the newborn, including spontaneous abortions [3], stillbirths [4,5], preterm birth [6,7], and low birthweight [8,9]. Despite these outcomes, our understanding of in utero alcohol's detrimental effects on the developing lung remain limited. Studies to further understand the mechanisms underlying the risks of in utero alcohol exposure for the premature newborn, specifically in pulmonary health, are needed.

Using experimental animal models, we showed that ethanol (ETOH) exposure results in chronic oxidant stress in both the neonatal and the adult lung, which resulted in impaired cellular immune defenses of the resident alveolar macrophage (AM) against infectious agents. It is the decreased availability of glutathione (GSH), the primary antioxidant in the airspace, that is one key to the ETOH-induced immune dysfunctions of both newborn and adult AM [10,11]. Over time, the ETOH-induced decreases in GSH and subsequent chronic oxidant stress promote the release of immune suppressors like transforming growth factor β1 (TGFβ1) that subsequently result in compromised immune cells that are critical in determining disease outcomes [12]. These studies also demonstrated that strategies to improve GSH availability subsequently improved the immune responses of newborn and adult AM, including both viral and bacterial clearance.

For infants, respiratory syncytial virus (RSV) is the most common etiologic agent for acute respiratory infections, infections in the lower respiratory tract such as bronchiolitis and pneumonia, and the leading cause of hospitalization for infants less than 2 years of age [13–16]. Newborns at the greatest risk for severe infections include the premature newborn and infants with chronic respiratory diseases such as bronchopulmonary dysplasia related to prematurity [17]. When compared to adults, the newborn's antioxidant status is low, but the generation of reactive oxygen species and lipid peroxidation products that occurs during a viral infection exacerbates their limited antioxidant availability, results in chronic oxidant stress, and contributes to the pathogenesis [18]. In premature newborns, their immature immune responses [19] also contribute to their risk of increased incidence and severity of viral infections such as RSV [20,21]. Indeed, lower respiratory tract infections with viruses such as RSV result in extended hospital stays, readmission to the pediatric intensive care unit, an increased need for oxygen and mechanical ventilation, and increased mortality [22]. Although airway epithelial cells are the primary site for RSV infection, innate immune responses by the mononuclear phagocytic system, including phagocytosis and the release of cytokines, chemokines, and other immune mediators, are critical for the resolution of pulmonary viral diseases [19,23]. As the resident mononuclear phagocytic cell in the lung and airspace, the AM is the key modulator of the local pulmonary immune response that is essential for RSV clearance [23,24].

The key role of antioxidant availability as a modulator of the pathogenic process was demonstrated in mouse models where antioxidant treatments attenuated RSV-induced oxidant stress as well as the viral burden [25,26]. In adult animal models, chronic alcohol exposure impairs multiple arms of immune defense mechanisms against respiratory viruses such as RSV [27,28]. In fetal lambs, in utero ETOH exposure predisposes the developing lung to RSV by altering host defenses via deranged surfactant proteins [29,30]. We recently demonstrated in our established fetal ETOH mouse model that in utero ETOH exposure impaired AM cellular capacity to defend against experimental RSV from the lung [31]. This impairment may potentially be modulated through the upregulation of the immunosuppressant TGFβ1, since we also observed that TGFβ1 directly impairs the capacity of AM to clear RSV [12].

Supplementation with an oral liposomal formulation of GSH (LGSH) has been shown to improve intracellular delivery of GSH and decrease oxidant stress in HIV subjects and patients with type 2 diabetes [32–34]. In addition, these studies demonstrated that LGSH treatments improved the immune responses of peripheral blood monocytes cells when treated in vitro with *Mycobacterium tuberculosis*, including improved bacterial clearance [33,34]. In ventilated preterm infants, a single intratracheal dose of LGSH increased pulmonary GSH pools and decreased oxidative stress [35]. The possibility that in vivo LGSH administration to the neonate exposed to ETOH in utero may improve immune defenses against viruses such as RSV is attractive, but has not been investigated. In the current study, we hypothesized that strategies such as LGSH treatment would augment the antioxidant GSH in the neonatal lung exposed to ETOH in utero, decrease AM oxidant stress, attenuate cellular immunosuppression, and improve neonatal AM immune defenses against an experimental RSV infection. The goals of the current study were to use an established mouse model of in utero ETOH exposure to (1) determine if a clinically relevant intervention, such as enteral LGSH, could protect against an experimental pulmonary RSV infection and (2) explore the potential mechanisms by which LGSH improved AM innate immune defenses against RSV.

2. Materials and Methods

Mouse model of fetal ethanol (ETOH) exposure. Our established mouse model of fetal ETOH exposure uses a continuous exposure of ETOH during pregnancy through a maternal Lieber-DeCarli liquid diet containing ETOH (BioServ, Frenchtown, NJ, USA) [31]. Female C57BL/6 mice were shipped from the vendor (Charles River, Burlington, MA, USA) and acclimated in the Emory Pediatrics animal facilities for one week. After breeding, the experimental liquid diet was started one day after visualization of the vaginal mucus plug. Pregnant dams were randomized to receive an isocaloric liquid diet ±25% ETOH-derived calories. For the ETOH group, the ETOH content of the diet was incrementally ramped up over a 1-week timeframe from 0% to 12.5% and then 25% of ETOH-derived calories. Food consumption was recorded daily and the liquid food was changed daily. The control group was pair-fed to the ETOH group with 25% of the calories obtained from maltose–dextrin. The only access to food during the experimental period was the assigned experimental liquid diet. The diet was continued throughout pregnancy and after spontaneous term delivery. Pups were kept with their respective dams and allowed to nurse *ad libitum*. All animals were used with protocols reviewed and approved by the Emory University Institutional Animal Care Committee (PROTO201800128) in accordance with NIH guidelines (Guide for the Care and Use of Laboratory Animals).

Respiratory syncytial virus (RSV) and RSV Plaque Assay. RSV clone rA2 | 19F was a generous gift from Martin L. Moore, PhD [36]. As we previously reported [12,31], RSV was propagated in Hep-2 (CCL-23) cells in minimum essential medium supplemented with 10% fetal bovine serum, penicillin (100 U/mL) and streptomycin (100 μg/mL) (Sigma, St. Louis, MO, USA). RSV was then harvested 6–7 days after the Hep-2 cell inoculation and sonicated on ice and centrifuged (500× *g*, 10 min at 4 °C). Plaque assays were performed to determine RSV titers (plaque forming units (PFUs)) by serially diluting the supernatant, infecting 24-well plates of Hep-2 cells (6 days at 37 °C and 5% CO_2) and visualizing immunostaining.

Liposomal glutathione (LGSH) treatments and experimental Respiratory syncytial virus (RSV) in the neonatal mouse. To investigate a potential therapeutic role for LGSH during an in vivo RSV infection, we incorporated LGSH into our established in vivo mouse model of inhaled RSV [31]. LGSH (ReadiSorb Liposomal Glutathione) was a generous gift from Dr. Frederick T. Guilford (Your Energy Systems, Palo Alto, CA, USA). This preparation of LGSH contained reduced GSH (422.7 g/5 mL) plus purified water, glycerin, lecithin, and potassium sorbate. On day of life 10 (P10), mice pups (±in utero ETOH exposure) were treated with an oral gavage containing either LGSH (20 μL, 1.7 mg of L-glutathione) or saline (20 μL). Mouse pups were then given intranasal injections of RSV (Nanoliter Injector; 20 μL; each nasal nare; 2×10^5 PFU) before they were returned to their respective dams. After 24 h, pups received an additional dose of ±LGSH (or saline) by oral gavage. All pups were then euthanized for analyses after 48 h.

Alveolar macrophage (AM) isolation. After euthanasia with intraperitoneal sodium pentobarbital, the pup trachea was identified under a dissecting microscope and cannulated with a 27 G catheter. The lungs were then serially lavaged via the catheter with 40 μL sterile saline (5 times) to remove the bronchoalveolar fluid lining the airspace. The initial lavage from each pup in a litter was pooled and centrifuged (402× *g*; 8 min) and the cell-free supernatant lavage fluid was saved for further analysis (noted below). The subsequent bronchoalveolar lavages (BALs) from each pup within the same litter were also pooled and similarly centrifuged. The cell pellets obtained from the initial and the subsequent lavages were resuspended in media (RPMI 1640 1 time) containing 10% fetal bovine serum and 1% antibiotics before they were pooled. The pooled cells from each litter represents an *n* of 1. Cell viability and cell count were determined with trypan blue stain (0.4%; Life Technologies, Grand Island, NY, USA). Pup AMs were cultured on slides, fixed with 3.7% paraformaldehyde and permeabilized with ice-cold methanol.

Plasma collection for systemic biomarkers of oxidant stress. After euthanasia with intraperitoneal sodium pentobarbital, blood samples were obtained from all pups via cardiac puncture and the samples pooled per experimental group and litter. Samples were spun and stored at −80 °C until batch analyses. Total plasma antioxidant capacity (AOC) was measured via colorimetric assay (MAK187, Sigma-Aldrich, St. Louis, MO, USA) and 8-hydroxyguanosine (8-OHdG) was measured by ELISA (DNA Damage Competitive ELISA kit, Life Technologies Corporation, Carlsbad, CA, USA).

Determination of the Respiratory syncytial virus (RSV) burden in the bronchoalveolar lavage fluid (BAL) and whole lung. All pups were euthanized for analyses 48 h after intranasal delivery of RSV. The lungs were then serially lavaged via the catheter with 40 µL sterile saline (5 times) to remove the fluid and cells from the airspace. To determine the RSV burden in the airways and alveolar space, the multiple BALs from pups within the same litter were pooled. The pooled BAL was then serially diluted in phosphate-buffered saline and plated for determination of RSV growth via the Hep-2 cell plaque assay, as we have previously described [31]. After the pup was euthanized, the right upper lobe of the lung was flash-frozen in liquid nitrogen and stored at −80 °C until batch analysis. The lobe was weighed, sterile phosphate-buffered saline (10 times the tissue weight) was added, and then the lobe homogenized on ice. After the samples were centrifuged (2000× g; 10 min; 4 °C), the supernatants were serially diluted in phosphate-buffered saline and then 100 µL of the homogenate dilution was similarly plated for determination of RSV growth. For the whole lung, RSV is presented as percentage of control of pfu/g lung tissue. The remaining isolated neonatal lung lobes were also flash frozen in liquid nitrogen and stored at −80 °C until batch analyses (see below).

Cellular immunostaining. The freshly isolated AMs were plated and fixed with 3.7% paraformaldehyde before permeabilization with ice-cold methanol. For assessment of RSV phagocytosis by the AM in vivo, we evaluated whole-cell RSV content via fluorescent immunostaining (1 h; a 1:100 dilution; Santa Cruz Biotechnology, Inc., Santa Cruz, CA, USA). Given the critical role of the antioxidant GSH in the AM immune phenotype, we evaluated whole-cell GSH via fluorescent immunostaining (1 h; 1:100 dilution; Abcam, Inc®, Boston, MA, USA). Since fetal ETOH exposure increases AM expression of the immunosuppressant TGFβ1 [11,31] and arginase-1 (Arg-1) [11], we also used immunostaining to evaluate whole-cell TGFβ1 and Arg-1 in the neonatal AM. Immunostaining for the neutrophil cell surface marker (Gr-1) was used to differentiate AM from polymorphonuclear leukocytes (PMNs). Cells were incubated with the respective primary antibody in a 1:100 dilution (Santa Cruz Biotechnology, Inc., Santa Cruz, CA, USA) for 1 h. After the slides were washed three times with phosphate-buffered saline over 5 min, and the secondary antibody (anti-goat IgG, ThermoFisher, Waltham, MA, USA) was added in a 1:200 dilution and further incubated for 45 min. Cellular fluorescence was quantified using fluorescence microscopy via ImagePro Plus for Windows, Version 4.5 and presented as mean relative fluorescence units per cell (RFUS/cell) ± S.E.M. as tallied from at least 25 cells/litter. To correct for autofluorescence, the background fluorescence of unstained AMs was subtracted from the RFUs obtained for each analysis.

Whole-lung analyses. In addition to determining the tissue RSV, the flash-frozen neonatal lung tissue was evaluated for the inflammatory PMN marker myeloperoxidase (MPO, ng/mL) with a commercially available ELISA (product # MBS2702122, MyBioSource, Inc., San Diego, CA, USA). In parallel experiments, the neonatal right upper lobe lung samples were weighed (designated wet weight) and then reweighed after desiccation by overnight incubation at 70 °C (designated dry weight) and the lung wet/dry weight ratio was determined. MPO concentrations (ng/mL) were normalized to lung sample wet/dry weight.

Statistical analysis. SigmaPlot software (Systat Software 14.5; San Jose, CA, USA) was used for statistical analysis and graph generation. ANOVA or Kruskal–Wallis ANOVA on ranks was used when appropriate to detect overall differences between groups. Student–Newman–Keuls or Dunn's post hoc analysis was conducted for group comparisons as indicated. A value of $p \leq 0.05$ was deemed statistically significant. Data are presented as mean ± S.E.M, where each n represents one litter.

3. Results

Fetal ethanol (ETOH) exposure decreased alveolar macrophage (AM) viral clearance and exacerbated the experimental lung Respiratory syncytial virus (RSV) infection in the exposed pup. In agreement with our previous study [31], in utero ETOH exposure impaired AM phagocytosis of inhaled RSV by 40% in the neonatal mouse pup when compared to the control pup (Figure 1). With the ETOH-induced decrease in AM in virus phagocytosis, there was an accompanying increase in the burden of RSV growth in the pup lung. Both the BAL (Figure 2A) and neonatal whole lung (Figure 2B) demonstrated a ~twofold increase in RSV growth in the ETOH pups when compared to the control pups.

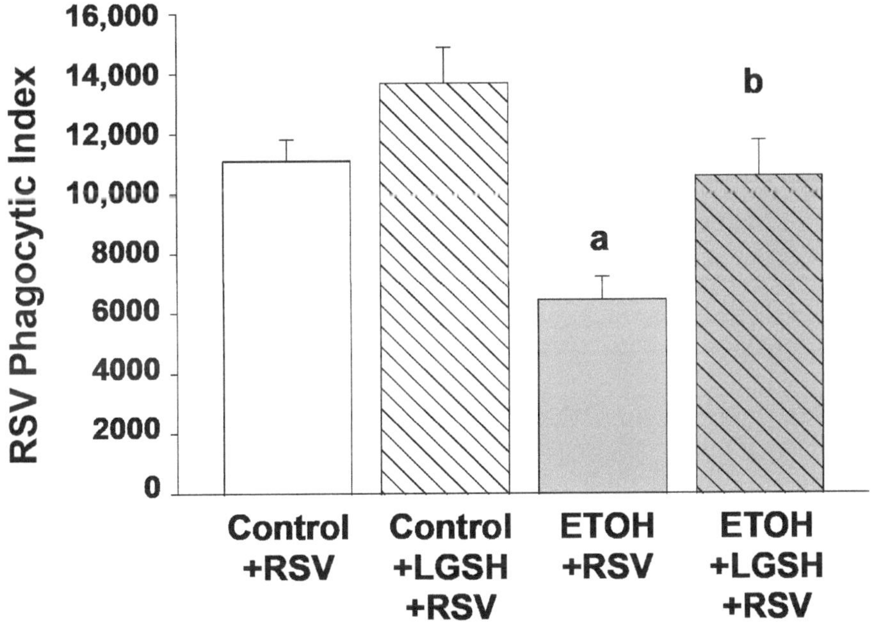

Figure 1. In utero ethanol (ETOH) exposure impaired in vivo alveolar macrophage (AM) phagocytosis of respiratory syncytial virus (RSV) but restored by enteral liposomal glutathione (LGSH) treatments. After breeding, the female mice were randomized to an experimental isocaloric liquid diet that contained either 25% ethanol (ETOH)-derived calories or 25% maltose–dextrin-derived calories. The diet was continued throughout pregnancy and after spontaneous term delivery. Pups were kept with their respective dams and allowed to nurse *ad libitum*. On postnatal day 10 (P10), pups from the control-fed dam and the ETOH-fed dam were then randomized to an oral gavage containing either liposomal glutathione (LGSH) (20 µL, 1.7 mg of L-glutathione) or saline (20 µL). All pups were then given intranasal injections of respiratory syncytial virus (RSV) (Nanoliter Injector; 20 µL; each nasal nare; 2×10^5 PFU) before they were returned to their respective dams. After 24 h, pups received an additional dose of ±LGSH (or saline) by gavage. All pups were then euthanized for analyses after 48 h, and the alveolar macrophages (AMs) from the pups were isolated from the multiple BALs and pooled per experimental group and litter. For assessment of RSV phagocytosis by the AM in vivo, we evaluated whole-cell RSV via fluorescent immunostaining (1 h; a 1:100 dilution; Santa Cruz Biotechnology, Inc., Santa Cruz, CA, USA). Background fluorescence of unstained AMs was used to account for autofluorescence and subtracted from the RFUs obtained for the RSV phagocytosis. The RSV burden was calculated relative to the RSV fluorescence for AMs from the control + RSV group. N = 5 litters for each group. ^{a} $p = 0.05$ when compared to the control + RSV group; ^{b} $p \leq 0.05$ when compared to the ETOH + RSV group.

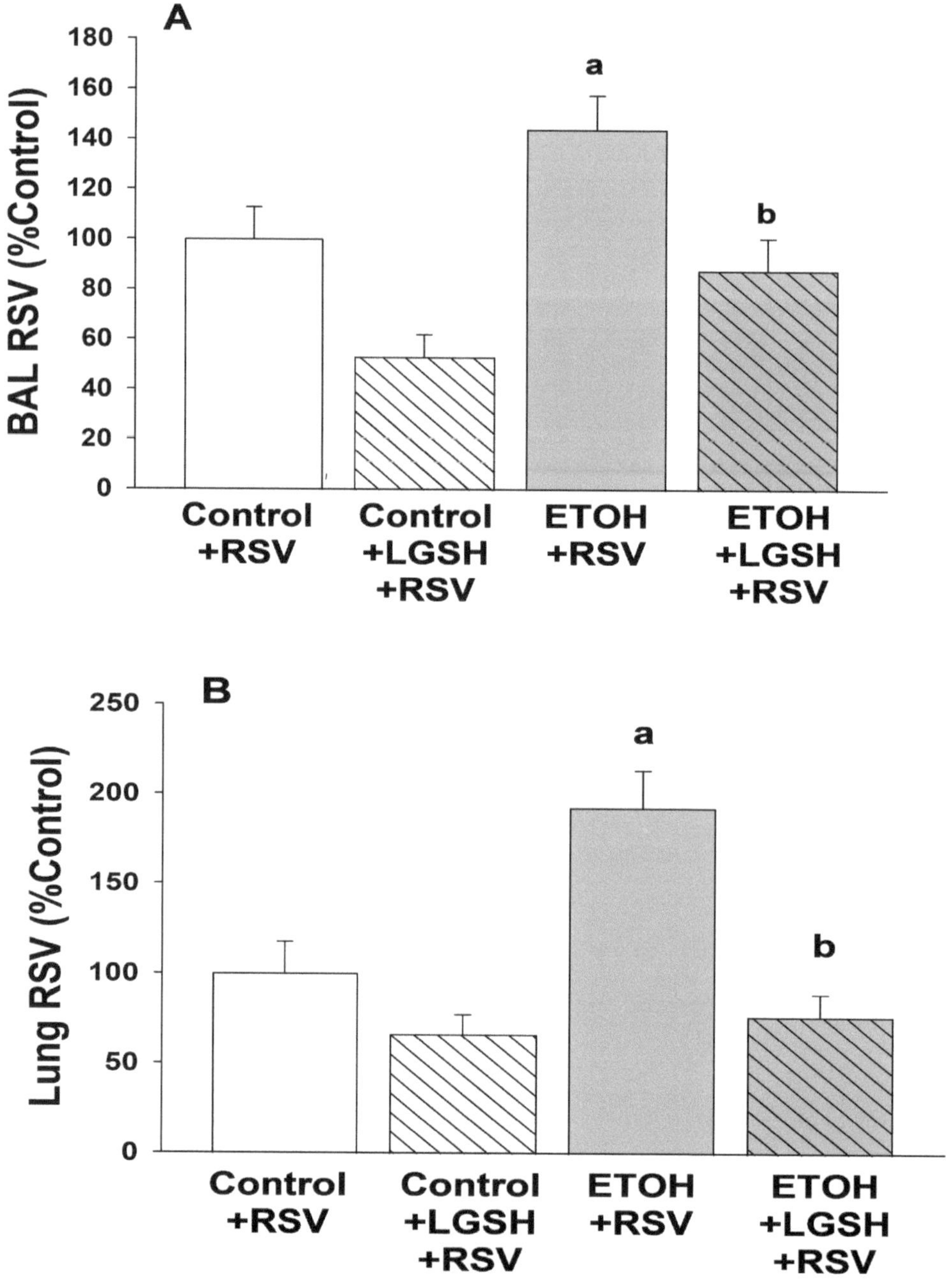

Figure 2. In utero ethanol (ETOH) exposure increased the respiratory syncytial virus (RSV) burden in the bronchoalveolar lavage (BAL, **A**) and lung tissue (**B**), but both were normalized by oral liposomal glutathione (LGSH) treatments. All pups were euthanized for analyses after 48 h of RSV delivery and the lungs were then serially lavaged via the catheter with 40 μL sterile saline (5 times) to remove the fluid and cells from the airspace. To determine the RSV burden in the airways and alveolar space, the multiple BALs from pups were pooled per experimental group and litter. The pooled BAL was serially

diluted in phosphate-buffered saline, plated for determination of RSV growth. The RSV burden in the BAL (**A**) is presented as percentage of control plaque forming units/mL (PFU/mL). To determine the RSV burden in the lung, the frozen right upper lung lobe was weighed, sterile phosphate-buffered saline (10 times the tissue weight) was added, and then the lobe homogenized on ice. After the samples were centrifuged (2000× *g*; 10 min; 4 °C), the supernatants were serially diluted in phosphate-buffered saline and plated for determination of RSV growth. For the whole lung (**B**), RSV is presented as percentage of control of PFU/g lung tissue. N = 5 litters for each group. [a] $p = 0.05$ when compared to the control + RSV group; [b] $p \leq 0.05$ when compared to the ETOH + RSV group.

Increased acute lung injury after Respiratory syncytial virus (RSV) inhalation in pups with in utero ethanol (ETOH) exposure. We next determined if the greater lung RSV burden associated with fetal ETOH exposure resulted in greater lung injury. With the lung wet/dry weight ratio as an indirect marker of lung edema, there was ~threefold increase in the ETOH + RSV group when compared to the control + RSV group (Figure 3A). Since pulmonary neutrophil infiltration can be associated with acute lung injury, the MPO content of the lung was used as a granulocyte marker. There was a ~fourfold increase in the MPO content of the lung in the ETOH + RSV group compared to the control + RSV group (Figure 3B). Using Gr-1+ as a marker of PMNs, we determined the percentage of cells that were neutrophils in the first lavage. In the ETOH + RSV group, there was ~twofold increase in neutrophils (GR-1+) in the alveolar space when compared to the control + RSV group (Figure 4). Taken together, these results suggest that compared to the control pup, fetal ETOH exposure resulted in a significant increase in these three indirect markers of acute lung injury after RSV inhalation.

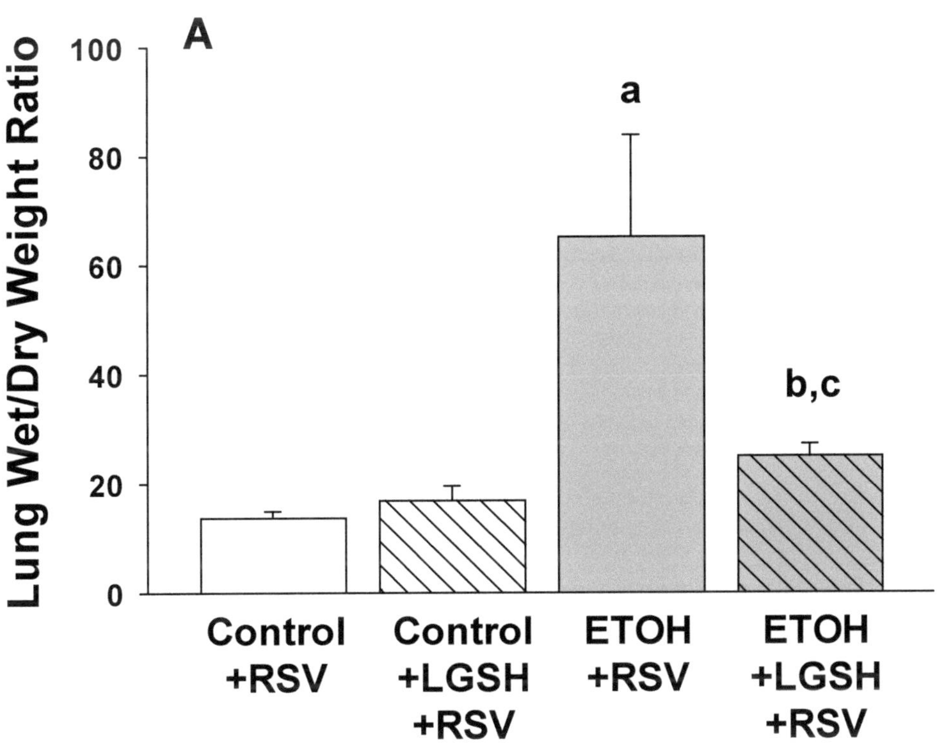

Figure 3. *Cont.*

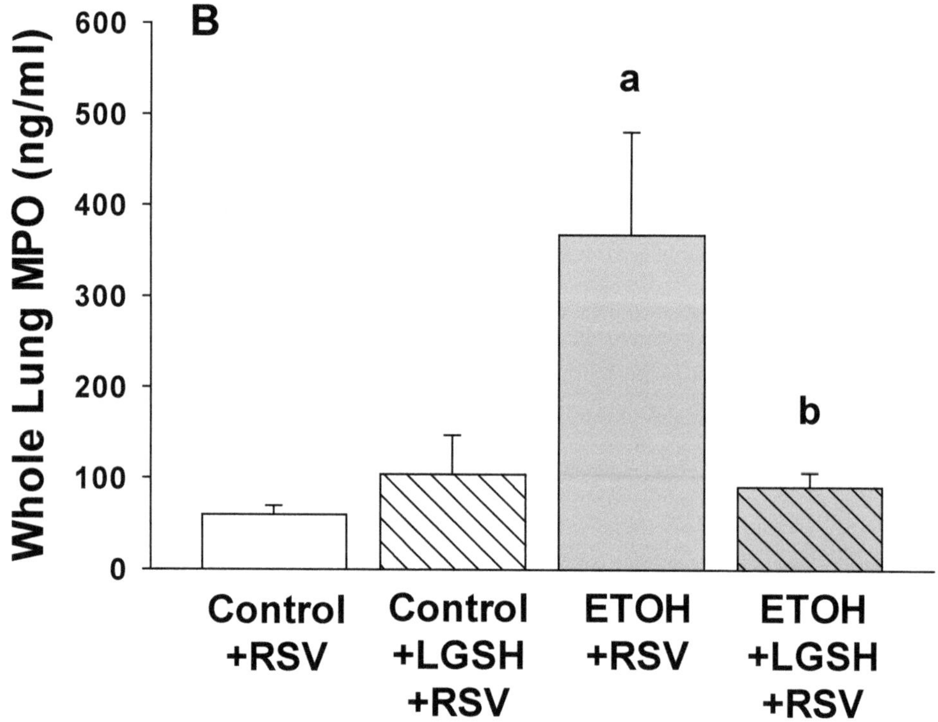

Figure 3. The amplified Respiratory syncytial virus (RSV) infection in the pup after in utero ethanol (ETOH) exposure resulted in increased acute lung injury, as defined by lung wet/dry weight ratio (**A**) and whole-lung myeloperoxidase (MPO) concentration (**B**), but acute lung injury was normalized by oral liposomal glutathione (LGSH) treatments. The flash-frozen right upper lobe lung samples from the neonatal pups were weighed (designated wet weight) and then reweighed after desiccation by overnight incubation at 70 °C (designated dry weight). The lung wet/dry weight ratio was determined as a marker of acute lung injury (**A**). The flash-frozen lung tissue was also evaluated for the inflammatory PMN marker myeloperoxidase (MPO) with a commercially available ELISA (product # MBS2702122, MyBioSource, Inc., San Diego, CA, USA). MPO concentrations (ng/mL) were normalized to the corresponding lung sample wet/dry weight (**B**). N = 5 litters for each group. [a] $p = 0.05$ when compared to the control + respiratory syncytial virus (RSV) group; [b] $p \leq 0.05$ when compared to the ETOH + RSV group; [c] denotes $p \leq 0.05$ when compared to the control + LGSH + RSV group.

Oxidant stress was exacerbated when the fetal ethanol (ETOH)-exposed pup was challenged with Respiratory syncytial virus (RSV). In the ETOH + RSV group, the plasma antioxidant capacity (AOC) was decreased ~35% when compared to the control + RSV group (Figure 5A). With plasma 8OH-dG as a marker of DNA oxidant damage, there was ~threefold increase in the ETOH + RSV group when compared to the control + RSV group (Figure 5B). When compared to the control + RSV group, fetal ETOH exposure superimposed on inhaled RSV also exacerbated AM oxidant stress, as shown by ~50% decrease in the cellular antioxidant GSH pool (Figure 6).

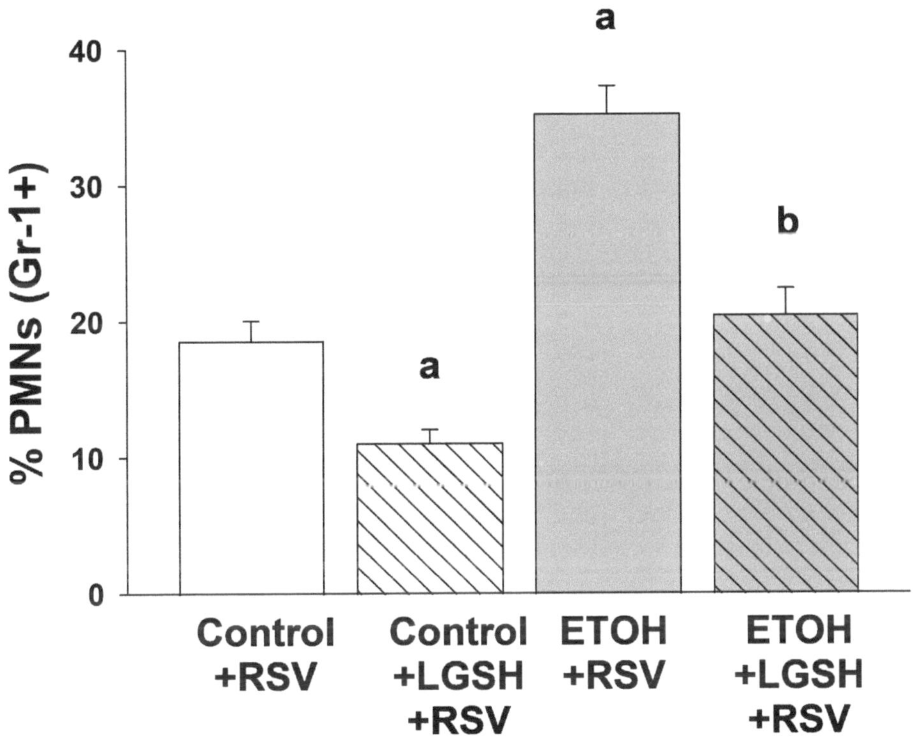

Figure 4. Respiratory syncytial virus (RSV) superimposed on in utero ethanol (ETOH) exposure increased polymorphonuclear leukocyte (PMN) migration into the alveolar space, but was attenuated by oral liposomal glutathione (LGSH) treatments. Immunostaining for the cell surface marker (Gr-1) was used to differentiate AM from the polymorphonuclear leukocytes (PMNs) that had migrated into the alveolar space. Cells retrieved from the lavage were plated, fixed, and incubated with the GR-1 primary antibody in a 1:100 dilution (Santa Cruz Biotechnology, Inc., Santa Cruz, CA, USA) for 1 h. After the slides were washed three times with phosphate-buffered saline over 5 min, the secondary antibody (anti-goat IgG) was added in a 1:200 dilution and further incubated for 45 min. Cellular fluorescence was quantified using fluorescence microscopy via ImagePro Plus for Windows Version 4.5 and presented as mean relative fluorescence units per cell (RFUS/cell) ± S.E.M. tallied from at least 25 cells/litter. To correct for autofluorescence, the background fluorescence of unstained AMs was subtracted from the RFUs obtained for each analysis. N = 5 litters for each group. [a] $p = 0.05$ when compared to the control + respiratory syncytial virus (RSV) group; [b] $p \leq 0.05$ when compared to the ETOH + RSV group.

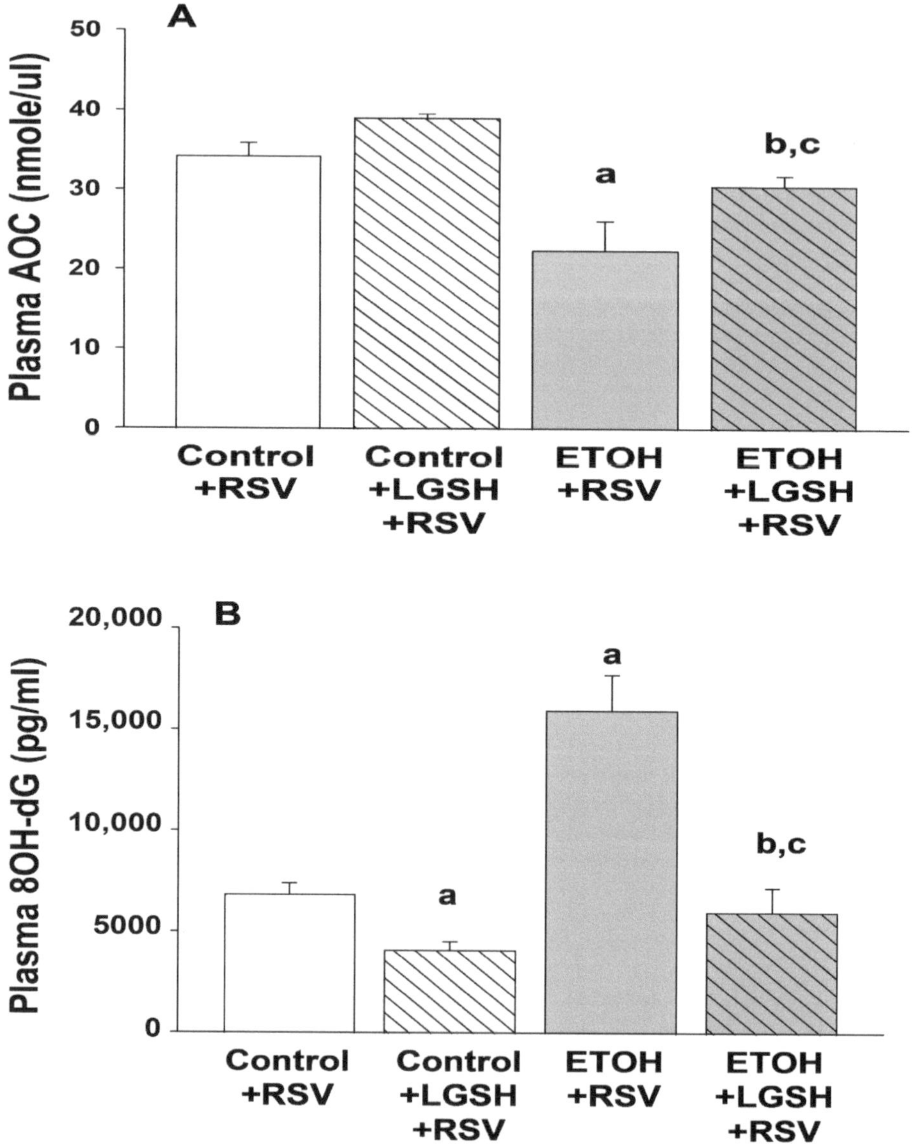

Figure 5. In utero ethanol (ETOH) exposure increased oxidant stress, as defined by a decrease in the plasma antioxidant capacity (AOC; **A**) and an increase in DNA oxidation (8OH-dG; **B**), but this oxidant stress was attenuated by oral liposomal glutathione (LGSH) treatments. After euthanasia with intraperitoneal sodium pentobarbital, blood samples were obtained from all pups via cardiac puncture and the samples pooled per experimental group and litter. Samples were spun and stored at −80 °C until batch analyses. Total plasma AOC (**A**) was measured via colorimetric assay (MAK187, Sigma-Aldrich, St. Louis, MO, USA) and 8-OHdG (**B**) was measured by ELISA (DNA Damage Competitive ELISA Kit, Life Technologies Corporation, Carlsbad, CA, USA). N = 5 litters for each group. ^{a} $p = 0.05$ when compared to the control + RSV group; ^{b} $p \leq 0.05$ when compared to the ETOH + respiratory syncytial virus (RSV) group; ^{c} $p \leq 0.05$ when compared to the control + LGSH + RSV group.

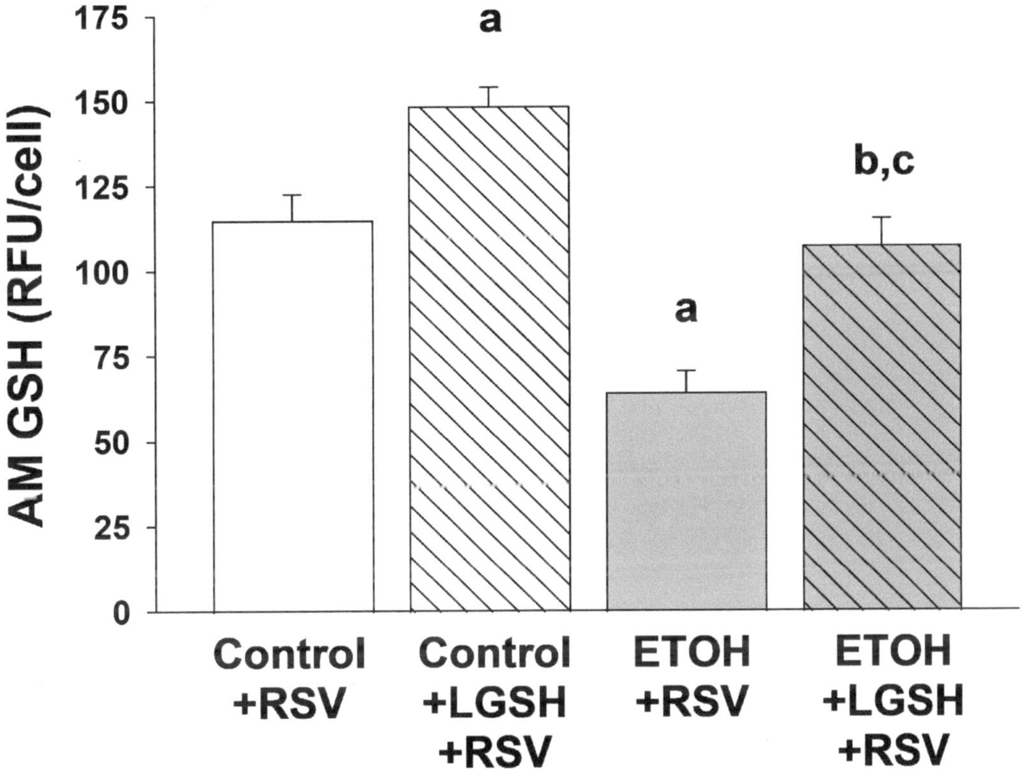

Figure 6. In utero ethanol (ETOH) exposure increased alveolar macrophage (AM) oxidant stress, as defined by a decrease in the glutathione (GSH) pool, but the GSH pool in the AM was restored by oral liposomal glutathione (LGSH) treatments. The freshly isolated AMs were plated and fixed with 3.7% paraformaldehyde before permeabilization with ice-cold methanol. The antioxidant GSH pool in the AMs was evaluated by whole-cell GSH via fluorescent immunostaining (1:100 dilution; Abcam, Inc®, Boston, MA, USA). Cellular fluorescence was quantified using fluorescence microscopy via ImagePro Plus for Windows version 4.5 [37] and is presented as mean relative fluorescence units per cell (RFUS/cell) ± S.E.M. as tallied from at least 25 cells/litter. To correct for autofluorescence, the background fluorescence of unstained AMs was subtracted from the RFUs obtained for each analysis. N = 5 litters for each group. [a] $p = 0.05$ when compared to the control + respiratory syncytial virus (RSV) group; [b] $p \leq 0.05$ when compared to the ETOH + RSV group; [c] $p \leq 0.05$ when compared to the control + LGSH + RSV group.

Increased markers of alveolar macrophage (AM) immune suppression with fetal ethanol (ETOH) exposure superimposed on experimental Respiratory syncytial virus (RSV). In the ETOH + RSV group, there was ~twofold increase in AM expression of the immunosuppressant TGFβ1 when compared to the control + RSV group (Figure 7A). Likewise, AM expression of Arg-1, an immunosuppressant, was increased ~twofold in the ETOH + RSV group when compared to the control + RSV group (Figure 7B).

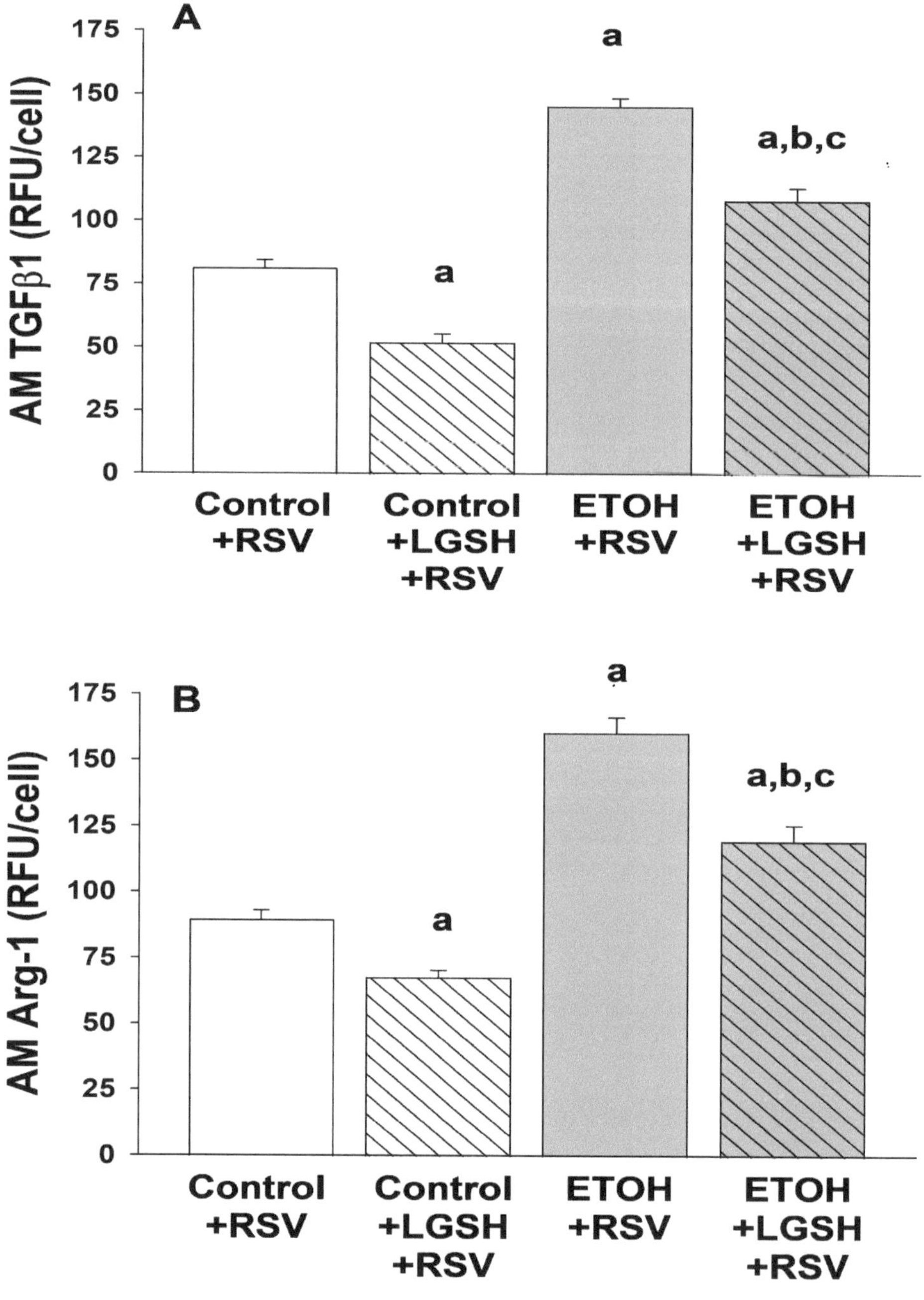

Figure 7. In utero ethanol (ETOH) exposure resulted in alveolar macrophage (AM) immunosuppression, as defined by an increase in transforming growth factor β1 (TGFβ1) (**A**) and arginase 1 (Arg-1) (**B**) expression, but oral liposomal glutathione (LGSH) treatments attenuated the AM immunosuppression. AM expression of TGFβ1 (**A**) and Arg-1 (**B**) were used as markers of immunosuppression. Cells were incubated with the primary antibody in a 1:100 dilution (Santa Cruz Biotechnology, Inc., Santa Cruz, CA, USA) for 1 h. After the slides were washed three times with

phosphate-buffered saline over 5 min, the secondary antibody (anti-goat IgG) was added in a 1:200 dilution and further incubated for 45 min. Cellular fluorescence was quantified using fluorescence microscopy via ImagePro Plus for Windows version 4.5 [38] and is presented as mean relative fluorescence units per cell (RFUS/cell) ± S.E.M. tallied from at least 25 cells/litter. To correct for autofluorescence, the background fluorescence of unstained AMs was subtracted from the RFUs obtained for each analysis. N = 5 litters for each group. ^{a} p = 0.05 when compared to the control + respiratory syncytial virus (RSV) group; ^{b} $p \leq 0.05$ when compared to the ETOH + RSV group; ^{c} $p \leq 0.05$ when compared to the control + LGSH + RSV group.

In vivo liposomal gluthaione (LGSH) treatments improved alveolar macrophage (AM) function and improved immune defense against Respiratory syncytial virus (RSV). To investigate the hypothesized therapeutic role of the antioxidant GSH in the setting of RSV, we evaluated whether LGSH administered to the pups by gavage would improve pulmonary defenses against experimental RSV. Enteral LGSH administered immediately before the intranasal delivery of RSV significantly improved the capacity of AMs to phagocytose RSV in the ETOH-exposed pup (Figure 1). Furthermore, RSV growth was significantly decreased in both the BAL (Figure 2A) and the whole lung (Figure 2B) of the ETOH-exposed pups gavaged with LGSH, approaching values observed in the control + RSV group.

In vivo liposomal glutathione (LGSH) treatments attenuated Respiratory syncytial virus (RSV)-induced acute lung injury. The diminished RSV growth seen in the ETOH-exposed pups gavaged with LGSH was accompanied by decreased markers of RSV-induced acute lung injury. Indeed, whole-lung wet/dry weight ratios (Figure 3A) and lung MPO levels (Figure 3B) were significantly decreased when the ETOH-exposed pup was treated with LGSH. Similarly, LGSH treatments also attenuated PMNs influx (Gr-1+ cells) into the airspace for both the RSV-treated control and the ETOH pups, suggestive of decreased acute lung injury (Figure 4).

Liposomal glutathione (LGSH) treatments also decreased the oxidant stress in pups challenged with Respiratory syncytial virus (RSV). In the ETOH + LGSH + RSV group, the plasma AOC was significantly improved when compared to the ETOH + RSV group (Figure 5A). Likewise, LGSH treatments resulted in a significant decrease in DNA oxidant damage for both the control + RSV group and the ETOH + RSV group (Figure 5B). Similar results were observed in the GSH pool for AM, as noted by improved GSH/cell for both the control + LGSH + RSV group and the ETOH + LGSH + RSV group (Figure 6).

Liposomal glutathione (LGSH) treatments improved alveolar macrophage (AM) immune functions in pups challenged with Respiratory syncytial virus (RSV). In the control + RSV + LGSH group, there was a significant decrease in AM expression of the immunosuppression markers TGFβ1 (Figure 7A) and Arg-1 (Figure 7B) when compared to AMs from the control + RSV group. In the ETOH + LGSH + RSV group, there was also a significant decrease in the TGFβ1 (Figure 7A) and Arg-1 (Figure 7B) immunosuppression markers in the AMs when compared to the ETOH + RSV group. However, these AM markers of immunosuppression remained significantly elevated in the ETOH + LGSH + RSV group when compared to the AMs from the control + RSV group and the control + LGSH + RSV group.

4. Discussion

Fetal alcohol exposure is well known to adversely affect the developing newborn, most notably the developing brain [39]. However, evidence continues to mount that in utero alcohol also adversely alters multiple organs in the developing fetus [40,41], including the developing lung. As we and others have described, multiple cell types within the ETOH-exposed developing lung are at risk of alcohol-induced injury and altered pulmonary immune function [42–44].

Using small-animal models of in utero ETOH exposure, we have previously shown that ETOH alters immune defenses in the both the premature and the term lung, diminishes the lung's major antioxidant GSH, and alters the immune phenotype of the resident

AMs [1]. Specifically, in utero ETOH exposure induces an immunosuppressed AM phenotype, characterized by increased cellular TGFβ1, an important immunosuppressant that decreases bacterial as well as viral phagocytosis by the AM [11,12,31].

In these models, the detrimental effects of in utero ETOH exposure were modulated by the availability of the antioxidant GSH. We have also demonstrated that experimental interventions to diminish oxidant stress by maintaining or replenishing GSH status in the neonatal ETOH-exposed lung results in decreased neonatal AM TGFβ1, and more importantly, improved AM immune functions after in utero ETOH exposure. Indeed, beneficial interventional strategies such as dietary S-adenosyl methionine, a GSH precursor, during maternal ETOH ingestion [45] or exogenous inhaled GSH to the neonatal pup improved neonatal defenses against experimental bacterial infection. These studies suggest a potential therapeutic role for exogenous GSH to improve pulmonary immune defenses in the vulnerable ETOH-exposed neonate.

In adult studies, enteral LGSH diminishes markers of systemic oxidant stress and improved immune function [46]. Furthermore, a randomized control trial in healthy adults demonstrated that oral LGSH increases body stores of GSH and improves immune cell function [47]. To expand on our previous findings of ETOH-induced immune dysfunction and neonatal infection risk with either bacteria or viruses [31], the current study evaluated the potential therapeutic role for enteral LGSH in the setting of an experimental neonatal viral infection with RSV. In our neonatal mouse model, in utero ETOH-exposed pups demonstrated increased pulmonary RSV infection, accentuated markers of acute lung injury, and diminished AM phagocytosis of RSV when compared to control + RSV pups. The ETOH-induced derangements in immune defenses against RSV were accompanied by decreased GSH in the AM and increased AM immunosuppression, as demonstrated by increased TGFβ1 and Arg-1 immunostaining. Excitingly, enteral LGSH administered immediately prior to experimental RSV delivery successfully improved the ETOH-exposed pup's pulmonary defense against RSV, as evidenced by diminished RSV growth; decreased markers of acute lung injury (lung wet/dry weight ratio, whole-lung MPO, and PMN count); decreased AM expression of immunosuppression markers (TGFβ1 and Arg-1); and restored AM in vivo phagocytosis of RSV. The role of LGSH in supplying the critical antioxidant GSH and decreasing ETOH-induced oxidant stress was demonstrated by normalization of the systemic antioxidant capacity, systemic DNA oxidation, and the AM pool of GSH.

The current study is important because newborns, particularly former premature newborns, are at risk of significant morbidity due to viral infections such as RSV [20,21]. Although we have demonstrated that in utero alcohol exposure is reported in a third of premature babies [2], the risk and severity of subsequent RSV infection in alcohol-exposed newborns remains poorly described. While RSV infections have been reported to be increased clinically in alcohol-exposed male babies [48], the mechanisms underlying the increased risk of RSV infection in the alcohol-exposed newborn requires further study. Furthermore, the potential for LGSH to augment immune defenses in the RSV-infected infant requires additional investigation.

The immune defense of airway epithelial cells is highly sensitive to alcohol and alcohol-induced oxidant stress [49,50], and these epithelial cells are the primary site for RSV infections. However, the first line of cellular defense against pathogens in the airspace is the AM, which plays a key role in engulfing and digesting pathogens, thereby minimizing the exposure of other airway cells to the pathogen. AMs also play a key role as a modulator of inflammation in the airspace through cytokine production and cellular interactions, thereby making AM immune functions critical for pulmonary defenses against RSV. Optimal AM defenses dictate effective viral clearance and significantly contribute to the resolution of pulmonary disease [23,24,51–54].

For RSV, a central role for reactive oxygen species and subsequent oxidant stress are demonstrated by significant alterations in oxidant response pathways and increases in markers of oxidative damage [55,56]. In addition to downregulation of the cellular antioxidant enzyme systems, RSV augments its own replication [57]. In contrast, antioxidant treatments attenuate both RSV-induced oxidant stress and that caused by the viral burden [25,26]. This suggests that modulation of oxidant stress represents a potential novel pharmacological approach to ameliorate RSV-induced acute lung inflammation and injury [57].

Availability of the endogenous antioxidant tripeptide GSH (glutamine, cysteine, and glycine) is critical for maintaining redox homeostasis and preventing the immune cell damage caused by reactive oxygen species. One strategy for GSH supplementation during oxidant stress is oral LGSH, which has been shown to decrease oxidant stress in HIV subjects and patients with type 2 diabetes [32–34]. In a mouse model of an active *Mycobacterium tuberculosis* infection, GSH depletion exacerbated the pathogen burden, but oral LGSH decreased both pulmonary oxidant stress and granuloma-promoting immune responses, resulting in decreases in the pulmonary bacterial infection [58]. Similar results were obtained in a diabetic mouse model with an active *Mycobacterium tuberculosis* infection, demonstrating that oral LGSH treatments can also improve bacterial clearance in an immunocompromised model [59]. In human clinical trials, enteral LGSH elevated antioxidant defenses and improved immune functions of natural killer cells and lymphocytes [46,47], however studies of enteral LGSH in the neonatal population are lacking.

Overall, the current study provides provocative evidence that in utero ETOH exposure results in chronic oxidant stress, which subsequently mediates dysregulation of the AM immune responses and the risk of a viral infection. Correspondingly, this increased the risk of acute lung injury associated with a viral infection. However, restoration of the GSH pool through enteral LGSH treatments improved the GSH pool in AMs and decreased the AM immunosuppression associated with in utero ETOH exposure. Similarly, the systemic oxidant stress as well as the exacerbation of the RSV infection and acute lung injury in the ETOH + RSV group were attenuated by the enteral LGSH therapy in the exposed pup. In this study, we focused on AMs and cannot rule out the possibility that other cell types also negatively affected by in utero ETOH ethanol exposure were positively impacted by the enteral LGSH therapy and contributed to the improved RSV clearance. Likewise, additional studies are needed to determine if these effects of fetal ethanol exposure on chronic oxidant stress or AM immune function improve with pulmonary maturation.

Additional studies are also needed to determine if there is a potential therapeutic role for LGSH in this vulnerable population, where timing of LGSH delivery before or at the time of the viral exposure may be critical. If these findings ultimately prove to be relevant in the clinical situation, then LGSH therapy may provide a strategy to improve AM immune responses and decrease the risk and severity of pulmonary infections in the highly vulnerable preterm infant with fetal alcohol exposure. Novel strategies such as LGSH supplements may also become particularly important in this era of antibiotic-resistant bacterial infections.

5. Conclusions

- Fetal ethanol exposure promoted chronic oxidant stress in alveolar macrophages and systemically.
- Chronic oxidant stress resulted in immunosuppression of alveolar macrophages.
- In utero ethanol exposure impaired the capacity of pup alveolar macrophages to clear viruses.
- Fetal ethanol exposure exacerbated lung respiratory syncytial virus infection and acute lung injury.
- Enteral treatments of the pup with liposomal glutathione normalized alveolar macrophage immune responses, lung viral infection, and acute lung injury.

Author Contributions: Conceptualization, L.A.S.B. and T.W.G.; Methodology, X.-D.P., F.L.H., L.A.S.B. and T.W.G.; Formal analysis, X.-D.P., F.L.H., L.A.S.B. and T.W.G.; Data curation, X.-D.P., F.L.H., L.A.S.B. and T.W.G.; Writing—original draft preparation, L.A.S.B. and T.W.G.; Writing—review and editing, L.A.S.B. and T.W.G.; Supervision, L.A.S.B. and T.W.G.; Funding acquisition, L.A.S.B. and T.W.G. All authors have read and agreed to the published version of the manuscript.

Funding: This work was supported by the NIAAA 5R01AA27020.

Institutional Review Board Statement: The animal study protocol was approved by the Institutional Review Board of Emory University (PROTO201800128, 26 February 2021).

Informed Consent Statement: Not applicable.

Data Availability Statement: Data is contained within the article.

Conflicts of Interest: The authors declare no conflict of interest.

Abbreviations

AM	alveolar macrophage
AOC	antioxidant capacity
Arg-1	arginase 1
BAL	bronchoalveolar lavage
ETOH	ethanol
GSH	glutathione
8-OHdG	8-hydroxyguanosine
LGSH	liposomal glutathione
PFU	plaque forming units
PI	phagocytic index
PMNs	polymorphonuclear leukocytes
RFU	relative fluorescence unit
RSV	respiratory syncytial virus
TGFβ1	transforming growth factor β1

References

1. Gauthier, T.W. Prenatal Alcohol Exposure and the Developing Immune System. *Alcohol Res. Curr. Rev.* **2015**, *37*, 279–285.
2. Gauthier, T.W.; Guidot, D.M.; Kelleman, M.S.; McCracken, C.E.; Brown, L.A. Maternal Alcohol Use During Pregnancy and Associated Morbidities in Very Low Birth Weight Newborns. *Am. J. Med. Sci.* **2016**, *352*, 368–375. [CrossRef] [PubMed]
3. Henriksen, T.B.; Hjollund, N.H.; Jensen, T.K.; Bonde, J.P.; Andersson, A.M.; Kolstad, H.; Ernst, E.; Giwercman, A.; Skakkebaek, N.E.; Olsen, J. Alcohol consumption at the time of conception and spontaneous abortion. *Am. J. Epidemiol.* **2004**, *160*, 661–667. [CrossRef] [PubMed]
4. Kesmodel, U.; Wisborg, K.; Olsen, S.F.; Henriksen, T.B.; Secher, N.J. Moderate alcohol intake during pregnancy and the risk of stillbirth and death in the first year of life. *Am. J. Epidemiol.* **2002**, *155*, 305–312. [CrossRef] [PubMed]
5. Kesmodel, U.; Wisborg, K.; Olsen, S.F.; Henriksen, T.B.; Secher, N.J. Moderate alcohol intake in pregnancy and the risk of spontaneous abortion. *Alcohol Alcohol.* **2002**, *37*, 87–92. [CrossRef]
6. Albertsen, K.; Andersen, A.M.; Olsen, J.; Gronbaek, M. Alcohol consumption during pregnancy and the risk of preterm delivery. *Am. J. Epidemiol.* **2004**, *159*, 155–161. [CrossRef] [PubMed]
7. Kesmodel, U.; Olsen, S.F.; Secher, N.J. Does alcohol increase the risk of preterm delivery? *Epidemiology* **2000**, *11*, 512–518. [CrossRef] [PubMed]
8. Patra, J.; Bakker, R.; Irving, H.; Jaddoe, V.W.; Malini, S.; Rehm, J. Dose-response relationship between alcohol consumption before and during pregnancy and the risks of low birthweight, preterm birth and small for gestational age (SGA)-a systematic review and meta-analyses. *BJOG* **2011**, *118*, 1411–1421. [CrossRef]
9. O'Callaghan, F.V.; O'Callaghan, M.; Najman, J.M.; Williams, G.M.; Bor, W. Maternal alcohol consumption during pregnancy and physical outcomes up to 5 years of age: A longitudinal study. *Early Hum. Dev.* **2003**, *71*, 137–148. [CrossRef]
10. Liang, Y.; Harris, F.L.; Jones, D.P.; Brown, L.A. Alcohol induces mitochondrial redox imbalance in alveolar macrophages. *Free Radic. Biol. Med.* **2013**, *65*, 1427–1434. [CrossRef]
11. Brown, S.D.; Brown, L.A. Ethanol (EtOH)-induced TGF-beta1 and reactive oxygen species production are necessary for EtOH-induced alveolar macrophage dysfunction and induction of alternative activation. *Alcohol Clin. Exp. Res.* **2012**, *36*, 1952–1962. [CrossRef] [PubMed]
12. Grunwell, J.R.; Yeligar, S.M.; Stephenson, S.; Ping, X.D.; Gauthier, T.W.; Fitzpatrick, A.M.; Brown, L.A.S. TGF-beta1 Suppresses the Type I IFN Response and Induces Mitochondrial Dysfunction in Alveolar Macrophages. *J. Immunol.* **2018**, *200*, 2115–2128. [CrossRef] [PubMed]
13. Bennett, M.V.; McLaurin, K.; Ambrose, C.; Lee, H.C. Population-based trends and underlying risk factors for infant respiratory syncytial virus and bronchiolitis hospitalizations. *PLoS ONE* **2018**, *13*, e0205399. [CrossRef] [PubMed]

14. Shi, T.; McAllister, D.A.; O'Brien, K.L.; Simoes, E.A.F.; Madhi, S.A.; Gessner, B.D.; Polack, F.P.; Balsells, E.; Acacio, S.; Aguayo, C.; et al. Global, regional, and national disease burden estimates of acute lower respiratory infections due to respiratory syncytial virus in young children in 2015: A systematic review and modelling study. *Lancet* **2017**, *390*, 946–958. [CrossRef] [PubMed]
15. Packnett, E.R.; Winer, I.H.; Larkin, H.; Oladapo, A.; Gonzales, T.; Wojdyla, M.; Goldstein, M.; Smith, V.C. RSV-related hospitalization and outpatient palivizumab use in very preterm (born at <29 wGA) infants: 2003-2020. *Hum. Vaccin Immunother.* **2022**, *18*, 2140533. [CrossRef] [PubMed]
16. Shi, T.; Balsells, E.; Wastnedge, E.; Singleton, R.; Rasmussen, Z.A.; Zar, H.J.; Rath, B.A.; Madhi, S.A.; Campbell, S.; Vaccari, L.C.; et al. Risk factors for respiratory syncytial virus associated with acute lower respiratory infection in children under five years: Systematic review and meta-analysis. *J. Glob. Health* **2015**, *5*, 020416. [CrossRef] [PubMed]
17. Odumade, O.A.; van Haren, S.D.; Angelidou, A. Implications of the Severe Acute Respiratory Syndrome Coronavirus 2 (SARS-CoV-2) Pandemic on the Epidemiology of Pediatric Respiratory Syncytial Virus Infection. *Clin. Infect. Dis.* **2022**, *75* (Suppl. S1), S130–S135. [CrossRef] [PubMed]
18. Tachibana, A.; Kimura, H.; Kato, M.; Nako, Y.; Kozawa, K.; Morikawa, A. Respiratory syncytial virus enhances the expression of CD11b molecules and the generation of superoxide anion by human eosinophils primed with platelet-activating factor. *Intervirology* **2002**, *45*, 43–51. [CrossRef]
19. Eddens, T.; Parks, O.B.; Williams, J.V. Neonatal Immune Responses to Respiratory Viruses. *Front. Immunol.* **2022**, *13*, 863149. [CrossRef]
20. Fonseca, W.; Lukacs, N.W.; Ptaschinski, C. Factors Affecting the Immunity to Respiratory Syncytial Virus: From Epigenetics to Microbiome. *Front. Immunol.* **2018**, *9*, 226. [CrossRef]
21. Paes, B. Respiratory Syncytial Virus in Otherwise Healthy Prematurely Born Infants: A Forgotten Majority. *Am. J. Perinatol.* **2018**, *35*, 541–544. [CrossRef] [PubMed]
22. Simoes, E.A.; Anderson, E.J.; Wu, X.; Ambrose, C.S. Effects of Chronologic Age and Young Child Exposure on Respiratory Syncytial Virus Disease among US Preterm Infants Born at 32 to 35 Weeks Gestation. *PLoS ONE* **2016**, *11*, e0166226. [CrossRef]
23. Bohmwald, K.; Espinoza, J.A.; Pulgar, R.A.; Jara, E.L.; Kalergis, A.M. Functional Impairment of Mononuclear Phagocyte System by the Human Respiratory Syncytial Virus. *Front. Immunol.* **2017**, *8*, 1643. [CrossRef] [PubMed]
24. Kolli, D.; Gupta, M.R.; Sbrana, E.; Velayutham, T.S.; Chao, H.; Casola, A.; Garofalo, R.P. Alveolar macrophages contribute to the pathogenesis of human metapneumovirus infection while protecting against respiratory syncytial virus infection. *Am. J. Respir. Cell Mol. Biol.* **2014**, *51*, 502–515. [CrossRef]
25. Castro, S.M.; Guerrero-Plata, A.; Suarez-Real, G.; Adegboyega, P.A.; Colasurdo, G.N.; Khan, A.M.; Garofalo, R.P.; Casola, A. Antioxidant treatment ameliorates respiratory syncytial virus-induced disease and lung inflammation. *Am. J. Respir. Crit. Care Med.* **2006**, *174*, 1361–1369. [CrossRef]
26. Cho, H.Y.; Imani, F.; Miller-DeGraff, L.; Walters, D.; Melendi, G.A.; Yamamoto, M.; Polack, F.P.; Kleeberger, S.R. Antiviral activity of Nrf2 in a murine model of respiratory syncytial virus disease. *Am. J. Respir. Crit. Care Med.* **2009**, *179*, 138–150. [CrossRef] [PubMed]
27. Wyatt, T.A.; Bailey, K.L.; Simet, S.M.; Warren, K.J.; Sweeter, J.M.; DeVasure, J.M.; Pavlik, J.A.; Sisson, J.H. Alcohol potentiates RSV-mediated injury to ciliated airway epithelium. *Alcohol* **2019**, *80*, 17–24. [CrossRef] [PubMed]
28. Warren, K.J.; Poole, J.A.; Sweeter, J.M.; DeVasure, J.M.; Wyatt, T.A. An association between MMP-9 and impaired T cell migration in ethanol-fed BALB/c mice infected with respiratory syncytial virus-2A. *Alcohol* **2019**, *80*, 25–32. [CrossRef]
29. Ackermann, M.R. Lamb model of respiratory syncytial virus-associated lung disease: Insights to pathogenesis and novel treatments. *ILAR J.* **2014**, *55*, 4–15. [CrossRef]
30. Lazic, T.; Wyatt, T.A.; Matic, M.; Meyerholz, D.K.; Grubor, B.; Gallup, J.M.; Kersting, K.W.; Imerman, P.M.; Almeida-De-Macedo, M.; Ackermann, M.R. Maternal alcohol ingestion reduces surfactant protein A expression by preterm fetal lung epithelia. *Alcohol* **2007**, *41*, 347–355. [CrossRef]
31. Johnson, J.K.; Harris, F.L.; Ping, X.D.; Gauthier, T.W.; Brown, L.A.S. Role of zinc insufficiency in fetal alveolar macrophage dysfunction and RSV exacerbation associated with fetal ethanol exposure. *Alcohol* **2019**, *80*, 5–16. [CrossRef] [PubMed]
32. Ly, J.; Lagman, M.; Saing, T.; Singh, M.K.; Tudela, E.V.; Morris, D.; Anderson, J.; Daliva, J.; Ochoa, C.; Patel, N.; et al. Liposomal Glutathione Supplementation Restores TH1 Cytokine Response to Mycobacterium tuberculosis Infection in HIV-Infected Individuals. *J. Interferon Cytokine Res.* **2015**, *35*, 875–887. [CrossRef] [PubMed]
33. To, K.; Cao, R.; Yegiazaryan, A.; Owens, J.; Nguyen, T.; Sasaninia, K.; Vaughn, C.; Singh, M.; Truong, E.; Medina, A.; et al. Effects of Oral Liposomal Glutathione in Altering the Immune Responses Against Mycobacterium tuberculosis and the Mycobacterium bovis BCG Strain in Individuals With Type 2 Diabetes. *Front. Cell. Infect. Microbiol.* **2021**, *11*, 657775. [CrossRef] [PubMed]
34. To, K.; Cao, R.; Yegiazaryan, A.; Owens, J.; Sasaninia, K.; Vaughn, C.; Singh, M.; Truong, E.; Sathananthan, A.; Venketaraman, V. The Effects of Oral Liposomal Glutathione and In Vitro Everolimus in Altering the Immune Responses against Mycobacterium bovis BCG Strain in Individuals with Type 2 Diabetes. *Biomol. Concepts* **2021**, *12*, 16–26. [CrossRef] [PubMed]
35. Cooke, R.W.; Drury, J.A. Reduction of oxidative stress marker in lung fluid of preterm infants after administration of intra-tracheal liposomal glutathione. *Biol. Neonate* **2005**, *87*, 178–180. [CrossRef]
36. Lukacs, N.W.; Moore, M.L.; Rudd, B.D.; Berlin, A.A.; Collins, R.D.; Olson, S.J.; Ho, S.B.; Peebles, R.S. Differential immune responses and pulmonary pathophysiology are induced by two different strains of respiratory syncytial virus. *Am. J. Pathol.* **2006**, *169*, 977–986. [CrossRef] [PubMed]

37. Brown, L.A.; Ping, X.D.; Harris, F.L.; Gauthier, T.W. Glutathione availability modulates alveolar macrophage function in the chronic ethanol-fed rat. *Am. J. Physiol. Lung Cell. Mol. Physiol.* **2007**, *292*, L824–L832. [CrossRef]
38. Ping, X.D.; Harris, F.L.; Brown, L.A.; Gauthier, T.W. In vivo dysfunction of the term alveolar macrophage after in utero ethanol exposure. *Alcohol Clin. Exp. Res.* **2007**, *31*, 308–316. [CrossRef]
39. Akison, L.K.; Kuo, J.; Reid, N.; Boyd, R.N.; Moritz, K.M. Effect of Choline Supplementation on Neurological, Cognitive, and Behavioral Outcomes in Offspring Arising from Alcohol Exposure During Development: A Quantitative Systematic Review of Clinical and Preclinical Studies. *Alcohol Clin. Exp. Res.* **2018**, *42*, 1591–1611. [CrossRef]
40. Gupta, K.K.; Gupta, V.K.; Shirasaka, T. An Update on Fetal Alcohol Syndrome-Pathogenesis, Risks, and Treatment. *Alcohol Clin. Exp. Res.* **2016**, *40*, 1594–1602. [CrossRef]
41. Caputo, C.; Wood, E.; Jabbour, L. Impact of fetal alcohol exposure on body systems: A systematic review. *Birth Defects Res. C Embryo Today* **2016**, *108*, 174–180. [CrossRef] [PubMed]
42. Wang, X.; Gomutputra, P.; Wolgemuth, D.J.; Baxi, L. Effects of acute alcohol intoxication in the second trimester of pregnancy on development of the murine fetal lung. *Am. J. Obs. Gynecol.* **2007**, *197*, 269.e1–269.e4. [CrossRef] [PubMed]
43. Sozo, F.; O'Day, L.; Maritz, G.; Kenna, K.; Stacy, V.; Brew, N.; Walker, D.; Bocking, A.; Brien, J.; Harding, R. Repeated ethanol exposure during late gestation alters the maturation and innate immune status of the ovine fetal lung. *Am. J. Physiol. Lung Cell. Mol. Physiol.* **2009**, *296*, L510–L518. [CrossRef] [PubMed]
44. Lazic, T.; Sow, F.B.; Van Geelen, A.; Meyerholz, D.K.; Gallup, J.M.; Ackermann, M.R. Exposure to ethanol during the last trimester of pregnancy alters the maturation and immunity of the fetal lung. *Alcohol* **2011**, *45*, 673–680. [CrossRef]
45. Gauthier, T.W.; Manar, M.H.; Brown, L.A.S. Is maternal alcohol use a risk factor for early-onset sepsis in the premature newborn? *Alcohol* **2004**, *33*, 139–145. [CrossRef] [PubMed]
46. Sinha, R.; Sinha, I.; Calcagnotto, A.; Trushin, N.; Haley, J.S.; Schell, T.D.; Richie, J.P., Jr. Oral supplementation with liposomal glutathione elevates body stores of glutathione and markers of immune function. *Eur. J. Clin. Nutr.* **2018**, *72*, 105–111. [CrossRef] [PubMed]
47. Richie, J.P., Jr.; Nichenametla, S.; Neidig, W.; Calcagnotto, A.; Haley, J.S.; Schell, T.D.; Muscat, J.E. Randomized controlled trial of oral glutathione supplementation on body stores of glutathione. *Eur. J. Nutr.* **2015**, *54*, 251–263. [CrossRef] [PubMed]
48. Libster, R.; Ferolla, F.M.; Hijano, D.R.; Acosta, P.L.; Erviti, A.; Polack, F.P.; Network, I.R. Alcohol during pregnancy worsens acute respiratory infections in children. *Acta Paediatr.* **2015**, *104*, e494–e499. [CrossRef]
49. Bailey, K.L.; Wyatt, T.A.; Katafiasz, D.M.; Taylor, K.W.; Heires, A.J.; Sisson, J.H.; Romberger, D.J.; Burnham, E.L. Alcohol and Cannabis Use Alter Pulmonary Innate Immunity. *Alcohol* **2018**, *80*, 131–138. [CrossRef]
50. Price, M.E.; Gerald, C.L.; Pavlik, J.A.; Schlichte, S.L.; Zimmerman, M.C.; Devasure, J.M.; Wyatt, T.A.; Sisson, J.H. Loss of cAMP-dependent stimulation of isolated cilia motility by alcohol exposure is oxidant dependent. *Alcohol* **2018**, *80*, 91–98. [CrossRef]
51. Ren, J.; Liu, G.; Go, J.; Kolli, D.; Zhang, G.; Bao, X. Human metapneumovirus M2-2 protein inhibits innate immune response in monocyte-derived dendritic cells. *PLoS ONE* **2014**, *9*, e91865. [CrossRef] [PubMed]
52. Kumagai, Y.; Takeuchi, O.; Kato, H.; Kumar, H.; Matsui, K.; Morii, E.; Aozasa, K.; Kawai, T.; Akira, S. Alveolar macrophages are the primary interferon-alpha producer in pulmonary infection with RNA viruses. *Immunity* **2007**, *27*, 240–252. [CrossRef] [PubMed]
53. Reed, J.L.; Brewah, Y.A.; Delaney, T.; Welliver, T.; Burwell, T.; Benjamin, E.; Kuta, E.; Kozhich, A.; McKinney, L.; Suzich, J.; et al. Macrophage impairment underlies airway occlusion in primary respiratory syncytial virus bronchiolitis. *J. Infect. Dis.* **2008**, *198*, 1783–1793. [CrossRef] [PubMed]
54. Eichinger, K.M.; Egana, L.; Orend, J.G.; Resetar, E.; Anderson, K.B.; Patel, R.; Empey, K.M. Alveolar macrophages support interferon gamma-mediated viral clearance in RSV-infected neonatal mice. *Respir. Res.* **2015**, *16*, 122. [CrossRef] [PubMed]
55. Garofalo, R.P.; Kolli, D.; Casola, A. Respiratory syncytial virus infection: Mechanisms of redox control and novel therapeutic opportunities. *Antioxid. Redox Signal.* **2013**, *18*, 186–217. [CrossRef] [PubMed]
56. El Saleeby, C.M.; Bush, A.J.; Harrison, L.M.; Aitken, J.A.; Devincenzo, J.P. Respiratory syncytial virus load, viral dynamics, and disease severity in previously healthy naturally infected children. *J. Infect. Dis.* **2011**, *204*, 996–1002. [CrossRef]
57. Hosakote, Y.M.; Liu, T.; Castro, S.M.; Garofalo, R.P.; Casola, A. Respiratory syncytial virus induces oxidative stress by modulating antioxidant enzymes. *Am. J. Respir. Cell Mol. Biol.* **2009**, *41*, 348–357. [CrossRef]
58. Kachour, N.; Beever, A.; Owens, J.; Cao, R.; Kolloli, A.; Kumar, R.; Sasaninia, K.; Vaughn, C.; Singh, M.; Truong, E.; et al. Liposomal Glutathione Helps to Mitigate Mycobacterium tuberculosis Infection in the Lungs. *Antioxidants* **2022**, *11*, 673. [CrossRef]
59. Beever, A.; Kachour, N.; Owens, J.; Sasaninia, K.; Kolloli, A.; Kumar, R.; Ramasamy, S.; Sisliyan, C.; Khamas, W.; Subbian, S.; et al. L-GSH Supplementation in Conjunction With Rifampicin Augments the Treatment Response to Mycobacterium tuberculosis in a Diabetic Mouse Model. *Front. Pharmacol.* **2022**, *13*, 879729. [CrossRef]

antioxidants

MDPI

Review

Acetic Acid: An Underestimated Metabolite in Ethanol-Induced Changes in Regulating Cardiovascular Function

Andrew D. Chapp [1,*], Zhiying Shan [2] and Qing-Hui Chen [2,*]

1 Department of Neuroscience, University of Minnesota, Minneapolis, MN 55455, USA
2 Kinesiology and Integrative Physiology, Michigan Technological University, Houghton, MI 49931, USA; zhiyings@mtu.edu
* Correspondence: achapp@umn.edu (A.D.C.); qinghuic@mtu.edu (Q.-H.C.)

Abstract: Acetic acid is a bioactive short-chain fatty acid produced in large quantities from ethanol metabolism. In this review, we describe how acetic acid/acetate generates oxidative stress, alters the function of pre-sympathetic neurons, and can potentially influence cardiovascular function in both humans and rodents after ethanol consumption. Our recent findings from in vivo and in vitro studies support the notion that administration of acetic acid/acetate generates oxidative stress and increases sympathetic outflow, leading to alterations in arterial blood pressure. Real-time investigation of how ethanol and acetic acid/acetate modulate neural control of cardiovascular function can be conducted by microinjecting compounds into autonomic control centers of the brain and measuring changes in peripheral sympathetic nerve activity and blood pressure in response to these compounds.

Keywords: ethanol; acetic acid; acetate; nitric oxide; sympathetic nerve activity; cardiovascular function

1. Introduction

Ethanol metabolism produces systemic maladaptive changes through oxidative stress. Ethanol metabolism in vivo generates reactive oxygen and nitrogen species, predominantly through oxidative pathways involving NADPH/$NADP^+$, NAD+/NADH, and/or H_2O_2 [1]. Ethanol is oxidized to acetic acid, which becomes acetate at physiological pH (Figure 1A) [2]. Acetate is then shuttled into the citric acid cycle, where additional oxidative stress is likely generated through electron transport chain electron leak [3]. Chronic oxidative stress from ethanol metabolism leads to liver [4–8], brain [9–13], and cardiovascular pathologies [14–20].

For many years, moderate ethanol consumption was thought to provide some benefit to cardiovascular health [21,22]. This was endorsed by the American Heart Association (AHA) [23]. However, this was at odds with the literature on alcohol use disorder suggesting that chronic use of ethanol leads to the development of hypertension and cardiovascular disease [24–26]. By 2023, after several research articles suggested that ethanol offers no cardiovascular benefit [27,28], the AHA reversed course, warning that even one ethanol-containing drink a day can increase cardiovascular disease risk [29].

A major gap in the ethanol research field is the limited investigation of the role of acetic acid/acetate in the pathologies associated with ethanol use. Acetic acid/acetate has been assumed to be a relatively benign compound [1], as acetate is a feedstock for the generation of ATP through the citric acid cycle [30] and for the generation of acetyl-CoA and acetylation reactions [31,32]. However, acetic acid is linked to several distinct disease states, including Alzheimer's disease [33,34], neurodegenerative disease [33,35], obesity [36], and gut–brain dysbiosis [33,35,36]. In this review article, we highlight how acetic acid/acetate may influence the regulation of cardiovascular function through several mechanisms, including the generation of reactive oxygen and nitrogen species (Figures 1 and 2).

check for updates

Citation: Chapp, A.D.; Shan, Z.; Chen, Q.-H. Acetic Acid: An Underestimated Metabolite in Ethanol-Induced Changes in Regulating Cardiovascular Function. *Antioxidants* **2024**, *13*, 139. https://doi.org/10.3390/antiox13020139

Academic Editor: Marco Fiore

Received: 20 December 2023
Revised: 13 January 2024
Accepted: 18 January 2024
Published: 23 January 2024

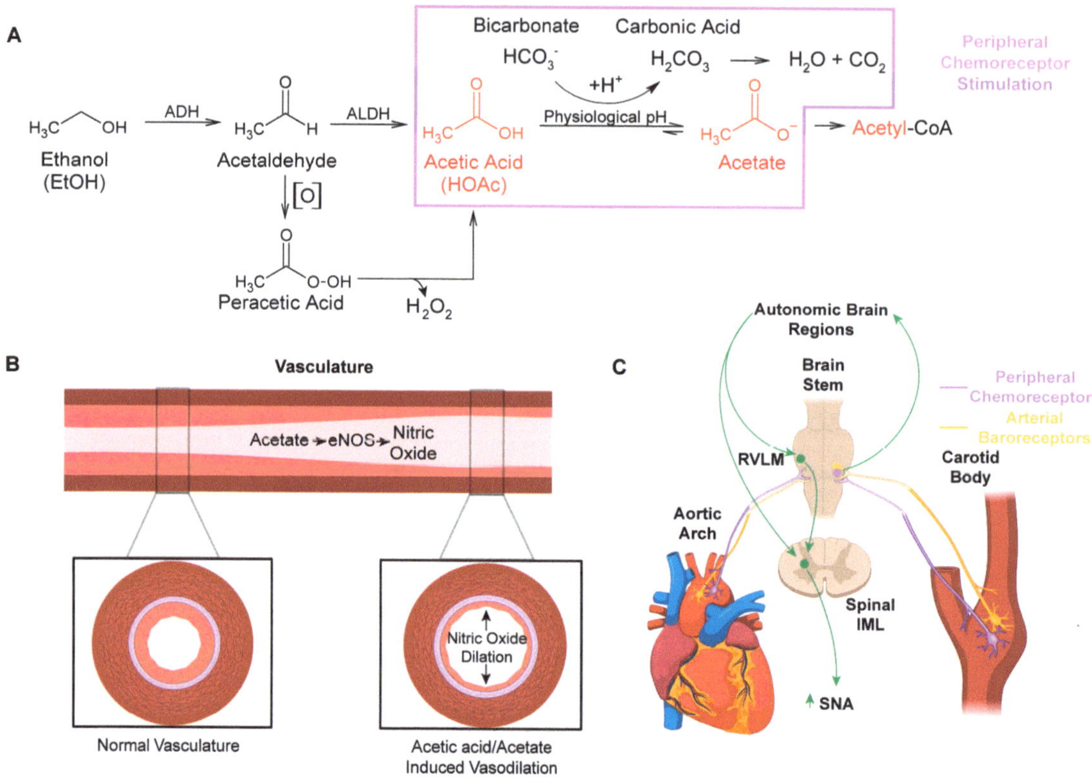

Figure 1. Impact of acetic acid/acetate on peripheral vasculature and the regulation of cardiovascular function. (**A**) Major metabolic pathway for ethanol metabolism (box in purple highlights potential chemoreceptor−stimulating pathways). Auto−oxidation (denoted [O]) of acetaldehyde, which can be increased by the presence of Fe^{3+} to form peracetic acid (peroxyacetic acid). Peracetic acid decomposes to acetic acid, generating hydrogen peroxide. (**B**) Acetate stimulates eNOS in the peripheral vasculature, which increases production of the powerful vasodilator, nitric oxide. (**C**) Schematic of the location of peripheral arterial baroreceptors and chemoreceptors, their innervation to the brainstem, and their relationship with neural control of autonomic function. Abbreviations: alcohol dehydrogenase (ADH), cytochrome P450 (CYP 450), aldehyde dehydrogenase (ALDH), endothelial nitric oxide synthase (eNOS), nitric oxide (NO), rostral ventrolateral medulla (RVLM), intermediolateral column (IML), sympathetic nerve activity (SNA).

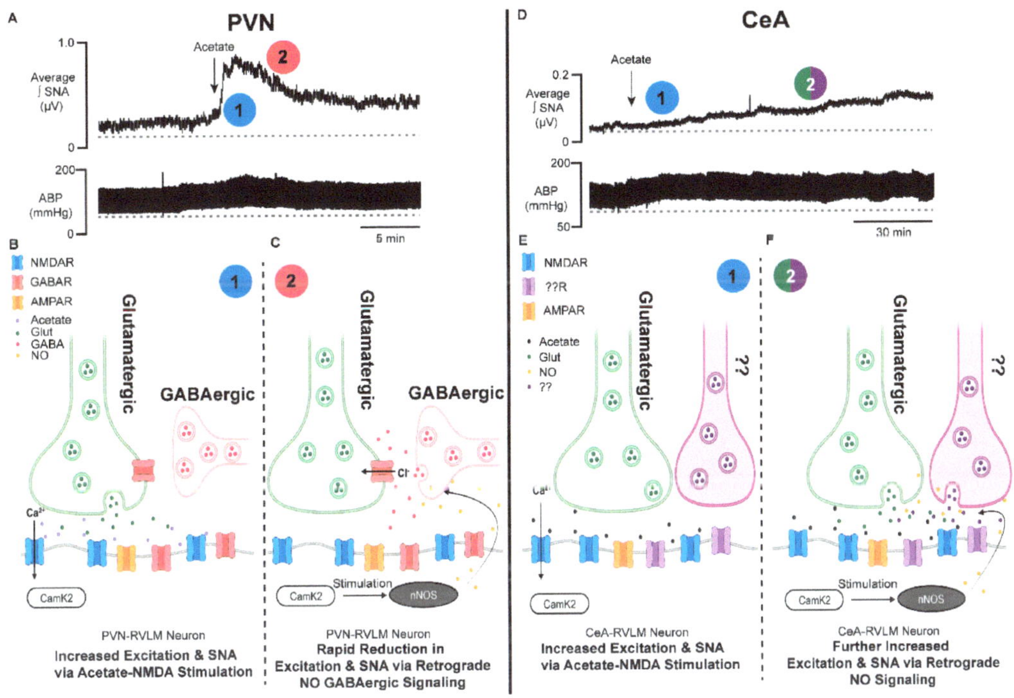

Figure 2. Brain-generated sympathoexcitatory responses to microinjections of acetate differ by autonomic region: implications for nitric oxide. (**A**) Representative sympathetic nerve recording and arterial blood pressure in response to PVN microinjected acetate. Note the rapid increase in sympathetic nerve activity (1, blue) and time of onset, followed by a rapid reduction in excitation (2, red). (**B**) Proposed schematic of the synaptic mechanisms contributing to the acetate-induced sympathoexcitatory response in PVN–RVLM neuronal circuitry. Note that (1, blue) corresponds to the response in (**A**). (**C**) Proposed schematic of NO retrograde synaptic signaling in PVN–RVLM neurons. Acetate activation of the NMDAR stimulates NO production, which increases presynaptic GABA release on glutamatergic synapses, reducing sympathoexcitation. Note that (2, red) corresponds to the response in (**A**). (**D**) Representative sympathetic nerve recording and arterial blood pressure in response to CeA–microinjected acetate. Note the steady increase in sympathetic nerve activity (1, blue) and the steady increase in excitation (2, green/purple). (**E**) Proposed schematic of the synaptic mechanisms contributing to the acetate-induced sympathoexcitatory response in CeA–RVLM neuronal circuitry. Note that (1, blue) corresponds to the response in (**D**). (**F**) Proposed schematic of NO retrograde synaptic signaling in CeA–RVLM neurons. Acetate activation of the NMDAR stimulates NO production, which increases the release of presynaptic glutamate and other neurotransmitters, increasing sympathoexcitation. Alternative inputs may also be propagating increased SNA. Note that (2, green/purple) corresponds to the response in (**D**). Abbreviations: paraventricular nucleus of the hypothalamus (PVN), rostral ventrolateral medulla (RVLM), neuronal nitric oxide synthase (nNOS), calcium calmodulin kinase 2 (CAMK2), nitric oxide (NO), calcium (Ca^{2+}), chloride (Cl^-), N-methyl-D-aspartate receptor (NMDAR), gamma aminobutyric acid receptor (GABAR), sympathetic nerve activity (SNA), glutamate (Glut), gamma aminobutyric acid (GABA), α-amino-3-hydroxy-5-methyl-4-isoxazolepropionic acid receptor (AMPAR).

2. Acetaldehyde and Peracetic Acid: All Roads Lead to Acetic Acid/Acetate

Acetaldehyde has been one of the most highly researched ethanol metabolites and has been implicated in the generation of reactive oxygen species, protein adduct formation [37–39], the increased production of salsolinol [40–46], and neuronal toxicity [47–50].

In individuals with aldehyde dehydrogenase (ALDH) 2 deficiencies, the buildup of acetaldehyde is thought to cause flushing, nausea, and headache [51–54]. Disulfiram, an ALDH inhibitor and alcohol use deterrent, also increases the buildup of acetaldehyde, which is thought to be the mechanism of action for reducing alcohol consumption [55–57].

Interestingly, blood acetaldehyde concentrations are not easily measured and have been reported in the range of 10–40 nM when blood ethanol concentrations are 10–20 mM [58,59]. Part of the prevailing theory on why blood acetaldehyde concentrations are so low is due to its rapid metabolism to acetic acid [1]. And while this may be true, there are other chemical factors and conversion routes for acetaldehyde that may be contributing to its low detection.

First, acetaldehyde is a volatile liquid that is soluble in aqueous solutions [60]. Its boiling point is 20.2 °C (68 °F) [60], much lower than the physiological temperature of 37 °C (98.6 °F). Thus, it is highly likely that any acetaldehyde that is not converted to acetic acid/acetate and remains in circulation passing to the lungs is exhaled in its gaseous state. Second, in a report published in 1929 by Bowen and Tietz, they noted that acetaldehyde shaken in the presence of oxygen (auto-oxidation) produced a peroxide, identified as peracetic acid [61]. These authors also noted that this type of reaction and similar reactions produced a "long chain mechanism" [61], which would be consistent with what is now a free radical-type mechanism. Indeed, follow-up work in 1950 by Bawn and Williamson further elucidated that aqueous solutions of acetaldehyde were also capable of auto-oxidation, producing the peracetic acid reported by Bowen and Tietz [61], and that these reactions could be increased and/or induced in the presence of metals, of note, iron III (Fe^{3+}) [62]. Interestingly, hemoglobin, the iron-containing protein that is responsible for oxygen and carbon dioxide transport in the blood, has an iron oxidation state of Fe^{3+} when oxygen is bound [63]. This close proximity of oxygen and Fe^{3+} with any circulating acetaldehyde may convert acetaldehyde into peracetic acid, and account for the low to undetectable circulating levels following ethanol consumption [59].

Peracetic acid (peroxyacetic acid) is a peroxide of acetic acid. It is utilized as a broad-spectrum disinfectant, displaying bactericidal, virucidal, fungicidal, and sporicidal properties [64]. In cell toxicity studies, peracetic acid was more cytotoxic than sodium hypochlorite (bleach) [65]. Moreover, given that peracetic acid is an organic peroxide, it undergoes similar homolysis reactions as other peroxides do [66]. As such, the spontaneous or metal-catalyzed homolysis reactions of peroxides generate free radicals [66,67] and these free radicals are thought to induce cellular and tissue damage [68,69]. Hydrogen peroxide is a classic example of a peroxide used as a source of oxygen-derived free radicals in research studies [70]. Additionally, hydrogen peroxide also interacts with iron in the well-known Fenton reaction to generate additional sources of free radicals that can lead to tissue damage [71]. Similar to the Fenton reactions involving hydrogen peroxide, peracetic acid was also found to undergo such Fenton reactions with Fe^{3+}, generating peroxyl ($CH_3C(O)OO\bullet$), alkoxyl ($CH_3C(O)\bullet$), and hydroperoxyl ($HO_2\bullet$) radicals [72]. In light of this information, ethanol and alcohol use disorder (AUD) researchers should consider the potential that many of the deleterious effects attributed from acetaldehyde might be due to its downstream metabolite, peracetic acid (Figure 1A).

3. Active Metabolites of Ethanol

Among the plethora of research on ethanol and AUD, few studies have progressed past acetaldehyde in the ethanol metabolic pathway. This is not to say that ethanol and AUD research has not entertained the possibility that acetic acid/acetate may be involved in the effects of ethanol consumption. Rather, it was more likely assumed that since acetic acid/acetate directly fed into the citric acid cycle to generate ATP, it had to be rather benign [1]. Indeed, glucose and L-lactate are both feedstock for the generation of ATP [73] and at face value, would appear benign. Glucose is consumed and L-lactic acid/lactate can be released into circulation and rapidly cleared following exercise [74]. However, both are kept within tight tolerance in humans and drastic deviations from these homeostatic values (glucose 4.0–6.0 mM [75] and L-lactic acid/lactate <2.0 mM) [76] can produce profound effects on physiological function [75,76]. As an example, an individual with diabetes whose blood glucose levels drop too low can display symptoms including confusion, shaking, nausea, vision changes, loss of consciousness, seizure, and/or death [75]. Thus, as there are tight homeostatic tolerances for glucose, the same may perhaps apply to acetic acid/acetate.

To the best of our knowledge there is no "normal" definition of blood acetate levels in humans. Blood acetate concentrations reported in humans and rodents have been between 0.05 and 0.6 mM [77–82], and in individuals with AUD, 0.9 mM [82]. However, a majority of these studies utilized older techniques for quantifying acetate, such as volatile derivatization reactions followed by organic extractions and gas chromatography [81], which may underestimate the actual concentrations (see Chapp et al. [83]). We have recently adapted ion chromatography methodology to measure short-chain fatty acids including acetic acid/acetate in rodents, with potential applicability to humans [83].

In rats, we have measured baseline serum acetate concentrations of 0.23 ± 0.04 mM [83]. In rats administered a dose of ethanol (2 g/kg), equivalent to a blood alcohol concentration (BAC) of 0.2% the time of metabolism influenced serum acetate concentrations, with peak serum acetate concentrations measured at ~4.2 mM in male rats and ~3.9 mM in female rats (unpublished data). In C57BL/6J mice, the baseline serum acetate level in males was 0.63 ± 0.04 mM and in females was 0.56 ± 0.6 mM [2]. Similar to the rats, mice administered ethanol (2 g/kg) had serum acetate concentrations influenced by metabolism times, with peak values of 3.67 mM in males and 3.60 mM in females [2]. These data demonstrate that: (1) Baseline serum acetate levels can potentially differ between rats and mice (~0.1–0.2 mM in rats and ~0.5–0.6 mM in mice) and thus may also be different in humans if measured using the same technique [83]. (2) The same dose of ethanol seems to produce similar serum acetate concentrations, at least between rats and mice. Again, whether this highly intoxicating dose of ethanol (2 g/kg, BAC ~0.2%) or lower doses (1 g/kg) in humans produces the same serum acetate concentrations as it does in rodents remains to be determined. We have, however, shown that these concentrations of acetate in dopaminergic-like PC12 cells can increase cytosolic calcium in an NMDAR-dependent manner, as well as increase cytosolic reactive oxygen species [84].

4. Peripheral Actions of Acetic Acid/Acetate: Impact on Nitric Oxide Synthase (NOS) and Nitric Oxide (NO)

Ethanol in humans has been found to influence heart rate, cardiac output, stroke volume, and sympathetic nerve activity [85–91]. It causes both vasoconstriction and vasodilation [14,92,93]. In rodent models, ethanol has been shown to increase the bioavailability of nitric oxide [93–95]. However, these studies did not account for ethanol metabolism or the production of acetic acid/acetate.

In studies investigating the effect of buffering agents in hemodialysis solutions, it is observed that acetate-based solutions create greater cardiovascular and hemodynamic instability than bicarbonate-based solutions do [96,97]. These instabilities include vasodilation, hypoxemia, increased cardiac output, and angina pectoris [98–100]. Methodical investigation identified that these acetate-based hemodialysis solutions increased levels of nitric oxide synthase (NOS) and nitric oxide (NO) [101,102], a powerful radical which stimulates vasodilation [103–105]. Several studies have also identified acetate-induced elevations in tumor necrosis factor alpha (TNFa) and cyclic adenosine monophosphate (cAMP), both mediators of NOS [102]. This was seen in dialysis buffers containing as little as 4 mM acetate [102]. A likely mechanism through which ethanol produces vasodilation is through ethanol's metabolism to acetic acid/acetate, which stimulates production of NOS and NO (Figure 1B). This finding was also substantiated by Sakakibara S et al., who found that consumption of vinegar increased eNOS production with as little as 200 uM acetate in both early-phase detection (20 min) as well as late-phase detection (4 h) [106]. Furthermore, they found increased forearm vasodilation in adults who had consumed acetic acid compared to a control. This time course of action for acetic acid/acetate suggests a profound impact on peripheral eNOS even with minimal concentrations of and exposure to acetate [106].

5. Direct Effects of Ethanol and Acetic Acid/Acetate on Cardiac Function

The potential of ethanol consumption to influence cardiac function has been well documented. In the acute setting, ethanol can increase cardiac output [107] and reduce plasma potassium [108], which may contribute to ethanol-induced arrhythmia or "holiday heart" [109]. The acute effects of ethanol on contractility appear to be mixed, with some studies suggesting either no change [107,110,111] or a decrease [112,113] in cardiac contractility. Chronic ethanol use also contributes to cardiomyopathy [114,115] and arrhythmias [20].

Similar to the effects of acute ethanol, acute acetate is also capable of altering cardiac function. Acute acetate administration has been found to increase cardiac output [116,117], precipitate hemodynamic instability [97,101,116], and reduce myocardial contractility [118,119]. Interestingly, some of the direct cardiac effects of acetate were speculated to be tied to acetate-induced NO release [101]. Indeed, the direct cardiac effects of NO are bimodal, with lower concentrations of NO increasing the ionotropic effects [120] and high concentrations of NO reducing cardiac myocyte contractility [121]. Whether the mixed effects of ethanol on cardiac function are a result of acetate and/or acetate-induced NO release remains to be determined. Furthermore, the effect of chronically elevated acetate from chronic ethanol use and metabolism on cardiac function has yet to be investigated. It is provocative to postulate that the major compound driving alcohol-induced cardiomyopathy [122,123] may be acetate rather than ethanol, and as such, future studies of chronic acetate seem warranted.

6. Ethanol Metabolism to Acetic Acid Alters Acid/Base Homeostasis and Likely Engages Peripheral Chemoreceptors

Although ethanol itself is not acidic (pKa value ~16) [124], ethanol acts as an exogenous source of acidic hydrogens. Once ethanol is fully oxidized to acetic acid (pKa value ~4.78) [125], the labile acidic hydrogen becomes >95% dissociated at physiological pH [126]. pKa is a measure of hydrogen acidity, as a review of pKa (see *General Chemistry*, sixth edition, Julia Burdge). The higher the pKa value, the less acidic the hydrogen. At physiological pH, acetic acid likely reacts with the bicarbonate buffering system (i.e., the vinegar and baking soda reaction), forming acetate and carbonic acid (Figure 1A). Carbonic acid is then broken down into water and carbon dioxide. The acidic hydrogen on acetic acid and the excess carbon dioxide cause activation of (1) peripheral chemoreceptors located in the carotid sinus and aortic arches, and (2) neural control centers in the brain, which control sympathetic outflow and increase sympathetic nerve activity (SNA) (Figure 1C).

In the United States, humans are considered legally intoxicated at a blood alcohol concentration (BAC) of 0.08% or a blood serum concentration of ~17 mM ethanol [127]. The conversion of ethanol to acetic acid is nearly 1:1, with ~95% of ethanol converted to acetic acid [128]. Thus, for 17 mM of ethanol consumed, the predicted amount of acetic acid produced would be ~16–17 mM. To buffer or neutralize this concentration of protons, roughly 17 mM of bicarbonate would be consumed. The normal concentration of bicarbonate in the serum is ~23–28 mM [129,130]. This bicarbonate consumption creates a high acid load, generating compensatory mechanisms within the kidneys and lungs to maintain pH homeostasis [130–132].

7. Integration of Arterial Baroreceptors, Chemoreceptors, and NO Signaling in Response to Acetic Acid Generated by Ethanol Metabolism

The integration of the peripheral cardiovascular regulatory systems engaged by acetic acid likely explains the complex phenomenon of mixed vasodilation and vasoconstriction induced by ethanol. First, acetic acid increases NOS activity and NO production, which leads to vasodilation [98,101,102]. Arterial baroreceptors located in the carotid sinus and aortic arch [133] would relay the resultant drop in blood pressure to neural control centers, which regulate cardiovascular function [134] (Figure 1C). These neural control centers then increase SNA to constrict the vasculature in an attempt to maintain homeostasis [133]. Third, the acid load generated from the acetic acid and/or excess carbon dioxide would similarly be sensed by the peripheral chemoreceptors, also located in the carotid sinus and aortic arch. And, in the same type of neural feedback loop as the arterial baroreceptor activation, chemoreceptor activation would also increase SNA and constrict the peripheral vasculature [135–138]. Thus, the complex vasoconstriction and vasodilation effects of ethanol consumption [93] can be explained when looking at the effects being driven by acetic acid and not ethanol per se.

8. Acetic Acid-Induced Changes in Neural Control of Cardiovascular Regulation

Acute oral consumption of ethanol consistently increases arterial blood pressure (ABP) and SNA in humans [85,89–91] and rodents [139–141]. Likewise, chronic ethanol use has been shown to increase norepinephrine, SNA, and ABP [111,142–146]. In one acute human study, dexamethasone was found to blunt ethanol's sympathoexcitatory effect [90] through a reduction in corticotropin-releasing hormones (CRHs), suggesting that this response is at least partially mediated by central effects. In rodent studies, oral ethanol has been less consistent [92], with some studies showing an increase in SNA and ABP [143,147] and other studies showing decreases in ABP [108,110]. As some other reviews have noted, the timing of measurements in rodents may be important [24]. In one rodent study, Crandall et al. observed that ABP was normal at peak blood ethanol concentrations but was significantly elevated 24 h post ethanol dosing [148]. Interestingly, we now know from pharmacokinetic studies that this time point corresponds to when serum acetate is still elevated, typically remaining elevated for 12–24 h post ethanol metabolism [149,150]. As such, reasonable evidence exists that suggests acetate may be a lead compound in driving the ethanol-induced effects on cardiovascular function from a centrally mediated standpoint.

When attempting to elucidate the neuronal mechanisms of ethanol-induced changes in cardiovascular regulation, several brain regions that regulate autonomic function have been studied. The rostral ventrolateral medulla (RVLM) is a significant brain region involved in the integration of upstream brain regional sympathoexcitatory outputs [139,151–158]. The RVLM projects to the spinal intermediolateral column (IML), which is the first synapse in the central to peripheral sympathetic output pathway [158,159]. In direct application studies of the RVLM conducted by El-Mas and colleagues, ethanol microinjection dose-dependently increased RVLM norepinephrine and ABP in spontaneously hypertensive rats (SHRs) [141,160]. Additional follow-up studies by the same group suggested that this ethanol response was due to enhanced catabolism of ethanol to acetaldehyde [141]. They however indirectly speculated that acetate had little effect on increasing ABP levels [141]. Interestingly, when investigating catalase and ALDH enzyme activity in the RVLM between SHRs and non-hypertensive control (Wistar–Kyoto, WKY) rats, the SHRs had higher levels of catalase activity compared to the WKY rats, with no differences in ALDH activity, suggesting a potential genetic component in relation to differences in brain ethanol metabolism [141]. While this study described above did not detect differences in ALDH activity, a recent and exciting report has highlighted brain region differences in ALDH2 expression in mice [161]. This previous finding suggests that investigators must be cognizant of how brain regions and/or genetic differences in ethanol metabolism may influence behavior and/or cardiovascular function.

The RVLM integrates projections from the paraventricular nucleus (PVN-RVLM) and the central nucleus of the amygdala (CeA-RVLM) [139,151–158], two brain regions that regulate autonomic function [139,151–154,162]. We have previously shown that microinjection of 1.7 μmole ethanol into the CeA consistently increases SNA and ABP [139]. This ethanol-induced pressor response within the CeA was *N*-methyl-D-aspartate receptor (NMDAR)-dependent [139,151], countering the predominant theory in the alcohol research field of ethanol inhibiting NMDAR function [163]. Furthermore, inhibiting glutamatergic receptors in the RVLM blunted the effect of ethanol microinjected in the CeA, suggesting that the sympathoexcitatory response acted through CeA-RVLM circuitry [139]. Moreover, we speculated that the CeA–ethanol response may have been at least partially driven by local brain metabolism of ethanol to acetate. We microinjected 0.2 μmole sodium acetate in the CeA and were able to elicit responses in the SNA similar to those achieved with ethanol. This acetate-induced response was also blunted by the NMDAR antagonist, AP5 [139]. Our follow-up whole-cell electrophysiology studies in brain slices containing CeA-RVLM-projecting neurons demonstrated that (1) acetate increased CeA-RVLM neuronal excitability in a dose-dependent manner, (2) the excitation was driven by acetate and not sodium concentrations, (3) the acetate-induced increase in CeA-RVLM neuronal excitability was sensitive to NMDAR antagonists, AP5 and memantine, and (4) acetate directly stimulated NMDAR-sensitive currents and cytosolic calcium increases, which were also abolished by NMDAR antagonists [151].

While we have shown that acetate appears to activate NMDAR in the CeA of rats, this finding has also been replicated in C57BL/6J mice in the nucleus accumbens shell [2,164], a key node in the mammalian reward circuit. Surprisingly, the effect of acetic acid-induced activation of the NMDAR was greater in females compared to males, highlighting a potential mechanism contributing to females' accelerated development of alcohol use disorder and the greater neuronal degeneration associated with long-term use in females when compared to males [2].

9. NMDAR–NO Interactions

NO activation and NMDAR activation are intricately tied together [165]. An increase in brain NO has been shown to lead to NMDAR activation [166], and oppositely, NMDAR activation is capable of stimulating NO production [165–167]. NO, which is a soluble and diffusible gas [168], can then enhance presynaptic neurotransmitter release, such as the release of glutamate, and further potentiate an excitatory response through retrograde signaling [169,170]. In the central nervous system, neuronal nitric oxide synthase (nNOS), inducible nitric oxide synthase (iNOS), and endothelial nitric oxide synthase (eNOS) are all present [171]. In neurons, nNOS–NMDAR interactions are governed predominantly by cytosolic calcium and calcium calmodulin kinase II (CamK2) [172]. Increased NMDAR activation stimulates CamK2, resulting in increased nNOS, which stimulates the production of NO and cyclic guanosine monophosphate (cGMP), which then diffuses out into the synaptic cleft where it can modulate neuronal function [173].

In the RVLM, NO was found to be excitatory, and this was likely through NO-dependent glutamatergic activation [174–176]. In our studies investigating the impact of acetate on NMDAR and neuronal function, we had hypothesized that acetic acid/acetate was capable of mimicking the structures of glutamate and glycine [139,151], accounting for the direct effects seen at the NMDAR during in vivo and ex vivo experiments. However, it is possible that acetic acid/acetate may also be stimulating NMDAR and glutamatergic activation [169,170] via increased NO, which then may stimulate presynaptic glutamate release and/or postsynaptic NMDAR activity.

Thus, in addition to the effect of acetate generation of NO on the peripheral vasculature, there is the potential for acetate to stimulate NO production in the brain. NO has been shown to have direct neuromodulatory effects on several neurotransmitters and signaling pathways in the brain, and this includes NMDA [166,167,170,171], GABA [177–180], and glutamatergic signaling [174–176]. These ligand-gated ion channels and neurotransmitters are impacted in ethanol studies [181] and as such, the possibility of acetic acid/acetate-generated NO production may be a major driving force in altered neuronal function post ethanol exposure.

The existence of temporal sympathoexcitatory effects between PVN and CeA microinjections of acetate also lends credibility to an acetate–NMDAR–NO interaction (Figure 2). The PVN has been extensively studied in relation to neural control of cardiovascular function [145,152–157,162,177–180,182–188], while for the CeA, such research is much less common [139,151,189]. The PVN contains an abundance of GABAergic presynaptic terminals [155,190,191] and these terminals have been shown to be modulated by

NO [178–180,192,193], which can influence cardiovascular function. In an elegant study from Li, Mayhan and Patel, this group found that microinjection of NMDA into the PVN elicited increases in SNA, ABP, and HR [192]. To investigate the contribution of NO in this response, the group included a nitric oxide synthase inhibitor, N(G)-monomethyl-L-arginine (L-NMMA), with and without NMDA. PVN microinjections of L-NMMA and NMDA produced significantly larger pressor responses compared to NMDA alone [192]. Thus, the inhibition of NO production with L-NMMA was found to exacerbate NMDAR sympathoexcitation in the PVN. To put this another way, PVN NMDAR activation generated NO, which acted as an inhibitory break on sympathoexcitation. Finally, the group found that administration of the NMDAR antagonist AP5 blunted these effects, demonstrating that they were NMDAR-dependent. This same group (Li and Patel et al.) identified that the NMDAR–NO inhibitory break within the PVN was mediated by NMDAR activation, leading to NO modulation of the GABA release on presynaptic glutamate [192].

Sympathoexcitatory responses to the PVN microinjection of acetate, which are sensitive to NMDAR antagonists, display a quick on/off response relative to CeA-microinjected acetate (Figure 2A). As has been previously discussed above, the PVN contains strong GABAergic inputs on glutamatergic synapses, which have been demonstrated to be inhibited by a retrograde NMDAR–NO signaling mechanism [192]. We hypothesize that the quick excitatory acetate response to PVN microinjection (Figure 2A,B) followed by a rapid normalization to baseline (Figure 2A,C) is likely due to an acetate–NMDAR–NO–GABA release type of mechanism. This PVN response can be contrasted to those we have observed from CeA-microinjected acetate (Figure 2D).

Contrary to the abundance of GABAergic inputs on presynaptic glutamatergic inputs to the PVN, presynaptic inputs to the CeA appear as a heterogenous mixture of glutamatergic [194], GABAergic [195], serotonergic [196], and norepinephrinergic [197] projections. However, a recent review has also suggested the potential for heavy glutamatergic innervation to the CeA [198]. We have demonstrated in this brain region an acetate–NMDAR effect both in vivo [139] and also ex vivo [151]. Given the blend of presynaptic projections to the CeA and the limited to non-existent mapping on CeA-RVLM neurons, which we have studied the effects of acetate on [139,151], it is unclear as to what, if anything, may be presynaptically influencing this autonomic circuitry. Similar to the NMDAR–NO effect within the PVN [192], a potential NMDAR–NO effect in the CeA is also a strong possibility (Figure 2D–F). However, given the slower response time and maximum sustained pressor responses to CeA-microinjected acetate relative to the PVN (Figure 2A,D), we speculate a potential acetate–NMDAR–NO retrograde signaling response that is more aligned with a mixed-mechanism involving glutamatergic and G-protein-coupled receptor responses (Figure 2E,F).

10. Intersection of Alcohol Cardiovascular Research and AUD Research

While alcohol consumption and its associated cardiovascular sequelae are a component of AUD pathophysiology, continued alcohol use despite negative outcomes remains the major issue driving secondary pathologies [199–201]. Thus, understanding the key neurobiological mechanisms which contribute to the rewarding aspects of alcohol use and those which eventually lead to the development of AUD is crucial for the development of treatment options for individuals suffering from AUD. The investigation of how alcohol metabolism (the generation of acetic acid/acetate) influences neural control of cardiovascular function intersects directly with researchers investigating the neurobiology of AUD, as the key mechanisms driving both are likely translatable across brain regions. We have highlighted the effect of acetic acid/acetate and ethanol on different organs/organ systems in Table 1.

Indeed, we have directly tested this hypothesis and have shown that acetic acid at physiologically relevant concentrations applied to medium spiny neurons (MSNs) of the nucleus accumbens shell (NAcSh), a key node in the mammalian reward circuit, is capable of increasing presynaptic glutamate release and increasing neuronal excitability [164]. In a subsequent follow-up study, we identified sex differences in acetic acid-induced NMDAR responses (females > males) in this same brain region [2]. Whether the acetic acid-induced increase in presynaptic glutamate release is mediated by presynaptic acidification [202,203], presynaptic NMDAR activation, retrograde NMDAR–NO signaling, and/or a combination of all of these remains to be determined. However, the stark sex difference in acetic acid-induced NMDAR responses may potentially underlie the greater propensity for females to experience accelerated AUD and suffer greater alcohol-induced neurodegeneration compared to males.

Table 1. Impact of ethanol and acetic acid/acetate on different organs/organ systems.

Organ/Organ System	Acetic Acid/Acetate	Ethanol
Brain	↑ GABA [161,204], ↑ Glutamate [164,204], ↑ NMDAR [2,84,139,151], ↑ Dopamine [203], ↑ Neuropathology [34,35,84], ↑ Cerebral blood flow [205]	↑ GABA [195,206], ↓ Glutamate [207,208], ↑ Glutamate [206,209], ↑ Dopamine [209–211], ↑ Serotonin [211,212] ↓ NMDAR [163,213], ↑ Neuropathology [13,214,215], ↑ Cerebral blood flow [216,217]
Heart	↓ Contractility [119,218], ↑ O_2 consumption [116], ↑ Coronary flow [116], ↑ Cardiac output [116]	↓ Contractility [92,219], ↑ Coronary flow [220], Arrhythmia [221], ↑ Cardiac output [107]
Gastrointestinal	↑ Inflammation: Oral cavity [222,223], Esophagus [224], Stomach [225], Small intestine [226], Liver [8], Colon [227]	↑ Inflammation: Oral cavity [228], Esophagus [228], Stomach [229], Small intestine [230,231], Liver [8,232,233], Colon [230]
Vasculature	↑ Vasodilation [98,99,102,106,119], ↑ NOS [102,106], ↑ NO [102]	↑ Vasodilation [85,93,146,234], ↑ NOS [235,236], ↑ NO [93,95,236]

Note: This table is not an exhaustive list, ↑ = increase, ↓ = decrease.

11. Conclusions and Future Research Directions

Given the complex effects of alcohol on cardiovascular function and neural control of the autonomic nervous system, future studies are needed to determine the mechanisms of these effects. For instance, there is a substantial body of research in the literature on the impacts of alcohol consumption on the peripheral vasculature [93,234]. However, many of these studies have not explored how alcohol-induced changes in the peripheral vasculature can affect neural control of the sympathetic nervous system (e.g., alcohol peripheral vasculature dilation stimulating arterial baroreceptors and initiating neural sympathoexcitation to maintain homeostasis). This push–pull dynamic between the peripheral and central nervous systems may be a contributing factor to the development of alcohol-induced hypertension [15,24–26,90,142,145,199] and cardiomyopathy [122,123]. While one could make an argument that moderate ethanol consumption provides cardiovascular benefits (e.g., increased vasodilation and cardiac output), these effects are likely countered by the acetic acid/acetate-initiated peripheral chemoreceptor reflex, arterial baroreceptor reflex, and activation of pre-sympathetic neurons in the central nervous system, which would increase SNA. Chronic ethanol use, and elevated acetic acid/acetate from its metabolism, would be anticipated to contribute to end organ damage due to increased sympathetic outflow and is potentially further exacerbated by ethanol/acetate-induced dysregulation of NO signaling. Dysregulation of NO signaling is a major finding in essential hypertension [237–240]. Furthermore, since alcohol and acetic acid/acetate are capable of crossing the blood–brain barrier, the impact of the brain's metabolism of ethanol on neural control of cardiovascular function also needs to be studied.

Future studies using rodent models could investigate the following: (1) What is the net output of ethanol and/or its metabolism in the brain on SNA and ABP? (2) Are responses to locally microinjected acetate in the PVN and CeA blunted or exacerbated by manipulating NO signaling? (3) How might acetic acid/acetate-induced NO signaling in the brain influence AUD and alcohol-related neurodegeneration?

The neural signaling mechanisms influencing cardiovascular function listed above share potential overlapping mechanisms with the development and maintenance of AUD. Thus, in vivo microinjection studies offer a powerful pharmacological tool to evaluate the excitatory/inhibitory actions of ethanol and its metabolites in real time by measuring sympathetic nerve activity and arterial blood pressure. Future studies and findings within this area of research may be beneficial for the development of treatment strategies for not only alcohol-induced cardiovascular diseases, but also for AUD.

Author Contributions: All authors edited and revised the manuscript. A.D.C. and Q.-H.C. conceptualized and designed the research. All authors have read and agreed to the published version of the manuscript.

Funding: This study was supported by the National Institute of Health: NIHR15HL122952, NIHR01HL163159, and the American Heart Association, AHA 16PRE27780121.

Acknowledgments: We would like to thank Timothy W. Chapp and Andréa R. Collins for their proofreading and suggestions, and Mingjun Gu for excellent technical assistance. A portion of the figures were created in biorender.com.

Conflicts of Interest: The authors declare no conflict of interest.

References

1. Zakhari, S. Overview: How is alcohol metabolized by the body? *Alcohol. Res. Health* **2006**, *29*, 245–254. [PubMed]
2. Chapp, A.D.; Nwakama, C.A.; Collins, A.R.; Mermelstein, P.G.; Thomas, M.J. Physiological acetic acid concentrations from ethanol metabolism stimulate accumbens shell medium spiny neurons via NMDAR activation in a sex-dependent manner. *Neuropsychopharmacology* **2023**. [CrossRef] [PubMed]
3. Jastroch, M.; Divakaruni, A.S.; Mookerjee, S.; Treberg, J.R.; Brand, M.D. Mitochondrial proton and electron leaks. *Essays Biochem.* **2010**, *47*, 53–67. [CrossRef]
4. Fernández-Checa, J.C.; Kaplowitz, N.; Colell, A.; García-Ruiz, C. Oxidative stress and alcoholic liver disease. *Alcohol. Health Res. World* **1997**, *21*, 321–324.
5. Tan, H.K.; Yates, E.; Lilly, K.; Dhanda, A.D. Oxidative stress in alcohol-related liver disease. *World J. Hepatol.* **2020**, *12*, 332–349. [CrossRef] [PubMed]
6. Ambade, A.; Mandrekar, P. Oxidative Stress and Inflammation: Essential Partners in Alcoholic Liver Disease. *Int. J. Hepatol.* **2012**, *2012*, 853175. [CrossRef] [PubMed]
7. Cederbaum, A.I.; Lu, Y.; Wu, D. Role of oxidative stress in alcohol-induced liver injury. *Arch. Toxicol.* **2009**, *83*, 519–548. [CrossRef]
8. Kendrick, S.F.; O'Boyle, G.; Mann, J.; Zeybel, M.; Palmer, J.; Jones, D.E.; Day, C.P. Acetate, the key modulator of inflammatory responses in acute alcoholic hepatitis. *Hepatology* **2010**, *51*, 1988–1997. [CrossRef]
9. Comporti, M.; Signorini, C.; Leoncini, S.; Gardi, C.; Ciccoli, L.; Giardini, A.; Vecchio, D.; Arezzini, B. Ethanol-induced oxidative stress: Basic knowledge. *Genes. Nutr.* **2010**, *5*, 101–109. [CrossRef]
10. Sun, A.Y.; Sun, G.Y. Ethanol and oxidative mechanisms in the brain. *J. Biomed. Sci.* **2001**, *8*, 37–43. [CrossRef]
11. Crews, F.T. Alcohol-related neurodegeneration and recovery: Mechanisms from animal models. *Alcohol. Res. Health* **2008**, *31*, 377–388. [PubMed]
12. Kamal, H.; Tan, G.C.; Ibrahim, S.F.; Shaikh, M.F.; Mohamed, I.N.; Mohamed, R.M.P.; Hamid, A.A.; Ugusman, A.; Kumar, J. Alcohol Use Disorder, Neurodegeneration, Alzheimer's and Parkinson's Disease: Interplay Between Oxidative Stress, Neuroimmune Response and Excitotoxicity. *Front. Cell Neurosci.* **2020**, *14*, 282. [CrossRef] [PubMed]
13. Crews, F.T.; Nixon, K. Mechanisms of Neurodegeneration and Regeneration in Alcoholism. *Alcohol. Alcohol.* **2008**, *44*, 115–127. [CrossRef] [PubMed]
14. Piano, M.R. Alcohol's Effects on the Cardiovascular System. *Alcohol. Res.* **2017**, *38*, 219–241.
15. Biddinger, K.J.; Emdin, C.A.; Haas, M.E.; Wang, M.; Hindy, G.; Ellinor, P.T.; Kathiresan, S.; Khera, A.V.; Aragam, K.G. Association of Habitual Alcohol Intake With Risk of Cardiovascular Disease. *JAMA Netw. Open* **2022**, *5*, e223849. [CrossRef] [PubMed]
16. Larsson, S.C.; Burgess, S.; Mason, A.M.; Michaëlsson, K. Alcohol Consumption and Cardiovascular Disease. *Circ. Genom. Precis. Med.* **2020**, *13*, e002814. [CrossRef]

17. Toma, A.; Paré, G.; Leong, D.P. Alcohol and Cardiovascular Disease: How Much is Too Much? *Curr. Atheroscler. Rep.* **2017**, *19*, 13. [CrossRef]
18. Sutanto, H.; Cluitmans, M.J.M.; Dobrev, D.; Volders, P.G.A.; Bébarová, M.; Heijman, J. Acute effects of alcohol on cardiac electrophysiology and arrhythmogenesis: Insights from multiscale in silico analyses. *J. Mol. Cell Cardiol.* **2020**, *146*, 69–83. [CrossRef]
19. Kupari, M.; Koskinen, P. Alcohol, cardiac arrhythmias and sudden death. *Novartis Found. Symp.* **1998**, *216*, 68–79; discussion 79–85. [CrossRef]
20. Mustroph, J.; Baier, M.J.; Unsin, D.; Provaznik, Z.; Kozakov, K.; Lebek, S.; Tarnowski, D.; Schildt, S.; Voigt, N.; Wagner, S.; et al. Ethanol-Induced Atrial Fibrillation Results From Late I_{Na} and Can Be Prevented by Ranolazine. *Circulation* **2023**, *148*, 698–700. [CrossRef]
21. Agarwal, D.P. Cardioprotective effects of light–moderate consumption of alcohol: A review of putative mechanisms. *Alcohol. Alcohol.* **2002**, *37*, 409–415. [CrossRef] [PubMed]
22. Chen, C.H.; Gray, M.O.; Mochly-Rosen, D. Cardioprotection from ischemia by a brief exposure to physiological levels of ethanol: Role of epsilon protein kinase C. *Proc. Natl. Acad. Sci. USA* **1999**, *96*, 12784–12789. [CrossRef] [PubMed]
23. Pearson, T.A. AHA Science Advisory. Alcohol and heart disease. Nutrition Committee of the American Heart Association. *Am. J. Clin. Nutr.* **1997**, *65*, 1567–1569. [CrossRef] [PubMed]
24. Husain, K.; Ansari, R.A.; Ferder, L. Alcohol-induced hypertension: Mechanism and prevention. *World J. Cardiol.* **2014**, *6*, 245–252. [CrossRef] [PubMed]
25. Beilin, L.J.; Puddey, I.B. Alcohol and Hypertension. *Hypertension* **2006**, *47*, 1035–1038. [CrossRef]
26. Tsuruta, M.; Adachi, H.; Hirai, Y.; Fujiura, Y.; Imaizumi, T. Association Between Alcohol Intake and Development of Hypertension in Japanese Normotensive Men: 12-Year Follow-Up Study. *Am. J. Hypertens.* **2000**, *13*, 482–487. [CrossRef]
27. Roerecke, M.; Kaczorowski, J.; Tobe, S.W.; Gmel, G.; Hasan, O.S.M.; Rehm, J. The effect of a reduction in alcohol consumption on blood pressure: A systematic review and meta-analysis. *Lancet Public. Health* **2017**, *2*, e108–e120. [CrossRef]
28. Federico, S.D.; Filippini, T.; Whelton, P.K.; Cecchini, M.; Iamandii, I.; Boriani, G.; Vinceti, M. Alcohol Intake and Blood Pressure Levels: A Dose-Response Meta-Analysis of Nonexperimental Cohort Studies. *Hypertension* **2023**, *80*, 1961–1969. [CrossRef]
29. Association, A.H. Even Just 1 Alcoholic Drink a Day May Increase Blood Pressure. Available online: https://www.heart.org/en/news/2023/07/31/even-just-1-alcoholic-drink-a-day-may-increase-blood-pressure (accessed on 19 December 2023).
30. Zakhari, S. Alcohol metabolism and epigenetics changes. *Alcohol. Res.* **2013**, *35*, 6–16.
31. Pietrocola, F.; Galluzzi, L.; Bravo-San Pedro, J.M.; Madeo, F.; Kroemer, G. Acetyl coenzyme A: A central metabolite and second messenger. *Cell Metab.* **2015**, *21*, 805–821. [CrossRef]
32. Mews, P.; Egervari, G.; Nativio, R.; Sidoli, S.; Donahue, G.; Lombroso, S.I.; Alexander, D.C.; Riesche, S.L.; Heller, E.A.; Nestler, E.J.; et al. Alcohol metabolism contributes to brain histone acetylation. *Nature* **2019**, *574*, 717–721. [CrossRef] [PubMed]
33. Erny, D.; Dokalis, N.; Mezö, C.; Castoldi, A.; Mossad, O.; Staszewski, O.; Frosch, M.; Villa, M.; Fuchs, V.; Mayer, A.; et al. Microbiota-derived acetate enables the metabolic fitness of the brain innate immune system during health and disease. *Cell Metab.* **2021**, *33*, 2260–2276.e7. [CrossRef] [PubMed]
34. Colombo, A.V.; Sadler, R.K.; Llovera, G.; Singh, V.; Roth, S.; Heindl, S.; Sebastian Monasor, L.; Verhoeven, A.; Peters, F.; Parhizkar, S.; et al. Microbiota-derived short chain fatty acids modulate microglia and promote Aβ plaque deposition. *eLife* **2021**, *10*, e59826. [CrossRef] [PubMed]
35. Sampson, T.R.; Debelius, J.W.; Thron, T.; Janssen, S.; Shastri, G.G.; Ilhan, Z.E.; Challis, C.; Schretter, C.E.; Rocha, S.; Gradinaru, V.; et al. Gut Microbiota Regulate Motor Deficits and Neuroinflammation in a Model of Parkinson's Disease. *Cell* **2016**, *167*, 1469–1480.e12. [CrossRef] [PubMed]
36. Perry, R.J.; Peng, L.; Barry, N.A.; Cline, G.W.; Zhang, D.; Cardone, R.L.; Petersen, K.F.; Kibbey, R.G.; Goodman, A.L.; Shulman, G.I. Acetate mediates a microbiome-brain-β-cell axis to promote metabolic syndrome. *Nature* **2016**, *534*, 213–217. [CrossRef] [PubMed]
37. Setshedi, M.; Wands, J.R.; Monte, S.M. Acetaldehyde adducts in alcoholic liver disease. *Oxid. Med. Cell Longev.* **2010**, *3*, 178–185. [CrossRef]
38. Tuma, D.J.; Hoffman, T.; Sorrell, M.F. The chemistry of acetaldehyde-protein adducts. *Alcohol. Alcohol. Suppl.* **1991**, *1*, 271–276.
39. Rintala, J.; Jaatinen, P.; Parkkila, S.; Sarviharju, M.; Kiianmaa, K.; Hervonen, A.; Niemelä, O. Evidence of acetaldehyde–protein adduct formation in rat brain after lifelong consumption of ethanol. *Alcohol. Alcohol.* **2000**, *35*, 458–463. [CrossRef]
40. Ito, A.; Jamal, M.; Ameno, K.; Tanaka, N.; Takakura, A.; Kawamoto, T.; Kitagawa, K.; Nakayama, K.; Matsumoto, A.; Miki, T.; et al. Acetaldehyde administration induces salsolinol formation in vivo in the dorsal striatum of Aldh2-knockout and C57BL/6N mice. *Neurosci. Lett.* **2018**, *685*, 50–54. [CrossRef]
41. Peana, A.T.; Rosas, M.; Porru, S.; Acquas, E. From Ethanol to Salsolinol: Role of Ethanol Metabolites in the Effects of Ethanol. *J. Exp. Neurosci.* **2016**, *10*, 137–146. [CrossRef]
42. Bassareo, V.; Frau, R.; Maccioni, R.; Caboni, P.; Manis, C.; Peana, A.T.; Migheli, R.; Porru, S.; Acquas, E. Ethanol-Dependent Synthesis of Salsolinol in the Posterior Ventral Tegmental Area as Key Mechanism of Ethanol's Action on Mesolimbic Dopamine. *Front. Neurosci.* **2021**, *15*, 675061. [CrossRef] [PubMed]
43. Correa, M.; Acquas, E.; Salamone, J.D. The renaissance of acetaldehyde as a psychoactive compound: Decades in the making. *Front. Behav. Neurosci.* **2014**, *8*, 249. [CrossRef] [PubMed]

44. Kuriyama, K.; Ohkuma, S.; Taguchi, J.-I.; Hashimoto, T. Alcohol, acetaldehyde and salsolinol-induced alterations in functions of cerebral GABA/benzodiazepine receptor complex. *Physiol. Behav.* **1987**, *40*, 393–399. [CrossRef] [PubMed]
45. Deehan Jr, G.A.; Engleman, E.A.; Ding, Z.M.; McBride, W.J.; Rodd, Z.A. Microinjections of acetaldehyde or salsolinol into the posterior ventral tegmental area increase dopamine release in the nucleus accumbens shell. *Alcohol. Clin. Exp. Res.* **2013**, *37*, 722–729. [CrossRef] [PubMed]
46. Jamal, M.; Ameno, K.; Ameno, S.; Okada, N.; Ijiri, I. In vivo study of salsolinol produced by a high concentration of acetaldehyde in the striatum and nucleus accumbens of free-moving rats. *Alcohol. Clin. Exp. Res.* **2003**, *27*, 79S–84S. [CrossRef]
47. Zuddas, A.; Corsini, G.U.; Schinelli, S.; Barker, J.L.; Kopin, I.J.; di Porzio, U. Acetaldehyde directly enhances MPP+ neurotoxicity and delays its elimination from the striatum. *Brain Res.* **1989**, *501*, 11–22. [CrossRef]
48. Cui, J.; Liu, Y.; Chang, X.; Gou, W.; Zhou, X.; Liu, Z.; Li, Z.; Wu, Y.; Zuo, D. Acetaldehyde Induces Neurotoxicity In Vitro via Oxidative Stress- and Ca(2+) Imbalance-Mediated Endoplasmic Reticulum Stress. *Oxid. Med. Cell Longev.* **2019**, *2019*, 2593742. [CrossRef]
49. Yan, T.; Zhao, Y. Acetaldehyde induces phosphorylation of dynamin-related protein 1 and mitochondrial dysfunction via elevating intracellular ROS and Ca^{2+} levels. *Redox Biol.* **2020**, *28*, 101381. [CrossRef]
50. Peana, A.T.; Sánchez-Catalán, M.J.; Hipólito, L.; Rosas, M.; Porru, S.; Bennardini, F.; Romualdi, P.; Caputi, F.F.; Candeletti, S.; Polache, A.; et al. Mystic Acetaldehyde: The Never-Ending Story on Alcoholism. *Front. Behav. Neurosci.* **2017**, *11*, 81. [CrossRef]
51. Matsumura, Y.; Stiles, K.M.; Reid, J.; Frenk, E.Z.; Cronin, S.; Pagovich, O.E.; Crystal, R.G. Gene Therapy Correction of Aldehyde Dehydrogenase 2 Deficiency. *Mol. Ther. Methods Clin. Dev.* **2019**, *15*, 72–82. [CrossRef]
52. Goldman, D. Aldehyde Dehydrogenase Deficiency as Cause of Facial Flushing Reaction to Alcohol in Japanese. *Alcohol. Health Res. World* **1995**, *19*, 48–49. [PubMed]
53. Chen, C.H.; Kraemer, B.R.; Mochly-Rosen, D. ALDH2 variance in disease and populations. *Dis. Model. Mech.* **2022**, *15*, dmm049601. [CrossRef]
54. Ohsawa, I.; Kamino, K.; Nagasaka, K.; Ando, F.; Niino, N.; Shimokata, H.; Ohta, S. Genetic deficiency of a mitochondrial aldehyde dehydrogenase increases serum lipid peroxides in community-dwelling females. *J. Hum. Genet.* **2003**, *48*, 404–409. [CrossRef] [PubMed]
55. Stokes, M.; Abdijadid, S. Disulfiram. In *StatPearls*; StatPearls Publishing: St. Petersburg, FL, USA, 2023.
56. Rosman, A.S.; Waraich, A.; Baraona, E.; Lieber, C.S. Disulfiram treatment increases plasma and red blood cell acetaldehyde in abstinent alcoholics. *Alcohol. Clin. Exp. Res.* **2000**, *24*, 958–964. [CrossRef]
57. Kleczkowska, P.; Sulejczak, D.; Zaremba, M. Advantages and disadvantages of disulfiram coadministered with popular addictive substances. *Eur. J. Pharmacol.* **2021**, *904*, 174143. [CrossRef] [PubMed]
58. Eriksson, C.J. Measurement of acetaldehyde: What levels occur naturally and in response to alcohol? *Novartis Found. Symp.* **2007**, *285*, 247–255; discussion 256–260. [CrossRef]
59. Lindros, K.O. Human blood acetaldehyde levels: With improved methods, a clearer picture emerges. *Alcohol. Clin. Exp. Res.* **1983**, *7*, 70–75. [CrossRef]
60. Lide, D.R. (Ed.) *CRC Handbook of Chemistry and Physics*, 78th ed.; CRC Press: Boca Raton, FL, USA, 1997; p. 3.
61. Bowen, E.J.; Tietz, E.L. The Oxidation of Acetaldehyde by Oxygen. *Nature* **1929**, *124*, 914. [CrossRef]
62. Bawn, C.E.H.; Williamson, J.B. The oxidation of acetaldehyde in solution. Part I.—The chemistry of the intermediate stages. *Trans. Faraday Soc.* **1951**, *47*, 721–734. [CrossRef]
63. Sánchez, M.; Sabio, L.; Gálvez, N.; Capdevila, M.; Dominguez-Vera, J.M. Iron chemistry at the service of life. *IUBMB Life* **2017**, *69*, 382–388. [CrossRef]
64. Koivunen, J.; Heinonen-Tanski, H. Peracetic acid (PAA) disinfection of primary, secondary and tertiary treated municipal wastewaters. *Water Res.* **2005**, *39*, 4445–4453. [CrossRef] [PubMed]
65. Viola, K.S.; Rodrigues, E.M.; Tanomaru-Filho, M.; Carlos, I.Z.; Ramos, S.G.; Guerreiro-Tanomaru, J.M.; Faria, G. Cytotoxicity of peracetic acid: Evaluation of effects on metabolism, structure and cell death. *Int. Endod. J.* **2018**, *51*, e264–e277. [CrossRef]
66. Turrà, N.; Neuenschwander, U.; Hermans, I. Molecule-induced peroxide homolysis. *Chemphyschem* **2013**, *14*, 1666–1669. [CrossRef] [PubMed]
67. Hiatt, R.R.; Irwin, K.C.; Gould, C.W. Homolytic decompositions of hydroperoxides. IV. Metal-catalyzed decompositions. *J. Org. Chem.* **1968**, *33*, 1430–1435. [CrossRef]
68. Uttara, B.; Singh, A.V.; Zamboni, P.; Mahajan, R.T. Oxidative stress and neurodegenerative diseases: A review of upstream and downstream antioxidant therapeutic options. *Curr. Neuropharmacol.* **2009**, *7*, 65–74. [CrossRef] [PubMed]
69. Machlin, L.J.; Bendich, A. Free radical tissue damage: Protective role of antioxidant nutrients. *Faseb J.* **1987**, *1*, 441–445. [CrossRef]
70. Wu, M.-L.; Tsai, K.-L.; Wang, S.-M.; Wu, J.-C.; Wang, B.-S.; Lee, Y.-T. Mechanism of Hydrogen Peroxide and Hydroxyl Free Radical–Induced Intracellular Acidification in Cultured Rat Cardiac Myoblasts. *Circ. Res.* **1996**, *78*, 564–572. [CrossRef]
71. Winterbourn, C.C. Toxicity of iron and hydrogen peroxide: The Fenton reaction. *Toxicol. Lett.* **1995**, *82–83*, 969–974. [CrossRef]
72. Carlos, T.D.; Bezerra, L.B.; Vieira, M.M.; Sarmento, R.A.; Pereira, D.H.; Cavallini, G.S. Fenton-type process using peracetic acid: Efficiency, reaction elucidations and ecotoxicity. *J. Hazard. Mater.* **2021**, *403*, 123949. [CrossRef]
73. Hui, S.; Ghergurovich, J.M.; Morscher, R.J.; Jang, C.; Teng, X.; Lu, W.; Esparza, L.A.; Reya, T.; Le, Z.; Yanxiang Guo, J.; et al. Glucose feeds the TCA cycle via circulating lactate. *Nature* **2017**, *551*, 115–118. [CrossRef]

74. Goodwin, M.L.; Harris, J.E.; Hernández, A.; Gladden, L.B. Blood lactate measurements and analysis during exercise: A guide for clinicians. *J. Diabetes Sci. Technol.* **2007**, *1*, 558–569. [CrossRef] [PubMed]
75. Mathew, T.K.; Zubair, M.; Tadi, P. Blood Glucose Monitoring. In *StatPearls*; StatPearls Publishing: St. Petersburg, FL, USA, 2023.
76. Foucher, C.D.; Tubben, R.E. Lactic Acidosis. In *StatPearls*; StatPearls Publishing: St. Petersburg, FL, USA, 2023.
77. Hosios, A.M.; Vander Heiden, M.G. Acetate metabolism in cancer cells. *Cancer Metab.* **2014**, *2*, 27. [CrossRef] [PubMed]
78. Richards, R.H.; Dowling, J.A.; Vreman, H.J.; Feldman, C.; Weiner, M.W. Acetate levels in human plasma. *Proc. Clin. Dial. Transpl. Forum* **1976**, *6*, 73–79.
79. Davies, P.G.; Venkatesh, B.; Morgan, T.J.; Presneill, J.J.; Kruger, P.S.; Thomas, B.J.; Roberts, M.S.; Mundy, J. Plasma acetate, gluconate and interleukin-6 profiles during and after cardiopulmonary bypass: A comparison of Plasma-Lyte 148 with a bicarbonate-balanced solution. *Crit. Care* **2011**, *15*, R21. [CrossRef] [PubMed]
80. Pomare, E.W.; Branch, W.J.; Cummings, J.H. Carbohydrate fermentation in the human colon and its relation to acetate concentrations in venous blood. *J. Clin. Investig.* **1985**, *75*, 1448–1454. [CrossRef] [PubMed]
81. Tollinger, C.D.; Vreman, H.J.; Weiner, M.W. Measurement of acetate in human blood by gas chromatography: Effects of sample preparation, feeding, and various diseases. *Clin. Chem.* **1979**, *25*, 1787–1790. [CrossRef] [PubMed]
82. Nuutinen, H.; Lindros, K.; Hekali, P.; Salaspuro, M. Elevated blood acetate as indicator of fast ethanol elimination in chronic alcoholics. *Alcohol* **1985**, *2*, 623–626. [CrossRef]
83. Chapp, A.D.; Schum, S.; Behnke, J.E.; Hahka, T.; Huber, M.J.; Jiang, E.; Larson, R.A.; Shan, Z.; Chen, Q.H. Measurement of cations, anions, and acetate in serum, urine, cerebrospinal fluid, and tissue by ion chromatography. *Physiol. Rep.* **2018**, *6*, e13666. [CrossRef]
84. Chapp, A.D.; Behnke, J.E.; Driscoll, K.M.; Fan, Y.; Hoban, E.; Shan, Z.; Zhang, L.; Chen, Q.H. Acetate Mediates Alcohol Excitotoxicity in Dopaminergic-like PC12 Cells. *ACS Chem. Neurosci.* **2019**, *10*, 235–245. [CrossRef]
85. van de Borne, P.; Mark, A.L.; Montano, N.; Mion, D.; Somers, V.K. Effects of alcohol on sympathetic activity, hemodynamics, and chemoreflex sensitivity. *Hypertension* **1997**, *29*, 1278–1283. [CrossRef]
86. Morvai, V.; Nádházi, Z.; Molnár, G.Y.; Ungváry, G.Y.; Folly, G. Acute effects of low doses of alcohol on the cardiovascular system in young men. *Acta Med. Hung.* **1988**, *45*, 339–348. [PubMed]
87. Sagawa, Y.; Kondo, H.; Matsubuchi, N.; Takemura, T.; Kanayama, H.; Kaneko, Y.; Kanbayashi, T.; Hishikawa, Y.; Shimizu, T. Alcohol has a dose-related effect on parasympathetic nerve activity during sleep. *Alcohol. Clin. Exp. Res.* **2011**, *35*, 2093–2100. [CrossRef] [PubMed]
88. Julian, T.H.; Syeed, R.; Glascow, N.; Zis, P. Alcohol-induced autonomic dysfunction: A systematic review. *Clin. Auton. Res.* **2020**, *30*, 29–41. [CrossRef] [PubMed]
89. Greenlund, I.M.; Cunningham, H.A.; Tikkanen, A.L.; Bigalke, J.A.; Smoot, C.A.; Durocher, J.J.; Carter, J.R. Morning sympathetic activity after evening binge alcohol consumption. *Am. J. Physiol.-Heart Circ. Physiol.* **2021**, *320*, H305–H315. [CrossRef] [PubMed]
90. Randin, D.; Vollenweider, P.; Tappy, L.; Jéquier, E.; Nicod, P.; Scherrer, U. Suppression of Alcohol-Induced Hypertension by Dexamethasone. *N. Engl. J. Med.* **1995**, *332*, 1733–1738. [CrossRef] [PubMed]
91. Carter, J.R.; Stream, S.F.; Durocher, J.J.; Larson, R.A. Influence of acute alcohol ingestion on sympathetic neural responses to orthostatic stress in humans. *Am. J. Physiol. Endocrinol. Metab.* **2011**, *300*, E771–E778. [CrossRef] [PubMed]
92. Kawano, Y. Physio-pathological effects of alcohol on the cardiovascular system: Its role in hypertension and cardiovascular disease. *Hypertens. Res.* **2010**, *33*, 181–191. [CrossRef] [PubMed]
93. Tawakol, A.; Omland, T.; Creager, M.A. Direct effect of ethanol on human vascular function. *Am. J. Physiol.-Heart Circ. Physiol.* **2004**, *286*, H2468–H2473. [CrossRef]
94. Rezvani, A.H.; Grady, D.R.; Peek, A.E.; Pucilowski, O. Inhibition of nitric oxide synthesis attenuates alcohol consumption in two strains of alcohol-preferring rats. *Pharmacol. Biochem. Behav.* **1995**, *50*, 265–270. [CrossRef]
95. Baraona, E.; Zeballos, G.A.; Shoichet, L.; Mak, K.M.; Lieber, C.S. Ethanol Consumption Increases Nitric Oxide Production in Rats, and Its Peroxynitrite-Mediated Toxicity Is Attenuated by Polyenylphosphatidylcholine. *Alcohol. Clin. Exp. Res.* **2002**, *26*, 883–889. [CrossRef]
96. Hakim, R.M.; Pontzer, M.A.; Tilton, D.; Lazarus, J.M.; Gottlieb, M.N. Effects of acetate and bicarbonate dialysate in stable chronic dialysis patients. *Kidney Int.* **1985**, *28*, 535–540. [CrossRef] [PubMed]
97. Leunissen, K.M.L.; Hoorntje, S.J.; Fiers, H.A.; Dekkers, W.T.; Mulder, A.W. Acetate versus Bicarbonate Hemodialysis in Critically Ill Patients. *Nephron* **2008**, *42*, 146–151. [CrossRef] [PubMed]
98. Keshaviah, P.R. The role of acetate in the etiology of symptomatic hypotension. *Artif. Organs* **1982**, *6*, 378–387. [CrossRef]
99. Vinay, P.; Cardoso, M.; Tejedor, A.; Prud'homme, M.; Levelillee, M.; Vinet, B.; Courteau, M.; Gougoux, A.; Rengel, M.; Lapierre, L.; et al. Acetate metabolism during hemodialysis: Metabolic considerations. *Am. J. Nephrol.* **1987**, *7*, 337–354. [CrossRef] [PubMed]
100. Keshaviah, P.; Shapiro, F.L. A critical examination of dialysis-induced hypotension. *Am. J. Kidney Dis.* **1982**, *2*, 290–301. [CrossRef] [PubMed]
101. Noris, M.; Todeschini, M.; Casiraghi, F.; Roccatello, D.; Martina, G.; Minetti, L.; Imberti, B.; Gaspari, F.; Atti, M.; Remuzzi, G. Effect of acetate, bicarbonate dialysis, and acetate-free biofiltration on nitric oxide synthesis: Implications for dialysis hypotension. *Am. J. Kidney Dis.* **1998**, *32*, 115–124. [CrossRef] [PubMed]
102. Amore, A.; Cirina, P.; Mitola, S.; Peruzzi, L.; Bonaudo, R.; Gianoglio, B.; Coppo, R. Acetate intolerance is mediated by enhanced synthesis of nitric oxide by endothelial cells. *J. Am. Soc. Nephrol.* **1997**, *8*, 1431–1436. [CrossRef]

103. Furchgott, R.F.; Zawadzki, J.V. The obligatory role of endothelial cells in the relaxation of arterial smooth muscle by acetylcholine. *Nature* **1980**, *288*, 373–376. [CrossRef]
104. Moncada, S.; Higgs, E.A. The discovery of nitric oxide and its role in vascular biology. *Br. J. Pharmacol.* **2006**, *147* (Suppl. S1), S193–S201. [CrossRef]
105. Moncada, S.; Palmer, R.M.; Higgs, E.A. Biosynthesis of nitric oxide from L-arginine. A pathway for the regulation of cell function and communication. *Biochem. Pharmacol.* **1989**, *38*, 1709–1715. [CrossRef]
106. Sakakibara, S.; Murakami, R.; Takahashi, M.; Fushimi, T.; Murohara, T.; Kishi, M.; Kajimoto, Y.; Kitakaze, M.; Kaga, T. Vinegar intake enhances flow-mediated vasodilatation via upregulation of endothelial nitric oxide synthase activity. *Biosci. Biotechnol. Biochem.* **2010**, *74*, 1055–1061. [CrossRef] [PubMed]
107. Riff, D.P.; Jain, A.C.; Doyle, J.T. Acute hemodynamic effects of ethanol on normal human volunteers. *Am. Heart J.* **1969**, *78*, 592–597. [CrossRef] [PubMed]
108. Puddey, I.B.; Vandongen, R.; Beilin, L.J.; Rouse, I.L. Alcohol stimulation of renin release in man: Its relation to the hemodynamic, electrolyte, and sympatho-adrenal responses to drinking. *J. Clin. Endocrinol. Metab.* **1985**, *61*, 37–42. [CrossRef] [PubMed]
109. Jain, A.; Yelamanchili, V.S.; Brown, K.N.; Goel, A. Holiday Heart Syndrome. In *StatPearls*; StatPearls Publishing: St. Petersburg, FL, USA, 2023.
110. Kupari, M. Acute cardiovascular effects of ethanol A controlled non-invasive study. *Br. Heart J.* **1983**, *49*, 174–182. [CrossRef] [PubMed]
111. Iwase, S.; Matsukawa, T.; Ishihara, S.; Tanaka, A.; Tanabe, K.; Danbara, A.; Matsuo, M.; Sugiyama, Y.; Mano, T. Effect of oral ethanol intake on muscle sympathetic nerve activity and cardiovascular functions in humans. *J. Auton. Nerv. Syst.* **1995**, *54*, 206–214. [CrossRef] [PubMed]
112. Delbridge, L.M.; Connell, P.J.; Harris, P.J.; Morgan, T.O. Ethanol effects on cardiomyocyte contractility. *Clin. Sci.* **2000**, *98*, 401–407. [CrossRef]
113. Ahmed, S.S.; Levinson, G.E.; Regan, T.J. Depression of myocardial contractility with low doses of ethanol in normal man. *Circulation* **1973**, *48*, 378–385. [CrossRef]
114. Fernández-Solà, J. The Effects of Ethanol on the Heart: Alcoholic Cardiomyopathy. *Nutrients* **2020**, *12*, 572. [CrossRef]
115. Figueredo, V.M.; Chang, K.C.; Baker, A.J.; Camacho, S.A. Chronic alcohol-induced changes in cardiac contractility are not due to changes in the cytosolic Ca2+ transient. *Am. J. Physiol.* **1998**, *275*, H122–H130. [CrossRef]
116. Kiviluoma, K.T.; Karhunen, M.; Lapinlampi, T.; Peuhkurinen, K.J.; Hassinen, I.E. Acetate-induced changes in cardiac energy metabolism and hemodynamics in the rat. *Basic. Res. Cardiol.* **1988**, *83*, 431–444. [CrossRef]
117. Suokas, A.; Kupari, M.; Heikkilä, J.; Lindros, K.; Ylikahri, R. Acute Cardiovascular and Metabolic Effects of Acetate in Men. *Alcohol. Clin. Exp. Res.* **1988**, *12*, 52–58. [CrossRef] [PubMed]
118. Jiang, X.; Zhang, Y.; Zhang, H.; Zhang, X.; Yin, X.; Yuan, F.; Wang, S.; Tian, Y. Acetate suppresses myocardial contraction via the short-chain fatty acid receptor GPR43. *Front. Physiol.* **2022**, *13*, 1111156. [CrossRef] [PubMed]
119. Poll, B.G.; Xu, J.; Jun, S.; Sanchez, J.; Zaidman, N.A.; He, X.; Lester, L.; Berkowitz, D.E.; Paolocci, N.; Gao, W.D.; et al. Acetate, a Short-Chain Fatty Acid, Acutely Lowers Heart Rate and Cardiac Contractility Along with Blood Pressure. *J. Pharmacol. Exp. Ther.* **2021**, *377*, 39–50. [CrossRef] [PubMed]
120. Massion, P.B.; Feron, O.; Dessy, C.; Balligand, J.-L. Nitric Oxide and Cardiac Function. *Circ. Res.* **2003**, *93*, 388–398. [CrossRef]
121. Finkel, M.S.; Oddis, C.V.; Jacob, T.D.; Watkins, S.C.; Hattler, B.G.; Simmons, R.L. Negative inotropic effects of cytokines on the heart mediated by nitric oxide. *Science* **1992**, *257*, 387–389. [CrossRef]
122. Shaaban, A.; Gangwani, M.K.; Pendela, V.S.; Vindhyal, M.R. Alcoholic Cardiomyopathy. In *StatPearls*; StatPearls Publishing: St. Petersburg, FL, USA, 2023.
123. Andersson, C.; Schou, M.; Gustafsson, F.; Torp-Pedersen, C. Alcohol Intake in Patients With Cardiomyopathy and Heart Failure: Consensus and Controversy. *Circ. Heart Fail.* **2022**, *15*, e009459. [CrossRef]
124. Riddick, J.A.; Bunger, W.B.; Sakano, T.K. *Organic Solvents: Physical Properties and Methods of Purification*, 4th ed.; Wiley: Hoboken, NJ, USA, 1986.
125. Serjeant, E.P.; Dempsey, B. Ionisation constants of organic acids in aqueous solution. *IUPAC Chem. Data Ser.* **1979**, *23*, 160–190.
126. Bjarnsholt, T.; Alhede, M.; Jensen, P.; Nielsen, A.K.; Johansen, H.K.; Homøe, P.; Høiby, N.; Givskov, M.; Kirketerp-Møller, K. Antibiofilm Properties of Acetic Acid. *Adv. Wound Care* **2015**, *4*, 363–372. [CrossRef]
127. Koob, G.F.; Arends, M.A.; Le Moal, M. Chapter 6—Alcohol. In *Drugs, Addiction, and the Brain*; Koob, G.F., Arends, M.A., Le Moal, M., Eds.; Academic Press: San Diego, CA, USA, 2014; pp. 173–219. [CrossRef]
128. Gemma, S.; Vichi, S.; Testai, E. Individual susceptibility and alcohol effects:biochemical and genetic aspects. *Ann. Ist. Super. Sanita* **2006**, *42*, 8–16.
129. Raphael, K.L.; Murphy, R.A.; Shlipak, M.G.; Satterfield, S.; Huston, H.K.; Sebastian, A.; Sellmeyer, D.E.; Patel, K.V.; Newman, A.B.; Sarnak, M.J.; et al. Bicarbonate Concentration, Acid-Base Status, and Mortality in the Health, Aging, and Body Composition Study. *Clin. J. Am. Soc. Nephrol.* **2016**, *11*, 308–316. [CrossRef]
130. Burger, M.; Schaller, D.J. Metabolic Acidosis. In *StatPearls*; StatPearls Publishing: St. Petersburg, FL, USA, 2023.
131. Hopkins, E.; Sanvictores, T.; Sharma, S. Physiology, Acid Base Balance. In *StatPearls*; StatPearls Publishing: St. Petersburg, FL, USA, 2023.
132. Raphael, K.L. Metabolic Acidosis in CKD: Core Curriculum 2019. *Am. J. Kidney Dis.* **2019**, *74*, 263–275. [CrossRef] [PubMed]

133. Armstrong, M.; Kerndt, C.C.; Moore, R.A. Physiology, Baroreceptors. In *StatPearls*; StatPearls Publishing: St. Petersburg, FL, USA, 2023.
134. Suarez-Roca, H.; Mamoun, N.; Sigurdson, M.I.; Maixner, W. Baroreceptor Modulation of the Cardiovascular System, Pain, Consciousness, and Cognition. *Compr. Physiol.* **2021**, *11*, 1373–1423. [CrossRef]
135. Caverson, M.M.; Ciriello, J.; Calaresu, F.R. Chemoreceptor and baroreceptor inputs to ventrolateral medullary neurons. *Am. J. Physiol.* **1984**, *247*, R872–R879. [CrossRef] [PubMed]
136. Cooper, V.L.; Pearson, S.B.; Bowker, C.M.; Elliott, M.W.; Hainsworth, R. Interaction of chemoreceptor and baroreceptor reflexes by hypoxia and hypercapnia—A mechanism for promoting hypertension in obstructive sleep apnoea. *J. Physiol.* **2005**, *568*, 677–687. [CrossRef]
137. Guyenet, P.G.; Bayliss, D.A. Neural Control of Breathing and CO_2 Homeostasis. *Neuron* **2015**, *87*, 946–961. [CrossRef] [PubMed]
138. Toledo, C.; Andrade, D.C.; Lucero, C.; Schultz, H.D.; Marcus, N.; Retamal, M.; Madrid, C.; Del Rio, R. Contribution of peripheral and central chemoreceptors to sympatho-excitation in heart failure. *J. Physiol.* **2017**, *595*, 43–51. [CrossRef] [PubMed]
139. Chapp, A.D.; Gui, L.; Huber, M.J.; Liu, J.; Larson, R.A.; Zhu, J.; Carter, J.R.; Chen, Q.H. Sympathoexcitation and pressor responses induced by ethanol in the central nucleus of amygdala involves activation of NMDA receptors in rats. *Am. J. Physiol. Heart Circ. Physiol.* **2014**, *307*, H701–H709. [CrossRef]
140. Brizzolara, A.L.; Morris, D.G.; Burnstock, G. Ethanol affects sympathetic cotransmission and endothelium-dependent relaxation in the rat. *Eur. J. Pharmacol.* **1994**, *254*, 175–181. [CrossRef]
141. El-Mas, M.M.; Abdel-Rahman, A.A. Enhanced catabolism to acetaldehyde in rostral ventrolateral medullary neurons accounts for the pressor effect of ethanol in spontaneously hypertensive rats. *Am. J. Physiol. Heart Circ. Physiol.* **2012**, *302*, H837–H844. [CrossRef]
142. Chan, T.C.; Wall, R.A.; Sutter, M.C. Chronic ethanol consumption, stress, and hypertension. *Hypertension* **1985**, *7*, 519–524. [CrossRef]
143. Grassi, G.M.; Somers, V.K.; Renk, W.S.; Abboud, F.M.; Mark, A.L. Effects of alcohol intake on blood pressure and sympathetic nerve activity in normotensive humans: A preliminary report. *J. Hypertens. Suppl.* **1989**, *7*, S20–S21. [CrossRef] [PubMed]
144. Hering, D.; Kucharska, W.; Kara, T.; Somers, V.K.; Narkiewicz, K. Potentiated sympathetic and hemodynamic responses to alcohol in hypertensive vs. normotensive individuals. *J. Hypertens.* **2011**, *29*, 537–541. [CrossRef] [PubMed]
145. Resstel, L.B.; Scopinho, A.A.; Lopes da Silva, A.; Antunes-Rodrigues, J.; Corrêa, F.M. Increased circulating vasopressin may account for ethanol-induced hypertension in rats. *Am. J. Hypertens.* **2008**, *21*, 930–935. [CrossRef] [PubMed]
146. Tirapelli, C.R.; Leone, A.F.; Coelho, E.B.; Resstel, L.B.; Corrêa, F.M.; Lanchote, V.L.; Uyemura, S.A.; Padovan, C.M.; de Oliveira, A.M. Effect of ethanol consumption on blood pressure and rat mesenteric arterial bed, aorta and carotid responsiveness. *J. Pharm. Pharmacol.* **2007**, *59*, 985–993. [CrossRef] [PubMed]
147. Ireland, M.A.; Vandongen, R.; Davidson, L.; Beilin, L.J.; Rouse, I.L. Acute effects of moderate alcohol consumption on blood pressure and plasma catecholamines. *Clin. Sci.* **1984**, *66*, 643–648. [CrossRef] [PubMed]
148. Crandall, D.L.; Ferraro, G.D.; Lozito, R.J.; Cervoni, P.; Clark, L.T. Cardiovascular effects of intermittent drinking: Assessment of a novel animal model of human alcoholism. *J. Hypertens.* **1989**, *7*, 683–687. [CrossRef] [PubMed]
149. Pronko, P.S.; Velichko, M.G.; Moroz, A.R.; Rubanovich, N.N. Low-molecular-weight metabolites relevant to ethanol metabolism: Correlation with alcohol withdrawal severity and utility for identification of alcoholics. *Alcohol. Alcohol.* **1997**, *32*, 761–768. [CrossRef] [PubMed]
150. Jiang, L.; Gulanski, B.I.; De Feyter, H.M.; Weinzimer, S.A.; Pittman, B.; Guidone, E.; Koretski, J.; Harman, S.; Petrakis, I.L.; Krystal, J.H.; et al. Increased brain uptake and oxidation of acetate in heavy drinkers. *J. Clin. Investig.* **2013**, *123*, 1605–1614. [CrossRef]
151. Chapp, A.D.; Collins, A.R.; Driscoll, K.M.; Behnke, J.E.; Shan, Z.; Zhang, L.; Chen, Q.-H. Ethanol Metabolite, Acetate, Increases Excitability of the Central Nucleus of Amygdala Neurons through Activation of NMDA Receptors. *ACS Chem. Neurosci.* **2023**, *14*, 1278–1290. [CrossRef]
152. Chapp, A.D.; Wang, R.; Cheng, Z.J.; Shan, Z.; Chen, Q.H. Long-Term High Salt Intake Involves Reduced SK Currents and Increased Excitability of PVN Neurons with Projections to the Rostral Ventrolateral Medulla in Rats. *Neural Plast.* **2017**, *2017*, 7282834. [CrossRef]
153. Larson, R.A.; Chapp, A.D.; Gui, L.; Huber, M.J.; Cheng, Z.J.; Shan, Z.; Chen, Q.-H. High Salt Intake Augments Excitability of PVN Neurons in Rats: Role of the Endoplasmic Reticulum Ca^{2+} Store. *Front. Neurosci.* **2017**, *11*, 182. [CrossRef] [PubMed]
154. Larson, R.A.; Gui, L.; Huber, M.J.; Chapp, A.D.; Zhu, J.; LaGrange, L.P.; Shan, Z.; Chen, Q.H. Sympathoexcitation in ANG II-salt hypertension involves reduced SK channel function in the hypothalamic paraventricular nucleus. *Am. J. Physiol. Heart Circ. Physiol.* **2015**, *308*, H1547–H1555. [CrossRef]
155. Chen, Q.H.; Haywood, J.R.; Toney, G.M. Sympathoexcitation by PVN-injected bicuculline requires activation of excitatory amino acid receptors. *Hypertension* **2003**, *42*, 725–731. [CrossRef] [PubMed]
156. Chen, Q.H.; Toney, G.M. Excitability of paraventricular nucleus neurones that project to the rostral ventrolateral medulla is regulated by small-conductance Ca^{2+}-activated K^+ channels. *J. Physiol.* **2009**, *587*, 4235–4247. [CrossRef] [PubMed]
157. Chen, Q.H.; Toney, G.M. In vivo discharge properties of hypothalamic paraventricular nucleus neurons with axonal projections to the rostral ventrolateral medulla. *J. Neurophysiol.* **2010**, *103*, 4–15. [CrossRef]
158. Adams, J.M.; McCarthy, J.J.; Stocker, S.D. Excess dietary salt alters angiotensinergic regulation of neurons in the rostral ventrolateral medulla. *Hypertension* **2008**, *52*, 932–937. [CrossRef]

159. Kumagai, H.; Oshima, N.; Matsuura, T.; Iigaya, K.; Imai, M.; Onimaru, H.; Sakata, K.; Osaka, M.; Onami, T.; Takimoto, C.; et al. Importance of rostral ventrolateral medulla neurons in determining efferent sympathetic nerve activity and blood pressure. *Hypertens. Res.* **2012**, *35*, 132–141. [CrossRef]
160. El-Mas, M.M.; Fan, M.; Abdel-Rahman, A.A. Role of rostral ventrolateral medullary ERK/JNK/p38 MAPK signaling in the pressor effects of ethanol and its oxidative product acetaldehyde. *Alcohol. Clin. Exp. Res.* **2013**, *37*, 1827–1837. [CrossRef]
161. Jin, S.; Cao, Q.; Yang, F.; Zhu, H.; Xu, S.; Chen, Q.; Wang, Z.; Lin, Y.; Cinar, R.; Pawlosky, R.J.; et al. Brain ethanol metabolism by astrocytic ALDH2 drives the behavioural effects of ethanol intoxication. *Nat. Metab.* **2021**, *3*, 337–351. [CrossRef]
162. Jiang, E.; Chapp, A.D.; Fan, Y.; Larson, R.A.; Hahka, T.; Huber, M.J.; Yan, J.; Chen, Q.H.; Shan, Z. Expression of Proinflammatory Cytokines Is Upregulated in the Hypothalamic Paraventricular Nucleus of Dahl Salt-Sensitive Hypertensive Rats. *Front. Physiol.* **2018**, *9*, 104. [CrossRef]
163. Lovinger, D.M.; White, G.; Weight, F.F. Ethanol inhibits NMDA-activated ion current in hippocampal neurons. *Science* **1989**, *243*, 1721–1724. [CrossRef]
164. Chapp, A.D.; Mermelstein, P.G.; Thomas, M.J. The Ethanol Metabolite Acetic Acid Activates Mouse Nucleus Accumbens Shell Medium Spiny Neurons. *J. Neurophysiol.* **2021**, *125*, 620–627. [CrossRef] [PubMed]
165. Garthwaite, J.; Charles, S.L.; Chess-Williams, R. Endothelium-derived relaxing factor release on activation of NMDA receptors suggests role as intercellular messenger in the brain. *Nature* **1988**, *336*, 385–388. [CrossRef] [PubMed]
166. Brown, G.C. Nitric oxide and neuronal death. *Nitric Oxide* **2010**, *23*, 153–165. [CrossRef] [PubMed]
167. Ledo, A.; Barbosa, R.M.; Gerhardt, G.A.; Cadenas, E.; Laranjinha, J. Concentration dynamics of nitric oxide in rat hippocampal subregions evoked by stimulation of the NMDA glutamate receptor. *Proc. Natl. Acad. Sci. USA* **2005**, *102*, 17483–17488. [CrossRef]
168. Thiel, V.E.; Audus, K.L. Nitric oxide and blood-brain barrier integrity. *Antioxid. Redox Signal* **2001**, *3*, 273–278. [CrossRef]
169. Regehr, W.G.; Carey, M.R.; Best, A.R. Activity-dependent regulation of synapses by retrograde messengers. *Neuron* **2009**, *63*, 154–170. [CrossRef]
170. Cserép, C.; Szőnyi, A.; Veres, J.M.; Németh, B.; Szabadits, E.; de Vente, J.; Hájos, N.; Freund, T.F.; Nyiri, G. Nitric Oxide Signaling Modulates Synaptic Transmission during Early Postnatal Development. *Cereb. Cortex* **2011**, *21*, 2065–2074. [CrossRef]
171. Förstermann, U.; Sessa, W.C. Nitric oxide synthases: Regulation and function. *Eur. Heart J.* **2012**, *33*, 829–837, 837a–837d. [CrossRef]
172. Araki, S.; Osuka, K.; Takata, T.; Tsuchiya, Y.; Watanabe, Y. Coordination between Calcium/Calmodulin-Dependent Protein Kinase II and Neuronal Nitric Oxide Synthase in Neurons. *Int. J. Mol. Sci.* **2020**, *21*, 7997. [CrossRef]
173. Joca, S.R.L.; Sartim, A.G.; Roncalho, A.L.; Diniz, C.F.A.; Wegener, G. Nitric oxide signalling and antidepressant action revisited. *Cell Tissue Res.* **2019**, *377*, 45–58. [CrossRef] [PubMed]
174. Guo, Z.L.; Tjen, A.L.S.C.; Fu, L.W.; Longhurst, J.C. Nitric oxide in rostral ventrolateral medulla regulates cardiac-sympathetic reflexes: Role of synthase isoforms. *Am. J. Physiol. Heart Circ. Physiol.* **2009**, *297*, H1478–H1486. [CrossRef] [PubMed]
175. Martins-Pinge, M.C.; Baraldi-Passy, I.; Lopes, O.U. Excitatory Effects of Nitric Oxide Within the Rostral Ventrolateral Medulla of Freely Moving Rats. *Hypertension* **1997**, *30*, 704–707. [CrossRef]
176. Morimoto, S.; Sasaki, S.; Miki, S.; Kawa, T.; Nakamura, K.; Itoh, H.; Nakata, T.; Takeda, K.; Nakagawa, M.; Fushiki, S. Nitric oxide is an excitatory modulator in the rostral ventrolateral medulla in rats. *Am. J. Hypertens.* **2000**, *13*, 1125–1134. [CrossRef] [PubMed]
177. Sharma, N.M.; Patel, K.P. Post-translational regulation of neuronal nitric oxide synthase: Implications for sympathoexcitatory states. *Expert. Opin. Ther. Targets* **2017**, *21*, 11–22. [CrossRef] [PubMed]
178. Zhang, K.; Mayhan, W.G.; Patel, K.P. Nitric oxide within the paraventricular nucleus mediates changes in renal sympathetic nerve activity. *Am. J. Physiol.* **1997**, *273*, R864–R872. [CrossRef] [PubMed]
179. Zhang, K.; Li, Y.F.; Patel, K.P. Blunted nitric oxide-mediated inhibition of renal nerve discharge within PVN of rats with heart failure. *Am. J. Physiol. Heart Circ. Physiol.* **2001**, *281*, H995–H1004. [CrossRef]
180. Northcott, C.A.; Billecke, S.; Craig, T.; Hinojosa-Laborde, C.; Patel, K.P.; Chen, A.F.; D'Alecy, L.G.; Haywood, J.R. Nitric oxide synthase, ADMA, SDMA, and nitric oxide activity in the paraventricular nucleus throughout the etiology of renal wrap hypertension. *Am. J. Physiol. Heart Circ. Physiol.* **2012**, *302*, H2276–H2284. [CrossRef]
181. Crews, F.T.; Morrow, A.L.; Criswell, H.; Breese, G. Effects of ethanol on ion channels. *Int. Rev. Neurobiol.* **1996**, *39*, 283–367. [CrossRef]
182. Ferguson, A.V.; Latchford, K.J.; Samson, W.K. The paraventricular nucleus of the hypothalamus—A potential target for integrative treatment of autonomic dysfunction. *Expert. Opin. Ther. Targets* **2008**, *12*, 717–727. [CrossRef]
183. Gui, L.; LaGrange, L.P.; Larson, R.A.; Gu, M.; Zhu, J.; Chen, Q.H. Role of small conductance calcium-activated potassium channels expressed in PVN in regulating sympathetic nerve activity and arterial blood pressure in rats. *Am. J. Physiol. Regul. Integr. Comp. Physiol.* **2012**, *303*, R301–R310. [CrossRef] [PubMed]
184. Han, S.K.; Chong, W.; Li, L.H.; Lee, I.S.; Murase, K.; Ryu, P.D. Noradrenaline excites and inhibits GABAergic transmission in parvocellular neurons of rat hypothalamic paraventricular nucleus. *J. Neurophysiol.* **2002**, *87*, 2287–2296. [CrossRef]
185. Huber, M.J.; Fan, Y.; Jiang, E.; Zhu, F.; Larson, R.A.; Yan, J.; Li, N.; Chen, Q.H.; Shan, Z. Increased activity of the orexin system in the paraventricular nucleus contributes to salt-sensitive hypertension. *Am. J. Physiol. Heart Circ. Physiol.* **2017**, *313*, H1075–H1086. [CrossRef] [PubMed]
186. Li, D.P.; Yang, Q.; Pan, H.M.; Pan, H.L. Pre- and postsynaptic plasticity underlying augmented glutamatergic inputs to hypothalamic presympathetic neurons in spontaneously hypertensive rats. *J. Physiol.* **2008**, *586*, 1637–1647. [CrossRef]

187. Li, D.P.; Zhou, J.J.; Pan, H.L. Endogenous casein kinase-1 modulates NMDA receptor activity of hypothalamic presympathetic neurons and sympathetic outflow in hypertension. *J. Physiol.* **2015**, *593*, 4439–4452. [CrossRef] [PubMed]
188. Li, D.P.; Zhou, J.J.; Zhang, J.; Pan, H.L. CaMKII Regulates Synaptic NMDA Receptor Activity of Hypothalamic Presympathetic Neurons and Sympathetic Outflow in Hypertension. *J. Neurosci.* **2017**, *37*, 10690–10699. [CrossRef] [PubMed]
189. Sheng, Z.F.; Zhang, H.; Zheng, P.; Chen, S.; Gu, Z.; Zhou, J.J.; Phaup, J.G.; Chang, H.M.; Yeh, E.T.H.; Pan, H.L.; et al. Impaired Kv7 channel activity in the central amygdala contributes to elevated sympathetic outflow in hypertension. *Cardiovasc. Res.* **2022**, *118*, 585–596. [CrossRef] [PubMed]
190. Kalsbeek, A.; La Fleur, S.; Van Heijningen, C.; Buijs, R.M. Suprachiasmatic GABAergic inputs to the paraventricular nucleus control plasma glucose concentrations in the rat via sympathetic innervation of the liver. *J. Neurosci.* **2004**, *24*, 7604–7613. [CrossRef]
191. Park, J.B.; Jo, J.Y.; Zheng, H.; Patel, K.P.; Stern, J.E. Regulation of tonic GABA inhibitory function, presympathetic neuronal activity and sympathetic outflow from the paraventricular nucleus by astroglial GABA transporters. *J. Physiol.* **2009**, *587*, 4645–4660. [CrossRef]
192. Li, Y.F.; Mayhan, W.G.; Patel, K.P. NMDA-mediated increase in renal sympathetic nerve discharge within the PVN: Role of nitric oxide. *Am. J. Physiol. Heart Circ. Physiol.* **2001**, *281*, H2328–H2336. [CrossRef]
193. Zhang, K.; Patel, K.P. Effect of nitric oxide within the paraventricular nucleus on renal sympathetic nerve discharge: Role of GABA. *Am. J. Physiol.* **1998**, *275*, R728–R734. [CrossRef] [PubMed]
194. Herman, M.A.; Varodayan, F.P.; Oleata, C.S.; Luu, G.; Kirson, D.; Heilig, M.; Ciccocioppo, R.; Roberto, M. Glutamatergic transmission in the central nucleus of the amygdala is selectively altered in Marchigian Sardinian alcohol-preferring rats: Alcohol and CRF effects. *Neuropharmacology* **2016**, *102*, 21–31. [CrossRef] [PubMed]
195. Roberto, M.; Madamba, S.G.; Stouffer, D.G.; Parsons, L.H.; Siggins, G.R. Increased GABA Release in the Central Amygdala of Ethanol-Dependent Rats. *J. Neurosci.* **2004**, *24*, 10159–10166. [CrossRef] [PubMed]
196. Khom, S.; Wolfe, S.A.; Patel, R.R.; Kirson, D.; Hedges, D.M.; Varodayan, F.P.; Bajo, M.; Roberto, M. Alcohol Dependence and Withdrawal Impair Serotonergic Regulation of GABA Transmission in the Rat Central Nucleus of the Amygdala. *J. Neurosci.* **2020**, *40*, 6842–6853. [CrossRef] [PubMed]
197. Downs, A.M.; McElligott, Z.A. Noradrenergic circuits and signaling in substance use disorders. *Neuropharmacology* **2022**, *208*, 108997. [CrossRef] [PubMed]
198. Zhang, W.H.; Zhang, J.Y.; Holmes, A.; Pan, B.X. Amygdala Circuit Substrates for Stress Adaptation and Adversity. *Biol. Psychiatry* **2021**, *89*, 847–856. [CrossRef] [PubMed]
199. Clark, L.T. Alcohol-induced hypertension: Mechanisms, complications, and clinical implications. *J. Natl. Med. Assoc.* **1985**, *77*, 385–389.
200. Witkiewitz, K.; Litten, R.Z.; Leggio, L. Advances in the science and treatment of alcohol use disorder. *Sci. Adv.* **2019**, *5*, eaax4043. [CrossRef]
201. Patel, A.K.; Balasanova, A.A. Unhealthy Alcohol Use. *J. Am. Med. Assoc.* **2021**, *326*, 196. [CrossRef]
202. Harmata, G.I.S.; Chan, A.C.; Merfeld, M.J.; Taugher-Hebl, R.J.; Harijan, A.K.; Hardie, J.B.; Fan, R.; Long, J.D.; Wang, G.Z.; Dlouhy, B.J.; et al. Intoxicating effects of alcohol depend on acid-sensing ion channels. *Neuropsychopharmacology* **2022**, *48*, 806–815. [CrossRef]
203. Drapeau, P.; Nachshen, D.A. Effects of lowering extracellular and cytosolic pH on calcium fluxes, cytosolic calcium levels, and transmitter release in presynaptic nerve terminals isolated from rat brain. *J. Gen. Physiol.* **1988**, *91*, 305–315. [CrossRef] [PubMed]
204. Frost, G.; Sleeth, M.L.; Sahuri-Arisoylu, M.; Lizarbe, B.; Cerdan, S.; Brody, L.; Anastasovska, J.; Ghourab, S.; Hankir, M.; Zhang, S.; et al. The short-chain fatty acid acetate reduces appetite via a central homeostatic mechanism. *Nat. Commun.* **2014**, *5*, 3611. [CrossRef]
205. Tanabe, J.; Yamamoto, D.J.; Sutton, B.; Brown, M.S.; Hoffman, P.L.; Burnham, E.L.; Glueck, D.H.; Tabakoff, B. Effects of Alcohol and Acetate on Cerebral Blood Flow: A Pilot Study. *Alcohol. Clin. Exp. Res.* **2019**, *43*, 2070–2078. [CrossRef] [PubMed]
206. Zuo, W.; Wang, L.; Chen, L.; Krnjević, K.; Fu, R.; Feng, X.; He, W.; Kang, S.; Shah, A.; Bekker, A.; et al. Ethanol potentiates both GABAergic and glutamatergic signaling in the lateral habenula. *Neuropharmacology* **2017**, *113*, 178–187. [CrossRef] [PubMed]
207. Möykkynen, T.; Korpi, E.R. Acute Effects of Ethanol on Glutamate Receptors. *Basic. Clin. Pharmacol. Toxicol.* **2012**, *111*, 4–13. [CrossRef] [PubMed]
208. Roberto, M.; Schweitzer, P.; Madamba, S.G.; Stouffer, D.G.; Parsons, L.H.; Siggins, G.R. Acute and chronic ethanol alter glutamatergic transmission in rat central amygdala: An in vitro and in vivo analysis. *J. Neurosci.* **2004**, *24*, 1594–1603. [CrossRef]
209. Xiao, C.; Shao, X.M.; Olive, M.F.; Griffin, W.C.; Li, K.-Y.; Krnjević, K.; Zhou, C.; Ye, J.-H. Ethanol Facilitates Glutamatergic Transmission to Dopamine Neurons in the Ventral Tegmental Area. *Neuropsychopharmacology* **2009**, *34*, 307–318. [CrossRef]
210. Melendez, R.I.; Rodd-Henricks, Z.A.; McBride, W.J.; Murphy, J.M. Alcohol Stimulates the Release of Dopamine in the Ventral Pallidum but not in the Globus Pallidus: A Dual-Probe Microdialysis Study. *Neuropsychopharmacology* **2003**, *28*, 939–946. [CrossRef]
211. Yoshimoto, K.; McBride, W.J.; Lumeng, L.; Li, T.K. Alcohol stimulates the release of dopamine and serotonin in the nucleus accumbens. *Alcohol* **1992**, *9*, 17–22. [CrossRef]
212. Lovinger, D.M. Serotonin's role in alcohol's effects on the brain. *Alcohol. Health Res. World* **1997**, *21*, 114–120.
213. Woodward, J.J. Ethanol and NMDA receptor signaling. *Crit. Rev. Neurobiol.* **2000**, *14*, 69–89. [CrossRef]
214. Lovinger, D.M. Excitotoxicity and alcohol-related brain damage. *Alcohol. Clin. Exp. Res.* **1993**, *17*, 19–27. [CrossRef]
215. Crews, F.T.; Collins, M.A.; Dlugos, C.; Littleton, J.; Wilkins, L.; Neafsey, E.J.; Pentney, R.; Snell, L.D.; Tabakoff, B.; Zou, J. Alcohol-induced neurodegeneration: When, where and why? *Alcohol. Clin. Exp. Res.* **2004**, *28*, 350–364. [CrossRef] [PubMed]

216. Kuikka, J.; Tiihonen, J.; Hakola, P.; Paanila, J.; Airaksinen, J.; Eronen, M.; Hallikainen, T. Acute ethanol-induced changes in cerebral blood flow. *Radioact. Isot. Clin. Med. Res.* **1995**, *151*, 345–350.
217. Lyons, D.; Miller, M.D.; Hedgecock-Rowe, A.A.; Crane, A.M.; Porrino, L.J. Time-dependent effects of acute ethanol administration on regional cerebral blood flow in the rat. *Alcohol* **1998**, *16*, 213–219. [CrossRef] [PubMed]
218. Kirkendol, R.L.; Pearson, J.E.; Bower, J.D.; Holbert, R.D. Myocardial depressant effects of sodium acetate. *Cardiovasc. Res.* **1978**, *12*, 127–136. [CrossRef] [PubMed]
219. Leiris, J.d.; Lorgeril, M.d.; Boucher, F. Ethanol and cardiac function. *Am. J. Physiol.-Heart Circ. Physiol.* **2006**, *291*, H1027–H1028. [CrossRef]
220. Gazzieri, D.; Trevisani, M.; Tarantini, F.; Bechi, P.; Masotti, G.; Gensini, G.F.; Castellani, S.; Marchionni, N.; Geppetti, P.; Harrison, S. Ethanol dilates coronary arteries and increases coronary flow via transient receptor potential vanilloid 1 and calcitonin gene-related peptide. *Cardiovasc. Res.* **2006**, *70*, 589–599. [CrossRef]
221. Brunner, S.; Winter, R.; Werzer, C.; von Stülpnagel, L.; Clasen, I.; Hameder, A.; Stöver, A.; Graw, M.; Bauer, A.; Sinner, M.F. Impact of acute ethanol intake on cardiac autonomic regulation. *Sci. Rep.* **2021**, *11*, 13255. [CrossRef]
222. Pan, Y.; Lv, H.; Zhang, F.; Chen, S.; Cheng, Y.; Ma, S.; Hu, H.; Liu, X.; Cai, X.; Fan, F.; et al. Green tea extracts alleviate acetic acid-induced oral inflammation and reconstruct oral microbial balance in mice. *J. Food Sci.* **2023**, *88*, 5291–5308. [CrossRef] [PubMed]
223. Qi, X.; Lin, W.; Wu, Y.; Li, Q.; Zhou, X.; Li, H.; Xiao, Q.; Wang, Y.; Shao, B.; Yuan, Q. CBD Promotes Oral Ulcer Healing via Inhibiting CMPK2-Mediated Inflammasome. *J. Dent. Res.* **2022**, *101*, 206–215. [CrossRef] [PubMed]
224. Tsuji, H.; Fuse, Y.; Kawamoto, K.; Fujino, H.; Kodama, T. Healing process of experimental esophageal ulcers induced by acetic acid in rats. *Scand. J. Gastroenterol. Suppl.* **1989**, *162*, 6–10. [CrossRef] [PubMed]
225. Wang, J.Y.; Yamasaki, S.; Takeuchi, K.; Okabe, S. Delayed healing of acetic acid-induced gastric ulcers in rats by indomethacin. *Gastroenterology* **1989**, *96*, 393–402. [CrossRef] [PubMed]
226. Miller, M.J.; Zhang, X.J.; Gu, X.A.; Clark, D.A. Acute intestinal injury induced by acetic acid and casein: Prevention by intraluminal misoprostol. *Gastroenterology* **1991**, *101*, 22–30. [CrossRef] [PubMed]
227. Fabia, R.; Willén, R.; Ar'Rajab, A.; Andersson, R.; Ahrén, B.; Bengmark, S. Acetic Acid-Induced Colitis in the Rat: A Reproducible Experimental Model for Acute Ulcerative Colitis. *Eur. Surg. Res.* **2008**, *24*, 211–225. [CrossRef] [PubMed]
228. Dong, Y.J.; Peng, T.K.; Yin, S.J. Expression and activities of class IV alcohol dehydrogenase and class III aldehyde dehydrogenase in human mouth. *Alcohol* **1996**, *13*, 257–262. [CrossRef]
229. Andrade, M.C.; Menezes, J.S.; Cassali, G.D.; Martins-Filho, O.A.; Cara, D.C.; Faria, A.M. Alcohol-induced gastritis prevents oral tolerance induction in mice. *Clin. Exp. Immunol.* **2006**, *146*, 312–322. [CrossRef] [PubMed]
230. Elamin, E.E.; Masclee, A.A.; Dekker, J.; Jonkers, D.M. Ethanol metabolism and its effects on the intestinal epithelial barrier. *Nutr. Rev.* **2013**, *71*, 483–499. [CrossRef]
231. Bishehsari, F.; Magno, E.; Swanson, G.; Desai, V.; Voigt, R.M.; Forsyth, C.B.; Keshavarzian, A. Alcohol and Gut-Derived Inflammation. *Alcohol. Res.* **2017**, *38*, 163–171.
232. Lieber, C.S. Metabolism of Alcohol. *Clin. Liver Dis.* **2005**, *9*, 1–35. [CrossRef]
233. Lieber, C.S. Mechanism of ethanol induced hepatic injury. *Pharmacol. Ther.* **1990**, *46*, 1–41. [CrossRef] [PubMed]
234. Malpas, S.C.; Robinson, B.J.; Maling, T.J. Mechanism of ethanol-induced vasodilation. *J. Appl. Physiol.* **1990**, *68*, 731–734. [CrossRef] [PubMed]
235. Deng, X.-S.; Deitrich, R.A. Ethanol metabolism and effects: Nitric oxide and its interaction. *Curr. Clin. Pharmacol.* **2007**, *2*, 145–153. [CrossRef] [PubMed]
236. Venkov, C.D.; Myers, P.R.; Tanner, M.A.; Su, M.; Vaughan, D.E. Ethanol increases endothelial nitric oxide production through modulation of nitric oxide synthase expression. *Thromb. Haemost.* **1999**, *81*, 638–642. [CrossRef] [PubMed]
237. Ahmad, A.; Dempsey, S.K.; Daneva, Z.; Azam, M.; Li, N.; Li, P.L.; Ritter, J.K. Role of Nitric Oxide in the Cardiovascular and Renal Systems. *Int. J. Mol. Sci.* **2018**, *19*, 2605. [CrossRef]
238. Hermann, M.; Flammer, A.; Lüscher, T.F. Nitric oxide in hypertension. *J. Clin. Hypertens.* **2006**, *8*, 17–29. [CrossRef]
239. Node, K.; Kitakaze, M.; Yoshikawa, H.; Kosaka, H.; Hori, M. Reduced plasma concentrations of nitrogen oxide in individuals with essential hypertension. *Hypertension* **1997**, *30*, 405–408. [CrossRef]
240. Panza, J.A.; Quyyumi, A.A.; Brush, J.E., Jr.; Epstein, S.E. Abnormal endothelium-dependent vascular relaxation in patients with essential hypertension. *N. Engl. J. Med.* **1990**, *323*, 22–27. [CrossRef] [PubMed]

antioxidants

MDPI

Article

The Disease Model of Addiction: The Impact of Genetic Variability in the Oxidative Stress and Inflammation Pathways on Alcohol Dependance and Comorbid Psychosymptomatology

Evangelia Eirini Tsermpini [1], Katja Goričar [1], Blanka Kores Plesničar [2,3], Anja Plemenitaš Ilješ [4,*] and Vita Dolžan [1,*]

1 Pharmacogenetics Laboratory, Institute of Biochemistry and Molecular Genetics, Faculty of Medicine, University of Ljubljana, 1000 Ljubljana, Slovenia; evangelia.tsermpini@dal.ca (E.E.T.); katja.goricar@mf.uni-lj.si (K.G.)

2 University Psychiatric Clinic, 1000 Ljubljana, Slovenia; blanka.kores@psih-klinika.si

3 Faculty of Medicine, University of Ljubljana, 1000 Ljubljana, Slovenia

4 Department of Psychiatry, University Clinical Centre Maribor, 2000 Maribor, Slovenia

* Correspondence: anja.plemenitas@ukc-mb.si (A.P.I.); vita.dolzan@mf.uni-lj.si (V.D.)

Abstract: Oxidative stress and neuroinflammation are involved in the pathogenesis of alcohol addiction. However, little is known regarding the effect of genetic, behavioral, psychological, and environmental sources of origin on the inflammation and oxidative stress pathways of patients with alcohol addiction. Our study aimed to evaluate the impact of selected common functional single-nucleotide polymorphisms in inflammation and oxidative stress genes on alcohol addiction, and common comorbid psychosymptomatology. Our study included 89 hospitalized alcohol-addicted patients and 93 healthy individuals, all Slovenian males. Their DNA was isolated from peripheral blood and patients were genotyped for *PON1* rs705379, rs705381, rs854560, and rs662, *SOD2* rs4880, *GPX1* rs1050450, *IL1B* rs1143623, rs16944, and rs1071676, *IL6* rs1800795, *IL6R* rs2228145, and miR146a rs2910164. Kruskal–Wallis and Mann–Whitney tests were used for the additive and dominant genetic models, respectively. Our findings suggested the involvement of *IL6* rs1800795 in alcohol addiction. Moreover, our data indicated that the genetic variability of *SOD2* and *PON1*, as well as *IL1B* and *IL6R*, may be related to comorbid psychosymptomatology, revealing a potential indirect means of association of both the oxidative stress and inflammation pathways.

Keywords: alcohol addiction; alcohol-related psychosymptomatology; oxidative stress; inflammation; polymorphisms

check for updates

Citation: Tsermpini, E.E.; Goričar, K.; Kores Plesničar, B.; Plemenitaš Ilješ, A.; Dolžan, V. The Disease Model of Addiction: The Impact of Genetic Variability in the Oxidative Stress and Inflammation Pathways on Alcohol Dependance and Comorbid Psychosymptomatology. *Antioxidants* **2024**, *13*, 20. https://doi.org/10.3390/antiox13010020

Academic Editor: Marco Fiore

Received: 11 November 2023
Revised: 15 December 2023
Accepted: 19 December 2023
Published: 21 December 2023

1. Introduction

Alcohol addiction is a complex mental disorder characterized by an impaired control of alcohol use and it has a negative impact on overall health, disability, and death rates, worldwide [1].

Both genetic and environmental factors, such as age, sex, and socioeconomic variables, can influence alcohol consumption, possibly with combined effects [2]. A meta-analysis on heritability provided evidence that genetics is a crucial factor for alcohol use disorder, as it can be up to 50% heritable [3]. A recent study that included Europeans and Africans indicated that the genetic risk of alcohol abuse is transmitted from parents to offspring due to the influence of the home environment [4]. Other environmental and developmental risk factors include prenatal alcohol exposure and adverse childhood events, like abuse, neglect, or family dysfunction, although with modest effects. Moreover, drinking during the age of 18–25 years old, can potentially predict alcohol addiction in adulthood, which can lead to an impairment of psychosocial end points [2].

It is well-established that oxidative stress and neuroinflammation are involved in alcohol addiction. Preclinical and clinical studies have shown that alcohol metabolism is associated with oxidative stress. The low antioxidant activity of superoxide dismutase (SOD), catalase, and glutathione peroxidase (GPX) can lead to increased oxidative stress in alcohol addicted-patients, as well as in patients during detoxification treatment [5]. Paraoxonase-1 (PON1) is another antioxidant enzyme, which can prevent low-density lipoprotein oxidation and hydrolyses various substrates, including peroxides, thus acting as a defense against not only oxidative stress but also inflammation. PON1 also has an anti-inflammatory role, as it can also directly suppress macrophage pro-inflammatory responses [6].

Alcohol addiction can also be considered an inflammatory condition mostly due to the altered cytokine regulation found in many of the patients' organs, including the brain [7]. Alcohol addiction can lead to endotoxemia and the increased secretion of pro-inflammatory cytokines, like interleukin-1β (IL-1β) and interleukin-6 (IL-6), which can alter various signaling pathways and cause damage in the central nervous system and other organs [8]. Furthermore, the gene expression levels of microRNAs, known for their involvement in the inflammatory response, such as miR-146a [9], were reduced in the peripheral blood mononuclear cells from patients with cirrhosis, and in extracellular vesicles from alcohol-drinkers without liver injury, when compared with non-drinkers [10].

However, to our knowledge, there are no studies that focus on the genetic variability of *PON1*, *SOD2*, *GPX1*, *IL1B*, *IL6*, interleukin 6 receptor (*IL6R*), and miR146a related to the risk of alcohol addiction, and patients' comorbid psychosymptomatology. Aiming to fill this gap, we used genetic models to address the potential relationship between genetic factors and behavioral, psychological, and environmental factors in patients with alcohol addiction. More specifically, the present study investigated common functional genetic variations in the oxidative stress (*PON1*, *SOD2*, and *GPX1*) and inflammation (*IL1B*, *IL6*, *IL6R*, and miR146a) pathways and their potential association with alcohol addiction and alcohol-related comorbid psychosymptomatology, including obsessive–compulsive, social anxiety, depressive, anxious, and aggressive symptoms.

2. Materials and Methods

2.1. Study Population

This study was focused on two groups of participants: hospitalized patients receiving treatment for alcohol addiction, and healthy individuals with no alcohol addiction history. All enrolled participants were Slovenian males aged 18 to 66. Hospitalized alcohol-addicted patients were recruited at the University Clinical Centre, Maribor, and the University Psychiatric Clinic, Ljubljana. Their diagnosis was made by experienced psychiatrists, according to the DSM-IV [11] criteria of alcohol addiction, and patients presented no significant abstinence symptoms after hospitalization for at least two weeks. Alcohol-addicted patients who had a present or past diagnosis of dependence or abuse of other substances (except nicotine), bipolar disorder, major depression, schizophrenia, schizoaffective disorder, according to the DSM-IV, organic mental syndromes, head trauma, and neurological disorders, or other significant medical conditions potentially affecting the central nervous system, were excluded from the study. Healthy individuals were blood donors, matched for ethnicity and sex, with no DSM-IV axis I mental disorders, including alcohol addiction and a family history of alcohol dependence or other psychiatric disorders, as it was determined by the psychiatrists after a short clinical interview.

This study was approved by the Slovenian National Medical Ethics Committee (approval No. 117/06/10 and 148/02/1011), and all participants agreed and signed informed consent forms, following the Declaration of Helsinki, after being informed about the aims of the study.

Patients' demographic data, including age, residence, marital status, academic years, and smoking status, along with their clinical data, were recorded at baseline. All participants completed questionnaires to evaluate their drinking habits, the severity of alcohol use and addiction, as well as comorbid symptomatology, including depression and anxiety symptoms, social anxiety symptoms, obsessive–compulsive traits, and symptoms of aggression.

The Alcohol Use Disorders Identification Test (AUDIT) is a screening tool to identify hazardous drinkers or patients with alcohol abuse or addiction. It is a 10-item questionnaire that focuses on alcohol consumption, drinking behavior, and alcohol-related problems, and the higher the score, the more likely it is that the drinking is harmful and leads to alcohol addiction. The Obsessive Compulsive Drinking Scale (OCDS) is a scale that measures an individual's alcohol use and his attempts to control his drinking. On the other hand, the Zung depression scale is a screening tool to identify the presence of depressive symptoms, ranging from mild to moderate and severe depression. Similarly, the Zung anxiety scale measures the anxiety levels of individuals who have anxiety-related symptoms. The Brief Social Phobia Scale (BSPS) is designed to assess the characteristic symptoms of social phobia, evaluating commonly feared or avoided situations of autonomic distress. The Yale–Brown Obsessive Compulsive Scale (YBOCS) obsession scale screens for unwanted thoughts, the fear of losing important things, concerns with order, symmetry, or exactness that intrude on thinking against a person's wishes and efforts to resist them and usually involve themes of harm, risk, and danger. YBOCS compulsions are an assessment tool that can determine the severity of compulsions, i.e., repetitive, purposeful, intentional behaviors called rituals which urge people to do something to lessen feelings of anxiety or other discomfort. Finally, the Buss-Durkee Hostility Inventory (BDHI) is a screening test for measuring a person's level of hostility, by assessing verbal and physical aggression, and anger. More information about the scales can be found in our previous articles [12–14].

2.2. Molecular Genetic Analysis

DNA was extracted from whole blood using the QIAamp Blood Mini kit, according to the manufacturer's protocols (Qiagen GmbH, Hilden, Germany). Genotyping of all SNPs was performed with competitive allele-specific PCR (KASP), using the KASP Master mix and custom KASP genotyping assays (LGC, Teddington, UK) according to the manufacturer's instructions (KBiosciences, Herts, UK, and LGC Genomics, Teddington, UK). Thermal cycling conditions are presented in Supplementary Tables S16–S18.

2.3. Statistical Analysis

The statistical analysis was carried out using the 27.0 version of IBM SPSS Statistics (IBM Corporation, Armonk, NY, USA). The cut-off for the level of significance was set at 0.05. The deviation from Hardy–Weinberg equilibrium (HWE) in healthy donors and for all studied polymorphisms was evaluated using Pearson's chi-square test. Both additive and dominant genetic models were used for the analyses. In comparing the clinical characteristics of alcohol-addicted patients, Fisher's exact test was used for the distribution of categorical variables, while the nonparametric Mann–Whitney test was used for the continuous ones. The association of polymorphisms with alcohol addiction was evaluated using logistic regression analysis, and odds ratios (ORs) and 95% confidence intervals (CIs) were determined. In the multivariable logistic regression, age, residence place, marital status, academic years, and smoking status were considered covariates, and significant variables were used for adjustment in regression analysis. Finally, Kruskal–Wallis and Mann–Whitney non-parametric tests for additive and dominant genetic models were used to associate genotypes with psychosymptomatology scores.

3. Results

Regarding the demographic characteristics of our cohort, the median age of the hospitalized alcohol-addicted patients was significantly higher than the healthy controls ($p < 0.001$), while more controls were smokers ($p < 0.001$) and had a partner ($p = 0.002$) (Table 1).

Table 1. Cohorts' characteristics.

Characteristic		Hospitalized Alcohol-Addicted Patients (N = 89)	Healthy Controls (N = 93)	p *
Age	years, median (25–75%)	47 (39–54)	36 (26–44.5)	**<0.001**
Education	years, median (25–75%)	12 (11–12)	12 (12–12)	**<0.001**
Partnership	Single, N (%)	38 (42.7)	25 (26.9)	**0.002**
	Partnership, N (%)	51 (57.3)	68 (73.1)	
Environment	Rural, N (%)	46 (51.7)	37 (39.8)	0.137
	Urban, N (%)	43 (48.3)	56 (60.2)	
Smoking	No, N (%)	58 (65.2)	24 (25.8)	**<0.001**
	Yes, N (%)	31 (34.8)	69 (74.2)	

* Fisher's exact test for categorical variables, Mann–Whitney test for continuous variables. Statistically significant p-values are printed in bold.

Regarding the questionnaires used to evaluate psychosymptomatology, we observed statistically significant differences between the patients and controls in all questionnaires' scores, except the BSPS ($p = 0.063$) (Supplementary Table S1).

The genotypes of the controls were all in HWE for all studied polymorphisms (all $p > 0.05$) (Supplementary Table S2).

After a comparison of the genotype frequencies between patients and controls, a statistically significant difference was observed for *IL6* rs1800795 CC ($p = 0.038$), which remained significant after adjustment for age, education, smoking, environment, and partnership ($p = 0.043$) (Table 2). No statistically significant differences were observed for the rest of the studied polymorphisms (Supplementary Table S3). Table 2 shows the comparison of the genotype frequencies of genetic variants of the inflammation pathway between hospitalized alcohol-addicted patients and healthy controls, and Supplementary Table S3 shows the comparison of genotype frequencies of genetic variants of the oxidative stress pathway between hospitalized alcohol-addicted patients and healthy controls.

Regarding potential associations between the studied polymorphisms and the scores of the comorbid psychosymptomatology scales, we found that *SOD2* rs4880 CT + TT genotypes were associated with a higher YBOCS obsessions subtotal ($p = 0.016$) and YBOCS compulsions subtotal ($p = 0.046$) in alcohol-addicted patients. Similarly, *PON1* rs705381 CT + TT was associated with lower a YBOCS compulsions subtotal in healthy controls ($p = 0.027$) and with a lower BSPS in alcohol-addicted patients ($p = 0.041$). *PON1* rs705379 GG was associated with BSPS ($p = 0.001$), Zung depression ($p = 0.005$), Zung anxiety ($p = 0.002$), and BDHI scores ($p = 0.040$) in healthy controls. Associations were also found with *PON1* rs705379 GA + AA in healthy individuals, i.e., BSPS ($p = 0.002$), Zung depression ($p = 0.001$), Zung anxiety ($p = 0.001$), and BDHI scores ($p = 0.047$). *PON1* rs854560 AA was associated with a YBOCS obsessions subtotal ($p = 0.038$), BSPS ($p = 0.018$), and Zung anxiety ($p = 0.005$). The YBOCS obsession subtotal and BSPS scores were also associated with *PON1* rs854560 AT + TT ($p = 0.014$ and $p = 0.003$, respectively). *IL1B* rs1071676 GG was associated with AUDIT scores in alcohol-addicted patients ($p = 0.045$). *IL6R* rs2228145 AA was associated with a YBOCs compulsions subtotal in alcohol-addicted patients ($p = 0.033$), whereas in healthy individuals, *IL6R* rs2228145 AA and AC + CC were both associated with BDHI scores ($p = 0.014$ and $p = 0.004$, respectively) (Figures 1 and 2, Tables 3 and 4). No statistically significant associations were observed between the rest of the studied polymorphisms and the psychosymptomatology scores either in the hospitalized alcohol-addicted patients or the healthy individuals (Supplementary Tables S4–S15).

Table 2. Comparison of genotype frequencies of genetic variants of inflammation pathways between hospitalized alcohol-addicted patients and healthy controls.

Gene	SNP	Genotype	Patients (N = 89) N (%)	Controls (N = 93) N (%)	OR (95% CI)	*p*	OR (95% CI)$_{adj}$	P$_{adj}$
IL1B	rs1143623	GG	39 (43.8)	49 (53.3)	Reference		Reference	
		GC	45 (50.6)	36 (39.1)	1.57 (0.86–2.88)	0.145	1.15 (0.5–2.64)	0.739
		CC	5 (5.6)	7 (7.6)	0.9 (0.26–3.05)	0.862	0.53 (0.09–3.01)	0.476
		GC + CC	50 (56.2)	43 (46.7)	1.46 (0.81–2.62)	0.205	1.04 (0.47–2.33)	0.915
IL1B	rs16944	TT	9 (10.1)	9 (9.8)	Reference		Reference	
		TC	48 (53.9)	38 (41.3)	1.26 (0.46–3.49)	0.653	1.38 (0.34–5.66)	0.656
		CC	32 (36)	45 (48.9)	0.71 (0.25–1.99)	0.516	1.05 (0.25–4.4)	0.942
		TC + CC	80 (89.9)	83 (90.2)	0.96 (0.36–2.55)	0.941	1.22 (0.31–4.74)	0.777
IL1B	rs1071676	GG	54 (60.7)	45 (48.9)	Reference		Reference	
		GC	31 (34.8)	37 (40.2)	0.7 (0.38–1.3)	0.256	0.57 (0.25–1.34)	0.200
		CC	4 (4.5)	10 (10.9)	0.33 (0.1–1.13)	0.079	0.34 (0.07–1.66)	0.184
		GC + CC	35 (39.3)	47 (51.1)	0.62 (0.34–1.12)	0.113	0.53 (0.23–1.18)	0.118
MIRN146A	rs2910164	GG	51 (57.3)	57 (61.3)	Reference		Reference	
		GC	31 (34.8)	31 (33.3)	1.12 (0.6–2.09)	0.727	0.92 (0.39–2.16)	0.855
		CC	7 (7.9)	5 (5.4)	1.56 (0.47–5.24)	0.468	1.35 (0.23–8)	0.742
		GC + CC	38 (42.7)	36 (38.7)	1.18 (0.65–2.13)	0.584	0.97 (0.43–2.19)	0.947
IL6	rs1800795	GG	37 (41.6)	31 (33.7)	Reference		Reference	
		GC	43 (48.3)	41 (44.6)	0.88 (0.46–1.67)	0.693	0.57 (0.23–1.38)	0.210
		CC	9 (10.1)	20 (21.7)	0.38 (0.15–0.95)	**0.038**	0.27 (0.07–0.96)	**0.043**
		GC + CC	52 (58.4)	61 (66.3)	0.71 (0.39–1.31)	0.275	0.47 (0.2–1.1)	0.081
IL6R	rs2228145	AA	38 (42.7)	35 (38.5)	Reference		Reference	
		AC	36 (40.4)	40 (44)	0.83 (0.44–1.58)	0.568	0.43 (0.17–1.1)	0.078
		CC	15 (16.9)	16 (17.5)	0.86 (0.37–2)	0.732	0.67 (0.2–2.3)	0.527
		AC + CC	51 (57.3)	56 (61.6)	0.84 (0.46–1.52)	0.563	0.49 (0.21–1.16)	0.105

P$_{adj}$: adjusted for age, education, smoking, environment, and partnership. Statistically significant *p*-values are printed in bold.

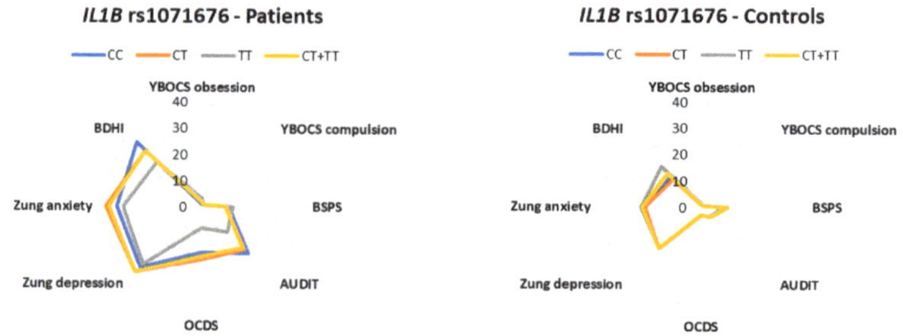

Figure 1. *Cont.*

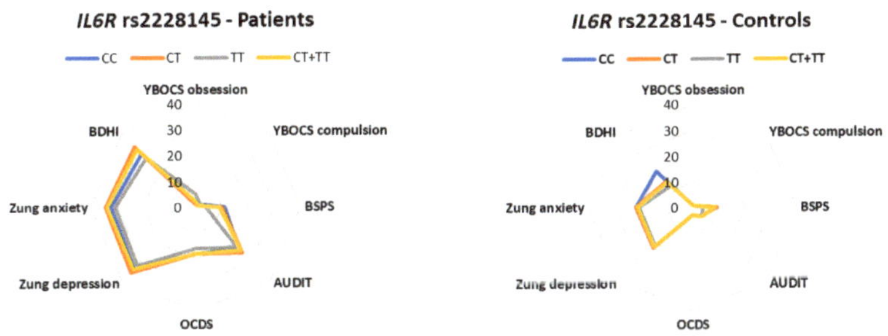

Figure 1. Polar diagrams that show the genetic variants of inflammation pathways that were statistically significant in at least one psychosymptomatology scale and in at least one group of either hospitalized alcohol-addicted patients or healthy individuals.

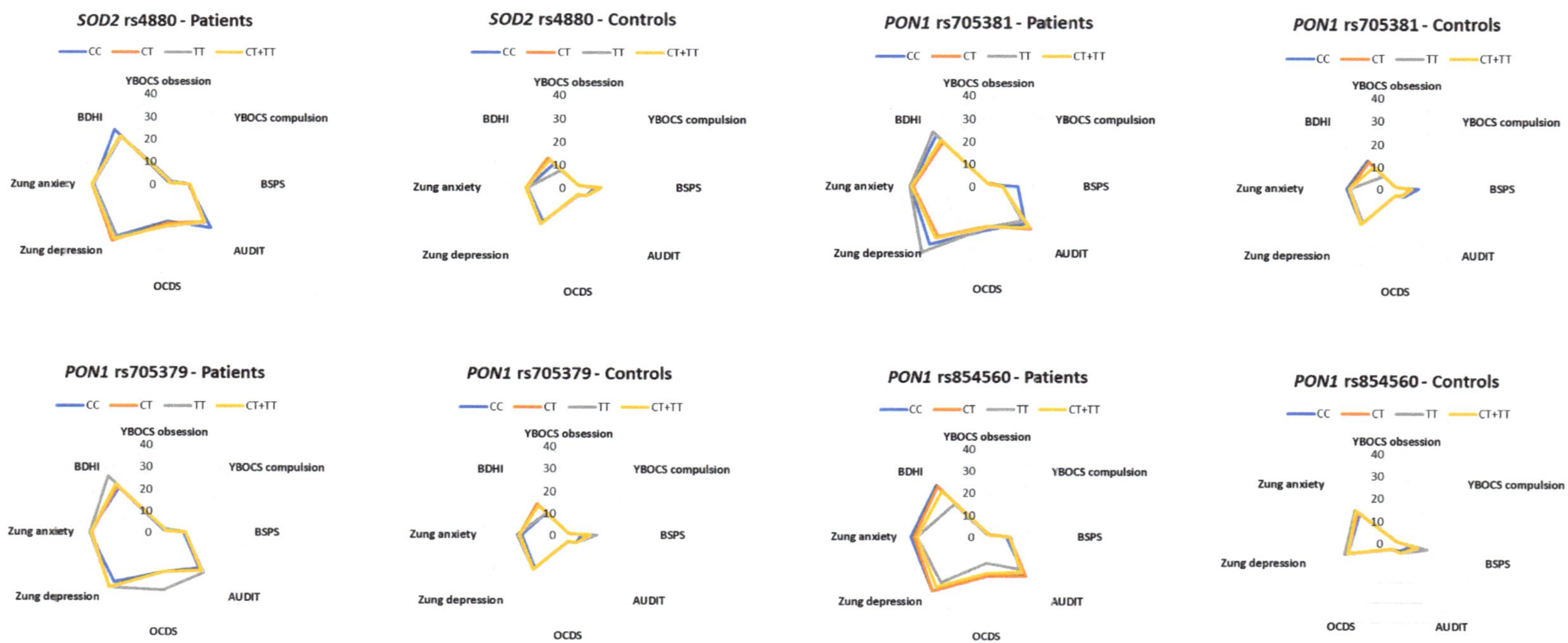

Figure 2. Polar diagrams that show the genetic variants of the oxidative stress pathways that were statistically significant in at least one psychosymptomatology scale and in at least one group of either hospitalized alcohol-addicted patients or healthy individuals.

Table 3. Associations between studied genes and polymorphisms and the scores of the selected psychosymptomatology scales in hospitalized patients with alcohol addiction.

Alcohol Addiction Comorbid Symptoms (Scale)	*SOD2*	*GPX1*	*PON1*				*IL1B*			*MIRN146A*	*IL6*	*IL6R*
	rs4880	**rs1050450**	**rs705381**	**rs705379**	**rs854560**	**rs662**	**rs1143623**	**rs16944**	**rs1071676**	**rs2910164**	**rs1800795**	**rs2228145**
Alcohol addiction severity (AUDIT)	$p = 0.591$	$p = 0.852$	$p = 0.315$	$p = 0.089$	$p = 0.779$	$p = 0.538$	$p = 0.865$	$p = 0.223$	$\mathbf{p = 0.045}$ [#]	$p = 0.705$	$p = 0.828$	$p = 0.749$
Obsessive-compulsive drinking symptoms (OCDS)	$p = 0.950$	$p = 0.709$	$p = 0.315$	$p = 0.502$	$p = 0.961$	$p = 0.227$	$p = 0.313$	$p = 0.688$	$p = 0.804$	$p = 0.891$	$p = 0.543$	$p = 0.542$
Obsessions (YBOCS)	$\mathbf{p = 0.016}$	$p = 0.899$	$p = 0.965$	$p = 0.570$	$p = 0.520$	$p = 0.951$	$p = 0.758$	$p = 0.834$	$p = 0.416$	$p = 0.236$	$p = 0.259$	$p = 0.467$
Compulsions (YBOCS)	$\mathbf{p = 0.046}$	$p = 0.742$	$p = 0.457$	$p = 0.437$	$p = 0.592$	$p = 0.376$	$p = 0.645$	$p = 0.282$	$p = 0.627$	$p = 0.580$	$p = 0.487$	$\mathbf{p = 0.033}$ [#]
Social phobia (BSPS)	$p = 0.591$	$p = 0.181$	$\mathbf{p = 0.041}$	$p = 0.555$	$p = 0.486$	$p = 0.935$	$p = 0.778$	$p = 0.624$	$p = 0.923$	$p = 0.826$	$p = 0.815$	$p = 0.104$
Depression (Zung)	$p = 0.915$	$p = 0.150$	$p = 0.250$	$p = 0.938$	$p = 0.415$	$p = 0.921$	$p = 0.451$	$p = 0.513$	$p = 0.626$	$p = 0.222$	$p = 0.382$	$p = 0.517$
Anxiety (Zung)	$p = 0.900$	$p = 0.753$	$p = 0.743$	$p = 0.866$	$p = 0.115$	$p = 0.438$	$p = 0.763$	$p = 0.723$	$p = 0.050$	$p = 0.842$	$p = 0.705$	$p = 0.558$
Aggression (BDHI)	$p = 0.685$	$p = 0.405$	$p = 0.188$	$p = 0.895$	$p = 0.165$	$p = 0.353$	$p = 0.449$	$p = 0.327$	$p = 0.380$	$p = 0.983$	$p = 0.832$	$p = 0.595$

SOD2: superoxide dismutase 2, GPX1: glutathione peroxidase 1, PON1: paraoxonase-11, IL1B: interleukin-1β, IL6: interleukin-6, IL6R: IL-6 receptor, YBOCS: Yale-Brown Obsessive Compulsive Scale, BSPS: Brief Social Phobia Scale, AUDIT: Alcohol Use Disorders Identification Test, OCDS: Obsessive Compulsive Drinking Scale, BDHI: Buss-Durkee Hostility Inventory. The p values are shown for the dominant genetic model, except for those marked with [#], which are for the additive model. Statistically significant p-values are printed in bold.

Table 4. Associations between studied genes and polymorphisms and the scores of the selected psychosymptomatology scales in healthy controls.

Alcohol Addiction Comorbid Symptoms (Scale)	*SOD2*	*GPX1*	*PON1*				*IL1B*			*MIRN146A*	*IL6*	*IL6R*
	rs4880	**rs1050450**	**rs705381**	**rs705379**	**rs854560**	**rs662**	**rs1143623**	**rs16944**	**rs1071676**	**rs2910164**	**rs1800795**	**rs2228145**
Alcohol addiction severity (AUDIT)	$p = 0.847$	$p = 0.254$	$p = 0.440$	$p = 0.489$	$p = 0.065$	$p = 0.229$	$p = 0.294$	$p = 0.775$	$p = 0.770$	$p = 0.518$	$p = 0.057$	$p = 0.083$
Obsessive-compulsive drinking symptoms (OCDS)	$p = 0.907$	$p = 0.564$	$p = 0.973$	$p = 0.795$	$p = 0.293$	$p = 0.895$	$p = 0.483$	$p = 0.822$	$p = 0.907$	$p = 0.626$	$p = 0.187$	$p = 0.269$
Obsessions (YBOCS)	$p = 0.338$	$p = 0.110$	$p = 0.445$	$p = 0.055$	$\mathbf{p = 0.038}$ [#]	$p = 0.414$	$p = 0.503$	$p = 0.082$	$p = 0.342$	$p = 0.478$	$p = 0.190$	$p = 0.439$
Compulsions (YBOCS)	$p = 0.196$	$p = 0.584$	$\mathbf{p = 0.027}$	$p = 0.097$	$p = 0.159$	$p = 0.976$	$p = 0.312$	$p = 0.167$	$p = 0.166$	$p = 0.315$	$p = 0.483$	$p = 0.204$
Social phobia (BSPS)	$p = 0.108$	$p = 0.133$	$p = 0.083$	$\mathbf{p = 0.002}$	$\mathbf{p = 0.014}$	$p = 0.151$	$p = 0.284$	$p = 0.921$	$p = 0.431$	$p = 0.382$	$p = 0.305$	$p = 0.313$
Depression (Zung)	$p = 0.073$	$p = 0.707$	$p = 0.873$	$\mathbf{p = 0.001}$	$p = 0.108$	$p = 0.751$	$p = 0.715$	$p = 0.584$	$p = 0.665$	$p = 0.630$	$p = 0.923$	$p = 0.096$
Anxiety (Zung)	$p = 0.373$	$p = 0.728$	$p = 0.271$	$\mathbf{p = 0.001}$	$\mathbf{p = 0.003}$	$p = 0.614$	$p = 0.883$	$p = 0.360$	$p = 0.662$	$p = 0.402$	$p = 0.260$	$p = 0.465$
Aggression (BDHI)	$p = 0.662$	$p = 0.735$	$p = 0.357$	$\mathbf{p = 0.047}$	$p = 0.331$	$p = 0.743$	$p = 0.959$	$p = 0.948$	$p = 0.664$	$p = 0.339$	$p = 0.990$	$\mathbf{p = 0.004}$

SOD2: superoxide dismutase 2, GPX1: glutathione peroxidase 1, PON1: paraoxonase-11, IL1B interleukin-1β, IL6: interleukin-6, IL6R: IL-6 receptor, YBOCS: Yale-Brown Obsessive Compulsive Scale, BSPS: Brief Social Phobia Scale, AUDIT: Alcohol Use Disorders Identification Test, OCDS: Obsessive Compulsive Drinking Scale, BDHI: Buss-Durkee Hostility Inventory. The *p* values are shown for the dominant genetic model, except for those marked with [#], which are for the additive model. Statistically significant *p*-values are printed in bold.

4. Discussion

In this pilot study, we used genetic models to investigate the relationship between the genetic variability in oxidative stress- and inflammation-related pathways and the behavioral, psychological, and environmental factors for patients with alcohol addiction. Thus, we focused on the role of *PON1* rs705379, rs705381, rs854560, and rs662, *SOD2* rs4880, *GPX1* rs1050450, *IL1B* rs1143623, rs16944, and rs1071676, *IL6* rs1800795, *IL6R* rs2228145, and miR146a rs2910164 in alcohol addiction, and patients' comorbid psychosymptomatology.

We found a statistically significant association of *IL6* rs1800795 with alcohol addiction after a comparison of the genotype frequencies of alcohol addicted patients and healthy individuals, which remained significant after adjustments for other parameters associated with alcohol addiction. No statistically significant associations with alcohol addiction were found for *PON1* rs705379, rs705381, rs854560, or rs662, *SOD2* rs4880, *GPX1* rs1050450, *IL1B* rs1143623, rs16944, or rs1071676, *IL6R* rs2228145, or miR146a rs2910164. We also observed statistically significant associations between *SOD2* rs4880 and obsessive–compulsive symptoms, *PON1* rs705381 and social phobia, *IL1B* rs1071676 and AUDIT scores, and *IL6R* rs2228145 and compulsions in alcohol-addicted patients. In the controls, *PON1* rs705381 was associated with compulsions, *PON1* rs705379 with social anxiety, depression, anxiety, and aggression, *PON1* rs854560 with obsessions, anxiety, and social phobia, and *IL6R* rs2228145 with aggression.

Cytokines are important markers of systemic inflammation that can induce neuroinflammation and further affect mood, cognition, and drinking habits [15]. Many of them, including IL-6, have demonstrated neurodegenerative and neuroprotective dynamics and have been associated with mental diseases [16]. Long-term alcohol abuse can cause multiple medical comorbidities including liver damage because of the inflammatory and oxidative stress induced by alcohol [17].

IL6 is located on chromosome 7p21 and rs1800795 is a promoter variant affecting IL-6 transcription levels. GG carriers have higher IL-6 serum levels in comparison to CC carriers [18]. Interestingly, this seems to be a gender-dependent genetic predisposition given that only male homozygotes for the G allele have higher IL-6 levels compared to carriers of the C allele [19]. Additionally, in healthy controls, the C allele was associated with significantly lower levels of plasma IL-6 [20]. To our knowledge, this is the first time that an association between *IL6* rs1800795 and alcohol addiction has been revealed. *IL6* rs1800795 was proven to be a risk factor for various diseases, including arteriovenous malformations of the brain [21].

Moreover, a whole-brain analysis revealed the neuroprotective role of rs1800795, given that GG carriers had significantly larger hippocampus gray matter volumes than heterozygous carriers. The hippocampus has a critical role in normal brain function, including learning, memory consolidation, and stress and the response to cytokines [18]. Furthermore, the hippocampus is related to various mental disorders, like depression, which does not cause permanent brain damage, but temporary neurotransmitter imbalances [22]. Interestingly, both human and animal studies have shown that heavy alcohol consumption is associated with episodes of alcohol-induced memory loss, blackouts, and functioning impairment, which are possibly linked to hippocampal brain volume loss. Also, a longitudinal magnetic resonance imaging study indicated that heavy drinkers had high rates of grey matter volume decline, which was further related to reduced memory and blackouts [23]. Alcohol consumption can also affect frontal regions and prefrontal white-matter pathways, which are linked to executive functions [24]. Alcohol consumption is associated with a reduced capacity to process new information, develop new skills, formulate, and execute plans and goals [25]. Moreover, IL6 has been associated with cognitive impairment proportionally, i.e., a high IL6 concentration is linked with a high rate of executive and memory function deficit in elderly participants. Furthermore, CC carriers had higher IL-6 concentrations and demonstrated a worse performance in the Stroop cognitive performance test in comparison with the GG carriers. The Stroop test is related to brain function, and measures selective attention capacity and skills, evaluating a person's overall executive processing abilities and examining the changes in neural activity [26].

Taken all together, the association revealed from our study regarding *IL6* rs1800795 and alcohol addiction is of great interest. Potentially, *IL6* rs1800795 influences IL6 expression, which can be further associated with cognitive impairment. Our finding comes as a possible explanation regarding the impairment of several cognitive function domains through brain morphology and connectivity in patients with alcohol addiction. At the same time, it can open the way to the diagnosis and prognosis of patients who are at risk of cognitive function deficits and empower therapeutic interventions to strengthen patients' ability to abstain from alcohol.

Our findings also indicated an association between *IL1B* rs1071676 and AUDIT scores. rs1071676 is located in the 3′-UTR region of *IL1B* but its exact function is still unknown. An in silico analysis showed that rs1071676 is located on the target site for has-miR-622 and thus potentially affects miRNA activity [27]. miR-622 has been associated with different types of cancer and cell proliferation, apoptosis, migration, and invasion [28].

Higher AUDIT scores are associated with alcohol-clustering conditions and more pronounced alcohol addiction. An AUDIT score equal to or higher than 20 has been associated with depression and anxiety, and the misuse of tobacco or other substances [29]. In our study, the median of patients homozygous for the G allele was five times higher than homozygous controls. This is the first time that an association between rs1071676 and AUDIT scores has been observed. However, a previous cross-sectional assessment of patients with depression aimed to elucidate the role of inflammation in light of previous alcohol use. Patients scoring above the AUDIT cutoff had higher IL-1β serum levels, but with no statistical significance [30].

Also, studies show that the heavier the drinking status, the heavier the smoking status [31]. Within our study, we also found an association between *IL6R* rs2228145 and compulsions in alcohol-addicted patients and aggressive symptomatology in healthy individuals. rs2228145 has been associated with decreased IL-6-induced C-reactive protein (CRP) levels [32] and depression severity [33]. Furthermore, high blood CRP and IL6 levels have been associated with greater depression symptom severity, obsessive compulsive disorder (OCD) anxiety and psychological distress [34]. Depressive symptoms were positively correlated with higher overall OCD severity, and higher obsession and compulsion levels. The higher the levels of depressive symptomatology, the higher the levels of counting compulsions, but there was no statistically significant association with specific categories of obsessions or compulsions [35]. Heavy, regular, and binge drinking may be associated with depression due to the brain chemical and homeostatic imbalance caused by alcohol. Alcohol consumption is considered a psychological stimulus that triggers a stress response [36]. Subsequently, this may result in immune system activation and the deregulation of pro-inflammatory cytokines, with CRP thought to be a good representative of this inflammatory response [34]. IL-6 was reported to be elevated in alcohol-addicted patients with alcohol-induced liver and/or pancreatic diseases, and the association between immune mediators and psychiatric comorbidity was noticed in those patients [17,37]. Combining our finding with the available scientific results, it is likely that rs2228145 is the mediator responsible for the compulsory symptoms in alcohol-addicted patients, possibly through a buffering system of IL6 and CRP levels. Genotyping patients for *IL6R* rs2228145 while measuring their IL6 and CRP levels could be useful tools as non-specific diagnostic markers of alcohol- addicted patients with comorbid psychiatric disorders.

Regarding the association, in controls, between *IL6R* rs2228145 and anger, hostility, and aggression, there are no similar findings.

We also indicated an association between *SOD2* rs4880 and obsessive–compulsive symptoms in alcohol-addicted patients and controls. Although rs4880 is the most studied genetic variant of SOD2, this is the first time that an association between obsessive–compulsive symptoms and alcohol addiction has been revealed.

ROS accumulation and oxidative damage can be affected by SOD2 genetic variability. rs4880 results in an amino acid change, which eventually causes changes in the secondary structure and function of the enzyme. rs4880 can lead to a redox status imbalance by altering enzyme localization and mitochondrial transportation. Individuals with the mutation have increased ROS production, which leads to mitochondrial dysfunction and increased oxidative stress vulnerability [38]. In the presence of rs4880, the translocation of SOD2 into the mitochondria is enhanced and the concentration of the active form of MnSOD is altered [39]. Homozygotes of the C allele have higher enzyme activity, whereas TT carriers have lower SOD2 enzyme activity, and a risk of higher ROS levels, probably due to the low efficiency of enzyme targeting from the cytoplasm to the mitochondria [40].

A case–control study using occupational stress as a variable to investigate the role of *SOD2* rs4880 in Han Chinese miners with dyslipidemia found that the condition was associated with occupational stress and genetic factors [41].

Moreover, the rs4880 T allele is associated with an increased risk of developing schizophrenia [42] and depression. Interestingly, sex was an influencing factor, given that the depression risk was only in males [40]. However, a large multi-site study investigating the impact of rs4880 on schizophrenia cognitive deficits, while measuring SOD2 activity, indicated that AA homozygotes had poorer attention performance. This allele was also associated with the scores of the Repeatable Battery for the Assessment of Neuropsychological Status scale, which measures cognitive functioning, while indexing, among others, immediate and delayed memory, visuospatial/constructional abilities, and language [43].

Additionally, rs4880 is associated with structural differences in the brain regions relevant to BD. More specifically, the caudal anterior cingulate cortex surface area and volume, as well as the prefrontal cortex volume, were lower in GG carriers. Also, GG BD patients had significant interaction effects in a temporal lobe brain region associated with smaller brain structures [42]. Chronic alcohol consumption has been associated with a reduction in gray and white matter volume, which leads to whole-brain volume decline. The shrinking of the brain volume subsequently causes cognitive function deficits in executive and occupational functioning, attention, working, and long-term memory, as well as the poor performance of daily living activities, social skills, and problem-solving in alcohol-addicted patients [44]. The brain is extremely vulnerable to oxidative stress, caused by the reduced concentration of antioxidant enzymes and increased oxygen metabolism [45]. Oxidative stress can damage brain white matter and lead to cognitive function impairment [46].

Alcohol consumption can potentially intensify ROS production in the brain, and induce alcohol-related organ damage by augmenting oxidative damage and neuronal cell death. Through the action of antioxidant enzymes, like SOD2, this imbalance can be shifted, and the production of active radicals can be reduced and further limit the damaging effects on cells. Notably, the low-activity Ala allele was a risk factor only in light and not in heavy drinkers, indicating a protective role of this polymorphism [45]. The involvement of *SOD2* rs4880 may be relevant to the brain changes related to alcohol consumption.

We found no association between *SOD2* rs4880 and alcohol addiction. However, we did observe an association between this genetic variant and obsessive–compulsive symptoms in alcohol-addicted patients.

OCD has been linked to increased oxidative stress [47]. Alcohol consumption may lead to an increase of ROS and may thus be connected with obsessive–compulsive symptoms [12]. Patients with anxiety, including OCD, are characterized by an oxidative stress imbalance, due to the dysregulation of free radicals, and DNA damage. Specifically, their antioxidant defense is reduced, and protein, lipid, and DNA damage is high, leading to the dysregulation of cell functions [48]. However, the findings regarding the role of SOD levels as an antioxidant marker in OCD are inconsistent [49–51]. SOD levels were statistically significant and higher in OCD patients in comparison with healthy individuals, all males, whose symptoms were assessed using the YBOCS scale. Given that SOD has a protective role against oxidative stress, the authors concluded that oxidative stress may be involved in the pathophysiology of OCD [49]. In contrast, a study investigating the serum antioxidant levels and DNA damage degree found no significant difference in the SOD levels between OCD patients and controls and no associations between the C-YBOCS and depression scores. Overall, OCD patients had increased antioxidant levels and oxidative DNA damage, supporting the involvement of oxidative stress in OCD [50]. In another study, SOD activity levels were higher in OCD patients with concurrent depression in comparison to controls [52].

As far as the involvement of *SOD2* rs4880 is concerned, a cross-sectional study demonstrated its potential involvement in OCD by increasing oxidative stress. Notably, rs4880 CT's frequency was lower and CC was higher in OCD patients than in controls [47].

In our case, *SOD2* rs4880 was associated with the YBOCS compulsions and obsessions subscales in both patients and healthy controls, which might be the reason for the inconsistent results in the available literature regarding SOD2 activity levels.

Our study also revealed an association between *PON1* rs705381 and social phobia. In the controls, *PON1* rs705381 was associated with compulsions, while rs705379 was associated with social phobia, aggression, anxiety, and depression and rs854560 with social phobia, obsessive symptoms, and anxiety.

PON1 is a calcium-dependent glycoprotein with anti inflammatory and antioxidant abilities. It reduces oxidized lipids in both low-density lipoprotein and high-density lipoprotein and inhibits their peroxidation [53]. The PON1 enzyme has also been shown to play an important role in aging [54]. PON1 activity can be altered due to various factors, including dietary and lifestyle habits, such as alcohol consumption [55]. Alcohol can inhibit serum PON1 activity [56], but also increases its activity levels, especially in men, possibly through the increase of high-density lipoprotein-cholesterol and apolipoprotein concentrations or the reduction of oxidative stress [57].

PON1 rs705381 is associated with increased serum PON1 levels. In fact, rs705381 is responsible for almost 3% of the variation in PON1 expression levels [53]. rs705381, combined with another variant of the acetylcholinesterase-paraoxonase 1 locus, significantly contributes to the reduced PON1 expression and trait-anxiety measures of healthy subjects [58]. Our study is the first to have investigated and revealed an association between *PON1* rs705381 and social phobia in patients with alcohol addiction and compulsion in healthy individuals. rs705379 contributes almost 23% of the PON1 expression variation, leading to increased serum and plasma PON1 levels [53]. rs705379 was also shown to influence PON1 enzyme activity [59].

In a case–control study on children with attention deficit/hyperactivity disorder (ADHD), GG carriers had low urinary dimethyl phosphate levels [60]. It was reported that one third of hospitalized patients with alcohol addiction have hypophosphatemia, caused by the impairment of phosphate absorption in the intestine due to alcohol consumption [61]. ADHD symptoms include difficulties in concentrating and focusing on organizing and other tasks, as well as hyperactivity and impulsiveness, which can lead to poor social interaction and psychological distress, including anxiety and depression. Although there are no studies investigating the potential association of rs705379 with psychological traits, we found an association between rs705379 and social phobia, aggression, anxiety, and depression in healthy controls.

rs854560 is a missense variation associated with reduced PON1 concentration, mRNA, and activity levels. Leu carriers have a higher PON1 activity and mRNA and protein levels than the Met homozygotes. Heterozygotes have intermediate levels of mRNA, protein, and gene activity [62]. rs854560's effect is probably due to the linkage disequilibrium of the SNP with rs705379 [63]. rs854560 has been extensively studied in patients with dyslipidemia, with inconclusive results [64]. Reviewing the available literature, our study is the only one that supports the association between rs854560 and social phobia, obsessive–compulsive, and anxiety symptoms in controls.

Finally, it should be noted that we found no association between *GPX1* rs1050450, *PON1* rs662, *IL1B* rs1143623, rs16944, and miR146a rs2910164 and alcohol addiction, as well as none with the psychological and behavioral traits examined.

One of the limitations of our study is its relatively small sample size. Thus, genetic variants with low penetrance which can potentially increase the risk of alcohol addiction might have been missed, despite the ethnically very homogenous genetic background. The alcohol-addicted patients were also older, more frequently smokers and single, less educated, and from rural environment, compared to healthy individuals. However, when we adjusted the genotype distribution for these clinical differences, we found no difference in the statistically significant findings. The other limitation is that the female population was not included in the study. Although the evidence from genetically informed studies suggests that the source and magnitude of genetic influences on alcohol outcomes are likely the same across sexes, there are sex differences in the subjective and neurobiological responses to alcohol intoxication and the genetic factors shared between AUD and endophenotypes such as alcohol sensitivity. Notably, Wigner et al. indicated that the rs4880 TT genotype is associated with an increased risk of depression only in male patients, and suggested that such a result might reflect differences in the regulation of SOD2 enzymatic activity between males and females [40]. By including only the male population, our results are also easier to compare with the other studies, as most clinical and preclinical studies have investigated alcohol exposure in males [65].

5. Conclusions

Our pilot study is the first investigating a genetic model of the inflammation- and oxidative stress-related pathways in patients with alcohol addiction and revealing associations with behavioral, psychological, and environmental factors. Our data revealed that *IL6* rs1800795 may play some role in the susceptibility to alcohol addiction, as well as that the genetic variation of both the oxidative stress and inflammation pathways potentially impact the psychosymptomatology of alcohol-addicted patients and healthy individuals. A better understanding of these underlying genetic factors and their interaction with psychosymptomatology may help in the clinical management of alcohol addiction.

Supplementary Materials: The following supporting information can be downloaded at https://www.mdpi.com/article/10.3390/antiox13010020/s1, Table S1: questionnaire traits and scores, Table S2: the investigated genetic variants, their minor allele frequencies, and Hardy–Weinberg Equilibrium in controls, Table S3: comparison of genotype frequencies of genetic variants of oxidative stress pathways between hospitalized alcohol-addicted patients and healthy controls, Table S4: associations between SOD2 rs4880 and the selected psychosymptomatology scale scores for hospitalized alcohol-dependent patients and healthy controls, Table S5: associations between GPX1 rs1050450 and the selected psychosymptomatology scale scores for hospitalized alcohol-dependent patients and healthy controls, Table S6: associations between PON1 rs705381 and the selected psychosymptomatology scale scores for hospitalized alcohol-dependent patients and healthy controls, Table S7: associations between PON1 rs705379 and the selected psychosymptomatology scale scores for hospitalized alcohol-dependent patients and healthy controls, Table S8: associations between PON1 rs854560 and the selected psychosymptomatology scale scores for hospitalized alcohol-dependent patients and healthy controls, Table S9: associations between PON1 rs662 and the selected psychosymptomatology scale scores for hospitalized alcohol-dependent patients and healthy controls, Table S10: associations between IL1B rs1143623 and the selected psychosymptomatology scale scores for hospitalized alcohol-dependent patients and healthy controls, Table S11: associations between IL1B rs16944 and the selected psychosymptomatology scale scores for hospitalized alcohol-dependent patients and healthy controls, Table S12: associations between IL1B rs1071676 and the selected psychosymptomatology scale scores for hospitalized alcohol-dependent patients and healthy controls, Table S13: associations between MIRN146A rs2910164 and the selected psychosymptomatology scale scores for hospitalized alcohol-dependent patients and healthy controls, Table S14: associations between IL6 rs1800795 and the selected psychosymptomatology scale scores for hospitalized alcohol-dependent patients and healthy controls, Table S15: associations between IL6R rs2228145 and the selected psychosymptomatology scale scores for hospitalized alcohol-dependent patients and healthy controls, Table S16: thermal cycling conditions used for GPX1 rs1050450, IL1β rs1143623 and rs16944, IL6 rs1800795, IL6R rs2228145, miR146a rs2910164, PON1 rs854560 and rs662, and SOD2 rs4880 genotyping, Table S17: thermal cycling conditions used for PON1 rs705379 and rs705381 genotyping, Table S18: thermal cycling conditions used for IL1β rs1071676 genotyping.

Author Contributions: E.E.T., A.P.I., B.K.P. and V.D.: conceptualization of the study. E.E.T.: formal experimental analysis, visualization, and writing—original draft preparation. K.G.: statistical analysis. E.E.T., A.P.I., B.K.P. and V.D.: review and editing of the manuscript. V.D.: supervision. V.D.: funding acquisition. All authors have read and agreed to the published version of the manuscript.

Funding: This study was financially supported by the Ministry of Education, Science and Sport of Slovenia (Grant No P1-0170).

Institutional Review Board Statement: This study was conducted according to the guidelines of the Declaration of Helsinki, and approved by the by the Slovenian National Medical Ethics Committee (approval No. 117/06/10 and 148/02/1011).

Informed Consent Statement: Informed consent was obtained from all subjects involved in the study.

Data Availability Statement: All the supporting data are reported in the Supplementary Files. All relevant raw data presented in this study are available on request from the corresponding authors. The raw data are not publicly available due to ethical restrictions.

Acknowledgments: We kindly acknowledge Savica Soldat from the Pharmacogenetics Laboratory for her expert technical assistance.

Conflicts of Interest: The authors declare no conflict of interest.

References

1. Griswold, M.G.; Fullman, N.; Hawley, C.; Arian, N.; Zimsen, S.R.M.; Tymeson, H.D.; Venkateswaran, V.; Tapp, A.D.; Forouzanfar, M.H.; Salama, J.S.; et al. Alcohol Use and Burden for 195 Countries and Territories, 1990–2016: A Systematic Analysis for the Global Burden of Disease Study 2016. *Lancet* **2018**, *392*, 1015–1035. [CrossRef] [PubMed]
2. MacKillop, J.; Agabio, R.; Feldstein Ewing, S.W.; Heilig, M.; Kelly, J.F.; Leggio, L.; Lingford-Hughes, A.; Palmer, A.A.; Parry, C.D.; Ray, L.; et al. Hazardous Drinking and Alcohol Use Disorders. *Nat. Rev. Dis. Primers* **2022**, *8*, 80. [CrossRef] [PubMed]
3. Verhulst, B.; Neale, M.C.; Kendler, K.S. The Heritability of Alcohol Use Disorders: A Meta-Analysis of Twin and Adoption Studies. *Psychol. Med.* **2015**, *45*, 1061–1072. [CrossRef] [PubMed]
4. Thomas, N.S.; Salvatore, J.E.; Kuo, S.I.C.; Aliev, F.; McCutcheon, V.V.; Meyers, J.M.; Bucholz, K.K.; Brislin, S.J.; Chan, G.; Edenberg, H.J.; et al. Genetic Nurture Effects for Alcohol Use Disorder. *Mol. Psychiatry* **2022**, *28*, 759–766. [CrossRef] [PubMed]
5. Tsermpini, E.E.; Plemenitaš Ilješ, A.; Dolžan, V. Alcohol-Induced Oxidative Stress and the Role of Antioxidants in Alcohol Use Disorder: A Systematic Review. *Antioxidants* **2022**, *11*, 1374. [CrossRef] [PubMed]
6. Milaciu, M.V.; Vesa, Ș.C.; Bocșan, I.C.; Ciumărnean, L.; Sâmpelean, D.; Negrean, V.; Pop, R.M.; Matei, D.M.; Pașca, S.; Răchișan, A.L.; et al. Paraoxonase-1 Serum Concentration and PON1 Gene Polymorphisms: Relationship with Non-Alcoholic Fatty Liver Disease. *J. Clin. Med.* **2019**, *8*, 2200. [CrossRef] [PubMed]
7. González-Reimers, E.; Santolaria-Fernández, F.; Martín-González, M.C.; Fernández-Rodríguez, C.M.; Quintero-Platt, G. Alcoholism: A Systemic Proinflammatory Condition. *World J. Gastroenterol.* **2014**, *20*, 14660–14671. [CrossRef]
8. Qin, L.; He, J.; Hanes, R.N.; Pluzarev, O.; Hong, J.S.; Crews, F.T. Increased Systemic and Brain Cytokine Production and Neuroinflammation by Endotoxin Following Ethanol Treatment. *J. Neuroinflamm.* **2008**, *5*, 10. [CrossRef]
9. Fan, W.; Liang, C.; Ou, M.; Zou, T.; Sun, F.; Zhou, H.; Cui, L. MicroRNA-146a Is a Wide-Reaching Neuroinflammatory Regulator and Potential Treatment Target in Neurological Diseases. *Front. Mol. Neurosci.* **2020**, *13*, 90. [CrossRef]
10. Eguchi, A.; Franz, N.; Kobayashi, Y.; Iwasa, M.; Wagner, N.; Hildebrand, F.; Takei, Y.; Marzi, I.; Relja, B. Circulating Extracellular Vesicles and Their MiR "Barcode" Differentiate Alcohol Drinkers with Liver Injury and Those without Liver Injury in Severe Trauma Patients. *Front. Med.* **2019**, *6*, 30. [CrossRef]
11. American Psychiatric Publishing. *Diagnostic and Statistical Manual of Mental Disorders DSM-5*, 5th ed.; American Psychiatric Publishing: Washington, DC, USA, 2013.
12. Plemenitas, A.; Kastelic, M.; Porcelli, S.; Serretti, A.; Rus Makovec, M.; Kores Plesnicar, B.; Dolžan, V. Genetic Variability in CYP2E1 and Catalase Gene among Currently and Formerly Alcohol-Dependent Male Subjects. *Alcohol Alcohol.* **2015**, *50*, 140–145. [CrossRef] [PubMed]
13. Ilješ, A.P.; Plesničar, B.K.; Dolžan, V. Associations of NLRP3 and CARD8 Gene Polymorphisms with Alcohol Dependence and Commonly Related Psychiatric Disorders: A Preliminary Study. *Arch. Ind. Hyg. Toxicol.* **2021**, *72*, 191–197. [CrossRef] [PubMed]
14. Tsermpini, E.E.; Goričar, K.; Kores Plesničar, B.; Plemenitaš Ilješ, A.; Dolžan, V. Genetic Variability of Incretin Receptors and Alcohol Dependence: A Pilot Study. *Front. Mol. Neurosci.* **2022**, *15*, 908948. [CrossRef] [PubMed]

15. Lanquetin, A.; Leclercq, S.; De Timary, P.; Segobin, S.; Naveau, M.; Coulbault, L.; MacCioni, P.; Lorrai, I.; Colombo, G.; Vivien, D.; et al. Role of Inflammation in Alcohol-Related Brain Abnormalities: A Translational Study. *Brain Commun.* **2021**, *3*, fcab154. [CrossRef] [PubMed]
16. Sasayama, D.; Hattori, K.; Wakabayashi, C.; Teraishi, T.; Hori, H.; Ota, M.; Yoshida, S.; Arima, K.; Higuchi, T.; Amano, N.; et al. Increased Cerebrospinal Fluid Interleukin-6 Levels in Patients with Schizophrenia and Those with Major Depressive Disorder. *J. Psychiatr. Res.* **2013**, *47*, 401–406. [CrossRef] [PubMed]
17. García-Marchena, N.; Maza-Quiroga, R.; Serrano, A.; Barrios, V.; Requena-Ocaña, N.; Suárez, J.; Chowen, J.A.; Argente, J.; Rubio, G.; Torrens, M.; et al. Abstinent Patients with Alcohol Use Disorders Show an Altered Plasma Cytokine Profile: Identification of Both Interleukin 6 and Interleukin 17A as Potential Biomarkers of Consumption and Comorbid Liver and Pancreatic Diseases. *J. Psychopharmacol.* **2020**, *34*, 1250–1260. [CrossRef] [PubMed]
18. Baune, B.T.; Konrad, C.; Grotegerd, D.; Suslow, T.; Birosova, E.; Ohrmann, P.; Bauer, J.; Arolt, V.; Heindel, W.; Domschke, K.; et al. Interleukin-6 Gene (IL-6): A Possible Role in Brain Morphology in the Healthy Adult Brain. *J. Neuroinflamm.* **2012**, *9*, 125. [CrossRef]
19. Bonafè, M.; Olivieri, F.; Cavallone, L.; Giovagnetti, S.; Marchegiani, F.; Cardelli, M.; Pieri, C.; Marra, M.; Antonicelli, R.; Lisa, R.; et al. A Gender--Dependent Genetic Predisposition to Produce High Levels of IL-6 Is Detrimental for Longevity. *Eur. J. Immunol.* **2001**, *31*, 2357–2361. [CrossRef]
20. Fishman, D.; Faulds, G.; Jeffey, R.; Mohamed-Ali, V.; Yudkin, J.S.; Humphries, S.; Woo, P. The Effect of Novel Polymorphisms in the Interleukin-6 (IL-6) Gene on IL-6 Transcription and Plasma IL-6 Levels, and an Association with Systemic-Onset Juvenile Chronic Arthritis. *J. Clin. Investig.* **1998**, *102*, 1369–1376. [CrossRef]
21. Chen, Y.; Pawlikowska, L.; Yao, J.S.; Shen, F.; Zhai, W.; Achrol, A.S.; Lawton, M.T.; Kwok, P.Y.; Yang, G.Y.; Young, W.L. Interleukin-6 Involvement in Brain Arteriovenous Malformations. *Ann. Neurol.* **2006**, *59*, 72–80. [CrossRef]
22. MacQueen, G.; Frodl, T. The Hippocampus in Major Depression: Evidence for the Convergence of the Bench and Bedside in Psychiatric Research? *Mol. Psychiatry* **2011**, *16*, 252–264. [CrossRef] [PubMed]
23. Meda, S.A.; Hawkins, K.A.; Dager, A.D.; Tennen, H.; Khadka, S.; Austad, C.S.; Wood, R.M.; Raskin, S.; Fallahi, C.R.; Pearlson, G.D. Longitudinal Effects of Alcohol Consumption on the Hippocampus and Parahippocampus in College Students. *Biol. Psychiatry Cogn. Neurosci. Neuroimaging* **2018**, *3*, 610–617. [CrossRef] [PubMed]
24. Crespi, C.; Galandra, C.; Canessa, N.; Manera, M.; Poggi, P.; Basso, G. Microstructural Damage of White-Matter Tracts Connecting Large-Scale Networks Is Related to Impaired Executive Profile in Alcohol Use Disorder. *Neuroimage Clin.* **2020**, *25*, 102141. [CrossRef] [PubMed]
25. Stephan, R.A.; Alhassoon, O.M.; Allen, K.E.; Wollman, S.C.; Hall, M.; Thomas, W.J.; Gamboa, J.M.; Kimmel, C.; Stern, M.; Sari, C.; et al. Meta-Analyses of Clinical Neuropsychological Tests of Executive Dysfunction and Impulsivity in Alcohol Use Disorder. *Am. J. Drug Alcohol Abus.* **2016**, *43*, 24–43. [CrossRef] [PubMed]
26. Mooijaart, S.P.; Sattar, N.; Trompet, S.; Lucke, J.; Stott, D.J.; Ford, I.; Jukema, J.W.; Westendorp, R.G.J.; de Craen, A.J.M. Circulating Interleukin-6 Concentration and Cognitive Decline in Old Age: The PROSPER Study. *J. Intern. Med.* **2013**, *274*, 77–85. [CrossRef]
27. Walczak, M.; Lykowska-Szuber, L.; Plucinska, M.; Stawczyk-Eder, K.; Zakerska-Banaszak, O.; Eder, P.; Krela-Kazmierczak, I.; Michalak, M.; Zywicki, M.; Karlowski, W.M.; et al. Is Polymorphism in the Apoptosis and Inflammatory Pathway Genes Associated with a Primary Response to Anti-TNF Therapy in Crohn's Disease Patients? *Front. Pharmacol.* **2020**, *11*, 1207. [CrossRef]
28. Lu, J.; Xie, Z.; Xiao, Z.; Zhu, D. The Expression and Function of MiR-622 in a Variety of Tumors. *Biomed. Pharmacother.* **2022**, *146*, 112544. [CrossRef]
29. Khan, M.R.; Young, K.E.; Caniglia, E.C.; Fiellin, D.A.; Maisto, S.A.; Marshall, B.D.L.; Edelman, E.J.; Gaither, J.R.; Chichetto, N.E.; Tate, J.; et al. Association of Alcohol Screening Scores with Adverse Mental Health Conditions and Substance Use Among US Adults. *JAMA Netw. Open.* **2020**, *3*, e200895. [CrossRef]
30. Toft, H.; Neupane, S.P.; Bramness, J.G.; Tilden, T.; Wampold, B.E.; Lien, L. The Effect of Trauma and Alcohol on the Relationship between Level of Cytokines and Depression among Patients Entering Psychiatric Treatment. *BMC Psychiatry* **2018**, *18*, 95. [CrossRef]
31. van Amsterdam, J.; van den Brink, W. The Effect of Alcohol Use on Smoking Cessation: A Systematic Review. *Alcohol* **2023**, *109*, 13–22. [CrossRef]
32. Garbers, C.; Monhasery, N.; Aparicio-Siegmund, S.; Lokau, J.; Baran, P.; Nowell, M.A.; Jones, S.A.; Rose-John, S.; Scheller, J. The Interleukin-6 Receptor Asp358Ala Single Nucleotide Polymorphism Rs2228145 Confers Increased Proteolytic Conversion Rates by ADAM Proteases. *Biochim. Biophys. Acta* **2014**, *1842*, 1485–1494. [CrossRef] [PubMed]
33. Khandaker, G.M.; Zammit, S.; Burgess, S.; Lewis, G.; Jones, P.B. Association between a Functional Interleukin 6 Receptor Genetic Variant and Risk of Depression and Psychosis in a Population-Based Birth Cohort. *Brain Behav. Immun.* **2018**, *69*, 264–272. [CrossRef] [PubMed]
34. Wium-Andersen, M.K.; Ørsted, D.D.; Nielsen, S.F.; Nordestgaard, B.G. Elevated C-Reactive Protein Levels, Psychological Distress, and Depression in 73,131 Individuals. *JAMA Psychiatry* **2013**, *70*, 176–184. [CrossRef] [PubMed]
35. Peris, T.S.; Bergman, R.L.; Asarnow, J.R.; Langley, A.; McCracken, J.T.; Piacentini, J. Clinical and Cognitive Correlates of Depressive Symptoms among Youth with Obsessive Compulsive Disorder. *J. Clin. Child Adolesc. Psychol.* **2010**, *39*, 616–626. [CrossRef] [PubMed]

36. Keyes, K.M.; Hatzenbuehler, M.L.; Grant, B.F.; Hasin, D.S. Stress and Alcohol: Epidemiologic Evidence. *Alcohol. Res.* **2012**, *34*, 391.
37. García-Marchena, N.; Araos, P.F.; Barrios, V.; Sánchez-Marín, L.; Chowen, J.A.; Pedraz, M.; Castilla-Ortega, E.; Romero-Sanchiz, P.; Ponce, G.; Gavito, A.L.; et al. Plasma Chemokines in Patients with Alcohol Use Disorders: Association of CCL11 (Eotaxin-1) with Psychiatric Comorbidity. *Front. Psychiatry* **2017**, *7*, 214. [CrossRef]
38. Pourvali, K.; Abbasi, M.; Mottaghi, A. Role of Superoxide Dismutase 2 Gene Ala16Val Polymorphism and Total Antioxidant Capacity in Diabetes and Its Complications. *Avicenna J. Med. Biotechnol.* **2016**, *8*, 48.
39. Schneider, A.; Rosendahl, J.; Bugert, P.; Weiss, C.; Unterschütz, H.; Kylanpää-Bäck, M.L.; Lempinen, M.; Kemppainen, E.; Diaconu, B.L.; Ebert, M.P.; et al. Genetic Variants in the Manganese Superoxide Dismutase 2 Gene and in the Catalase Gene Are Not Associated with Alcoholic Chronic Pancreatitis. *Alcohol Alcohol.* **2017**, *52*, 535–541. [CrossRef]
40. Wigner, P.; Czarny, P.; Synowiec, E.; Bijak, M.; Białek, K.; Talarowska, M.; Galecki, P.; Szemraj, J.; Sliwinski, T. Variation of Genes Involved in Oxidative and Nitrosative Stresses in Depression. *Eur. Psychiatry* **2018**, *48*, 38–48. [CrossRef]
41. Yang, Y.; Zheng, Z.; Chen, Y.; Wang, X.; Wang, H.; Si, Z.; Meng, R.; Wu, J. A Case Control Study on the Relationship between Occupational Stress and Genetic Polymorphism and Dyslipidemia in Coal Miners. *Sci. Rep.* **2023**, *13*, 2321. [CrossRef]
42. Zou, Y.; Kennedy, K.G.; Grigorian, A.; Fiksenbaum, L.; Freeman, N.; Zai, C.C.; Kennedy, J.L.; MacIntosh, B.J.; Goldstein, B.I. Antioxidative Defense Genes and Brain Structure in Youth Bipolar Disorder. *Int. J. Neuropsychopharmacol.* **2022**, *25*, 89–98. [CrossRef] [PubMed]
43. Zhang, X.Y.; Chen, D.C.; Xiu, M.H.; Yang, F.D.; Tan, Y.; Luo, X.; Zuo, L.; Kosten, T.A.; Kosten, T.R. Cognitive Function, Plasma MnSOD Activity, and MnSOD Ala-9Val Polymorphism in Patients with Schizophrenia and Normal Controls. *Schizophr. Bull.* **2014**, *40*, 592. [CrossRef] [PubMed]
44. Sullivan, E.V.; Harris, R.A.; Pfefferbaum, A. Alcohol's Effects on Brain and Behavior. *Alcohol. Res. Health* **2010**, *33*, 127. [PubMed]
45. Gitik, M.; Srivastava, V.; Hodgkinson, C.A.; Shen, P.H.; Goldman, D.; Meyerhoff, D.J. Association of Superoxide Dismutase 2 (SOD2) Genotype with Gray Matter Volume Shrinkage in Chronic Alcohol Users: Replication and Further Evaluation of an Addiction Gene Panel. *Int. J. Neuropsychopharmacol.* **2016**, *19*, pyw033. [CrossRef] [PubMed]
46. Salminen, L.E.; Schofield, P.R.; Pierce, K.D.; Bruce, S.E.; Griffin, M.G.; Tate, D.F.; Cabeen, R.P.; Laidlaw, D.H.; Conturo, T.E.; Bolzenius, J.D.; et al. Vulnerability of White Matter Tracts and Cognition to the SOD2 Polymorphism: A Preliminary Study of Antioxidant Defense Genes in Brain Aging. *Behav. Brain Res.* **2017**, *329*, 111–119. [CrossRef]
47. Orhan, N.; Kucukali, C.I.; Cakir, U.; Seker, N.; Aydin, M. Genetic Variants in Nuclear-Encoded Mitochondrial Proteins Are Associated with Oxidative Stress in Obsessive Compulsive Disorders. *J. Psychiatr. Res.* **2012**, *46*, 212–218. [CrossRef]
48. Fedoce, A.D.G.; Ferreira, F.; Bota, R.G.; Bonet-Costa, V.; Sun, P.Y.; Davies, K.J.A. The Role of Oxidative Stress in Anxiety Disorder: Cause or Consequence? *Free Radic. Res.* **2018**, *52*, 737–750. [CrossRef]
49. Behl, A.; Swami, G.; Sircar, S.S.; Bhatia, M.S.; Banerjee, B.D. Relationship of Possible Stress-Related Biochemical Markers to Oxidative/Antioxidative Status in Obsessive-Compulsive Disorder. *Neuropsychobiology* **2010**, *61*, 210–214. [CrossRef]
50. Şimşek, Ş.; Gençoğlan, S.; Yüksel, T. DNA Damage and Antioxidants in Treatment Naïve Children with Obsessive-Compulsive Disorder. *Psychiatry Res.* **2016**, *237*, 133–137. [CrossRef]
51. Ozdemir, E.; Cetinkaya, S.; Ersan, S.; Kucukosman, S.; Ersan, E.E. Serum Selenium and Plasma Malondialdehyde Levels and Antioxidant Enzyme Activities in Patients with Obsessive-Compulsive Disorder. *Prog. Neuropsychopharmacol. Biol. Psychiatry* **2009**, *33*, 62–65. [CrossRef]
52. Kuloglu, M.; Atmaca, M.; Tezcan, E.; Gecici, Ö.; Tunckol, H.; Ustundag, B. Antioxidant Enzyme Activities and Malondialdehyde Levels in Patients with Obsessive-Compulsive Disorder. *Neuropsychobiology* **2002**, *46*, 27–32. [CrossRef] [PubMed]
53. Vavlukis, M.; Vavlukis, A.; Krsteva, K.; Topuzovska, S. Paraoxonase 1 Gene Polymorphisms in Lipid Oxidation and Atherosclerosis Development. *Front. Genet.* **2022**, *13*, 966413. [CrossRef] [PubMed]
54. Oczos, J.; Grimm, C.; Barthelmes, D.; Sutter, F.; Menghini, M.; Kloeckener-Gruissem, B.; Berger, W. Regulatory Regions of the Paraoxonase 1 (PON1) Gene Are Associated with Neovascular Age-Related Macular Degeneration (AMD). *Age* **2013**, *35*, 1651–1662. [CrossRef] [PubMed]
55. Costa, L.G.; Cole, T.B.; Jarvik, G.P.; Furlong, C.E. Functional Genomic of the Paraoxonase (PON1) Polymorphisms: Effects on Pesticide Sensitivity, Cardiovascular Disease, and Drug Metabolism. *Annu. Rev. Med.* **2003**, *54*, 371–392. [CrossRef] [PubMed]
56. Debord, J.; Dantoine, T.; Bollinger, J.C.; Abraham, M.H.; Verneuil, B.; Merle, L. Inhibition of Arylesterase by Aliphatic Alcohols. *Chem. Biol. Interact.* **1998**, *113*, 105–115. [CrossRef]
57. Van Der Gaag, M.S.; Van Tol, A.; Scheek, L.M.; James, R.W.; Urgert, R.; Schaafsma, G.; Hendriks, H.F.J. Daily Moderate Alcohol Consumption Increases Serum Paraoxonase Activity; a Diet-Controlled, Randomised Intervention Study in Middle-Aged Men. *Atherosclerosis* **1999**, *147*, 405–410. [CrossRef]
58. Sklan, E.H.; Lowenthal, A.; Korner, M.; Ritov, Y.; Landers, D.M.; Rankinen, T.; Bouchard, C.; Leon, A.S.; Rice, T.; Rao, D.C.; et al. Acetylcholinesterase/Paraoxonase Genotype and Expression Predict Anxiety Scores in Health, Risk Factors, Exercise Training, and Genetics Study. *Proc. Natl. Acad. Sci. USA* **2004**, *101*, 5512–5517. [CrossRef]
59. Petrič, B.; Redenšek Trampuž, S.; Dolžan, V.; Gregorič Kramberger, M.; Trošt, M.; Maraković, N.; Goličnik, M.; Bavec, A. Investigation of Paraoxonase-1 Genotype and Enzyme-Kinetic Parameters in the Context of Cognitive Impairment in Parkinson's Disease. *Antioxidants* **2023**, *12*, 399. [CrossRef]

60. Chang, C.H.; Yu, C.J.; Du, J.C.; Chiou, H.C.; Hou, J.W.; Yang, W.; Chen, C.F.; Chen, H.C.; Chen, Y.S.; Hwang, B.; et al. The Associations among Organophosphate Pesticide Exposure, Oxidative Stress, and Genetic Polymorphisms of Paraoxonases in Children with Attention Deficit/Hyperactivity Disorder. *Sci. Total Environ.* **2021**, *773*, 145604. [CrossRef]
61. Farooq, A.; Richman, C.M.; Swain, S.M.; Shahid, R.A.; Vigna, S.R.; Liddle, R.A. The Role of Phosphate in Alcohol-Induced Experimental Pancreatitis. *Gastroenterology* **2021**, *161*, 982–995.e2. [CrossRef]
62. Belin, A.C.; Ran, C.; Anvret, A.; Paddock, S.; Westerlund, M.; Håkansson, A.; Nissbrandt, H.; Söderkvist, P.; Dizdar, N.; Ahmadi, A.; et al. Association of a Protective Paraoxonase 1 (PON1) Polymorphism in Parkinson's Disease. *Neurosci. Lett.* **2012**, *522*, 30–35. [CrossRef] [PubMed]
63. Brophy, V.H.; Jampsa, R.L.; Clendenning, J.B.; McKinstry, L.A.; Jarvik, G.P.; Furlong, C.E. Effects of 5′ Regulatory-Region Polymorphisms on Paraoxonase-Gene (PON1) Expression. *Am. J. Hum. Genet.* **2001**, *68*, 1428–1436. [CrossRef] [PubMed]
64. Luo, Z.; Li, S.; Muhammad, I.; Karim, M.R.; Song, Y. Associations of the PON1 Rs854560 Polymorphism with Plasma Lipid Levels: A Meta-Analysis. *Lipids Health Dis.* **2018**, *17*, 274. [CrossRef] [PubMed]
65. Erickson, E.K.; Grantham, E.K.; Warden, A.S.; Harris, R.A. Neuroimmune Signaling in Alcohol Use Disorder. *Pharmacol. Biochem. Behav.* **2019**, *177*, 34–60. [CrossRef]

antioxidants

MDPI

Article

Fenofibrate Decreases Ethanol-Induced Neuroinflammation and Oxidative Stress and Reduces Alcohol Relapse in Rats by a PPAR-α-Dependent Mechanism

Cristina Ibáñez [1,2,†], Tirso Acuña [3,†], María Elena Quintanilla [2,3], Diliana Pérez-Reytor [1], Paola Morales [2,3,4,*] and Eduardo Karahanian [1,2,*]

1 Institute of Biomedical Sciences, Faculty of Health Sciences, Universidad Autónoma de Chile, Santiago 8910060, Chile; cristina.ibanez@cloud.uautonoma.cl (C.I.); d.perez@uautonoma.cl (D.P.-R.)
2 Research Center for the Development of Novel Therapeutic Alternatives for Alcohol Use Disorders, Santiago 8910060, Chile; equintanilla@uchile.cl
3 Molecular and Clinical Pharmacology Program, Institute of Biomedical Sciences, Faculty of Medicine, Universidad de Chile, Santiago 8380453, Chile; tirsoacuna@ug.uchile.cl
4 Department of Neuroscience, Faculty of Medicine, Universidad de Chile, Santiago 8380453, Chile
* Correspondence: pmorales@uchile.cl (P.M.); eduardo.karahanian@uautonoma.cl (E.K.); Tel.: +56-996361996 (P.M.)
† These authors contributed equally to this work.

Abstract: High ethanol consumption triggers neuroinflammation, implicated in sustaining chronic alcohol use. This inflammation boosts glutamate, prompting dopamine release in reward centers, driving prolonged drinking and relapse. Fibrate drugs, activating peroxisome proliferator-activated receptor alpha (PPAR-α), counteract neuroinflammation in other contexts, prompting investigation into their impact on ethanol-induced inflammation. Here, we studied, in UChB drinker rats, whether the administration of fenofibrate in the withdrawal stage after chronic ethanol consumption reduces voluntary intake when alcohol is offered again to the animals (relapse-type drinking). Furthermore, we determined if fenofibrate was able to decrease ethanol-induced neuroinflammation and oxidative stress in the brain. Animals treated with fenofibrate decreased alcohol consumption by 80% during post-abstinence relapse. Furthermore, fenofibrate decreased the expression of the proinflammatory cytokines tumor necrosis factor-alpha (TNF-α) and interleukins IL-1β and IL-6, and of an oxidative stress-induced gene (heme oxygenase-1), in the hippocampus, nucleus accumbens, and prefrontal cortex. Animals treated with fenofibrate showed an increase M2-type microglia (with anti-inflammatory proprieties) and a decrease in phagocytic microglia in the hippocampus. A PPAR-α antagonist (GW6471) abrogated the effects of fenofibrate, indicating that they are dependent on PPAR-α activation. These findings highlight the potential of fenofibrate, an FDA-approved dyslipidemia medication, as a supplementary approach to alleviating relapse severity in individuals with alcohol use disorder (AUD) during withdrawal.

Keywords: alcohol use disorder; alcoholism; fibrates; neuroinflammation; PPAR alpha

Citation: Ibáñez, C.; Acuña, T.; Quintanilla, M.E.; Pérez-Reytor, D.; Morales, P.; Karahanian, E. Fenofibrate Decreases Ethanol-Induced Neuroinflammation and Oxidative Stress and Reduces Alcohol Relapse in Rats by a PPAR-α-Dependent Mechanism. *Antioxidants* **2023**, *12*, 1758. https://doi.org/10.3390/antiox12091758

Academic Editor: Marco Fiore

Received: 16 August 2023
Revised: 6 September 2023
Accepted: 8 September 2023
Published: 13 September 2023

1. Introduction

Alcohol use disorder (AUD) remains a significant public health concern, with available pharmacological interventions such as naltrexone and acamprosate, which primarily target reducing alcohol craving, help maintain abstinence, and reduce harmful drinking in the treatment phase of psychological dependence. However, they often fall short in providing sustained recovery, as most patients experience relapse in the short or middle term [1,2]. Throughout the detoxification phase, patients are typically prescribed benzodiazepines, such as diazepam, to alleviate symptoms of alcohol withdrawal syndrome [3]; however, their usage in cases of AUD should be limited to short durations due to the potential of substituting alcohol addiction with benzodiazepine dependence. Prolonged and excessive alcohol consumption leads to enduring changes in the brain, perpetuating addictive behavior in individuals with AUD. Ethanol-induced neuroinflammation in the brain has emerged as a pivotal factor in the neurobiological changes associated with persistent chronic alcohol abuse [4–6].

Neuroinflammation associated with ethanol consumption can be induced by two mechanisms: one of them involves the oxidation of ethanol by cytochrome P4502E1 (CYP2E1), generating reactive oxygen species (ROS) that activate nuclear factor kappa-light-chain-enhancer of activated B cells (NF-κB), leading to pro-inflammatory responses [7,8]. This activation also increases the expression of NADPH oxidase [9], which generates even more ROS, perpetuating the inflammatory response [5,6]. The second mechanism involves the activation of tumor necrosis factor-alpha (TNF-α) receptor in the brain, since ethanol consumption increases the permeability of the intestinal mucosa, which allows bacterial lipopolysaccharide (LPS) to diffuse into circulation [10]. LPS then triggers the release of TNF-α to the blood, which can cross the blood–brain barrier and activate microglial TNF receptors, inducing neuroinflammation [11,12]. The activation of either mechanism leads to the release of proinflammatory cytokines such as TNF-α, IL-1β, and IL-6 [13]. This glial-secreted TNF-α then binds to its TNF receptor in the same microglial cells, creating an activation loop that potentiates the initial neuroinflammation response [5]. The self-perpetuating positive feedback loops created by these mechanisms can maintain neuroinflammation for months after alcohol withdrawal, which increases the risk of relapse [9].

The establishment of alcohol addiction is a complex process, where the modulation of dopamine release in the brain's reward pathways plays a central role. This modulation involves interactions between various neurotransmitters' pathways, including γ-aminobutyric acid (GABA) and glutamate. Both neurotransmitters modulate dopamine release in the nucleus accumbens in response to ethanol. Activation of presynaptic GABA receptors by ethanol inhibits GABA release in the ventral tegmental area (VTA), which in turn stimulates dopamine release in the nucleus accumbens [14]. Dopamine release is closely linked to the motivational effects of alcohol, that is, to the positive reinforcement involved in the reward. However, a second behavioral aspect of AUD is negative reinforcing features such as hyper-anxious and hyperexcitability states, which lead the individual to continue drinking to obtain relief from this negative affective state associated with alcohol dependence. GABAergic mechanisms have been implicated in the neuroadaptations associated with the transition in humans from limited access to ethanol to chronic drinking [15]. Studies with pharmacological agonists and antagonists have implicated GABA systems in the anxiogenic effects of ethanol withdrawal, since GABA agonists decrease central nervous system hyperexcitability during ethanol withdrawal, whereas GABA antagonists exacerbate many of the symptoms of ethanol withdrawal [15]. Although there are no studies that demonstrate the direct relationship between ethanol-induced neuroinflammation and anxiogenic effects, studies have shown that the state of anxiety could be related to the release of proinflammatory cytokines [16]. Involvement of proinflammatory cytokines in anxiety has been demonstrated in transgenic mice lacking TNF-α receptors [17] and transgenic mice lacking IL-1β receptors [18]. In these animal models, the level of anxiety was found to be lower than that in the wild-type mice [17,18]. However, the involvement of GABAergic systems in anxiety triggered by neuroinflammation has not yet been demonstrated.

On the other hand, glutamate is a neurotransmitter strongly linked to the maintenance of chronic ethanol intake and relapse to alcohol and other substances of abuse [19]. Ethanol-induced neuroinflammation plays a pivotal role in heightening glutamatergic activity in the brain [20]. Ethanol is known to increase the extracellular glutamate levels within mesocorticolimbic structures. In AUD individuals, astrocytes exhibit modified glutamate, clearance due to the inadequate functioning of the brain's glutamate transporter 1 (GLT-1), also recognized as excitatory amino acid transporter 2 (EAAT2) or solute carrier family 1 member 2 (SLC1A2). This irregularity results in an increase in extracellular glutamate levels at the tripartite synapse [21]. Following chronic ethanol consumption and subsequent abstinence, environmental cues associated with alcohol usage can trigger drug-seeking behavior and the urge to re-administer the substance, leading to relapse and the perpetuation of alcohol intake [22]. This behavior cycle is powered by an amplified glutamate release into the nucleus accumbens through intricate circuit connecting hippocampus and prefrontal cortex [20].

Peroxisome proliferator-activated receptor alpha (PPAR-α) is a nuclear receptor with key functions in lipid metabolism [23]. Fibrate drugs such as fenofibrate, bezafibrate, gemfibrozil, ciprofibrate, and clofibrate are PPAR-α agonists [24], actively elevating the oxidation rate of fatty acids, and are commonly employed to treat hypertriglyceridemia [25]. Several studies have shown that fenofibrate administration to rodents that consume alcohol voluntarily leads to a reduction in alcohol intake [26–31]. Since PPAR-α activation in the brain decreases neuroinflammation [32–36], we hypothesized that fenofibrate exerts an anti-inflammatory effect even when administered only during the abstinence period following chronic alcohol intake. In line with this hypothesis, a previous study by our group reported that the administration of fenofibrate during the abstinence period following chronic alcohol consumption in rats was able to (i) reverse the ethanol-induced increase in glial acidic fibrillary protein (GFAP) levels, indicative of astrogliosis, (ii) decrease the deactivation of the NF-κB inhibitor (IκBα), and (iii) restore the diminished expression levels of GLT-1 caused by alcohol treatment [37].

In this study, our primary objective was to examine the potential effects of fenofibrate when administered exclusively during the abstinence stage following chronic alcohol consumption in rats, specifically focusing on its impact on alcohol intake in a relapse model. Additionally, we sought to evaluate the influence of fenofibrate administration during abstinence on the expression of proinflammatory cytokines, markers of oxidative stress in the brain, and the activation of microglia. We aimed to contribute to a deeper understanding of the potential therapeutic advantages that fenofibrate might offer in the management of alcohol use disorder and the corresponding neurobiological alterations.

2. Materials and Methods

2.1. Animals and Treatments

High-drinker UChB (University of Chile Bibulous) rats derived from the Wistar strain and bred selectively for their high alcohol intake [38] were used. Two-month-old female rats were housed in individual cages in temperature-controlled rooms under a regular 12-hour light/12-hour dark cycle (a total of 30 animals). For 46 days, rats were offered a free choice between 10% (v/v) ethanol solution and water. On day 47, rats were allowed concurrent three-bottle choice access to 10% and 20% (v/v) ethanol solutions and water for 14 additional days. This protocol was routinely adopted by our group, since, in a previous study, we observed that simultaneous access to 10% and 20% ethanol allowed the detection of a more marked effect in relapse-type alcohol consumption [39]. Food (Mardones rat formula, Alimentos Cisternas, Santiago, Chile) and water were always provided ad libitum, and the volume of water and ethanol was recorded daily. After these 60 days of alcohol consumption, the weight of the animals averaged 242.1 ± 18.9 g. The decision to utilize female rats in this study stems from compelling evidence in specific rat lines, such as high-alcohol-drinking-2 (HAD-2), and Sardinian alcohol-preferring (sP) lines, which were selectively bred for their elevated ethanol consumption. In these lines, females consistently demonstrated higher levels of ethanol consumption compared to males [40,41]. This pattern is found as well in the UChB rat line (unpublished results). Furthermore, the choice of female rats is supported by their relatively more stable body weight during the 2–3-month duration of the experiments conducted in this study.

After 60 days of continuous free choice between ethanol and water, animals were divided into 4 groups (n = 7 for groups I and IV, n = 8 for groups II and III, see below) and deprived of ethanol access for 14 days, keeping water and food consumption ad libitum. In the last five days of ethanol withdrawal, the four groups were, respectively, administered a daily dose of: Group I, Vehicle control: dimethyl sulfoxide (DMSO), i.p., plus water by gavage, which are the vehicles of GW6471 and of fenofibrate, respectively; Group II, Fenofibrate: micronized fenofibrate (Fibronil®, Royal Pharma, Santiago, Chile) 50 mg/kg/day per gavage, re-suspended in water (1 mL per each 150 g of body weight) [27] plus DMSO i.p. vehicle; Group III, Fenofibrate: 50 mg/kg/day + GW6471 1 mg/kg/day i.p. (a PPAR-α specific antagonist, MyBioSource, San Diego, CA, USA), dissolved in DMSO (0.1 mL per 100 g body weight) [42]; and Group IV, GW6471: 1 mg/kg/day i.p. plus water vehicle per gavage.

On day 75, after finishing the 14 days of abstinence (which included the 5 days of drug treatment), the animals were offered the 10% and 20% (v/v) ethanol solutions again for just 1 h, and the consumed volume was recorded.

Immediately after recording alcohol consumption during 1 h re-access, the animals were anesthetized with ketamine/xylazine (10:1 mg/kg of body weight, i.p.) and immediately decapitated to obtain brains samples.

2.2. Quantification of the Expression of Proinflammatory Cytokines and Oxidative Stress Markers

From one of the cerebral hemispheres, hippocampus, nucleus accumbens, and prefrontal cortex tissues were extracted and homogenized in RNA-Solv® Reagent (Omega Bio-tek, Inc., Norcross, GA, USA) with a mini Potter–Elvehjem pestle (Sigma-Aldrich, St. Louis, MO, USA). Total RNA was purified according to the manufacturer's instructions. For RT-qPCR analysis, 300 ng of RNA was treated with DNase I (New England Biolabs, Ipswich, MA, USA) and subjected to amplification by RT-qPCR using the following primers: TNF-α (forward) CAGCCGATTTGCCACTTCATA, TNF-α (reverse) TCCTTAGGGCAAGGGCTCTT, IL1-β (forward) AGGCTTCCTTGTGCAAGTGT, IL1-β (reverse) TGTCGAGATGCTGCTGTGAG, IL6 (forward) CCCAACTTCCAATGCTCTCCTAATG, IL6 (reverse) GCACACTAGGTTTGCCGAGTAGACC, β-Actin (forward) CTTGCAGCTCCTCCGTCGCC β-Actin (reverse) CTTGCTCTGGGCCTCGTCGC, HO-1 (forward) TCGACAACCCCACCAAGTTC, HO-1 (reverse) AGGTAGTATCTTGAACCAGGCT.

The corresponding GenBank accession numbers are: *TNF-α* X66539.1; *IL1-β* NM_031512.2; *IL6* NM_012589.2; *β-Actin* NM_031144.3; HO-1 NM_012580.2.

2.3. Determination of Oxidized Glutathione (GSSG) and Reduced Glutathione (GSH) Levels

A portion of the hippocampus was reserved for the determination of oxidized glutathione (GSSG) and reduced glutathione (GSH) levels using the GSH/GSSG Ratio Detection Assay Kit II (Abcam, Boston, MA, USA), following the manufacturer's instructions. Briefly, tissue samples were homogenized in ice-cold PBS/0.5% NP-40 with a mini Potter–Elvehjem pestle ((Sigma-Aldrich, St. Louis, MO, USA).), centrifuged for cell debris removal, and deproteinized to remove enzymes that could potentially metabolize glutathione (Deproteinizing Sample Preparation Kit—TCA, Abcam, Boston, MA, USA).

2.4. Determination of Microglia Immunoreactivity

The other cerebral hemisphere was used for immunofluorescence against the microglial marker ionized calcium-binding adaptor molecule 1 (Iba-1, 019-19741, Wako Chemicals, Richmond, VA, USA, 1:400 dilution) in coronal cryosections of the hippocampus (30 μm thick) as previously reported [43]. Nuclei were labeled with 4,6 diamino-2-phenylindol (DAPI), 0.02 M; 0.0125 mg/mL for nuclear labelling (Invitrogen, Carlsbad, CA, USA). Microphotographs were taken from the stratum radiatum of the hippocampus using a confocal microscope (Zeiss, LMS700, Oberkochen, Germany). The area analyzed for each stack was 0.04 mm^2, and the thickness (z-axis) was measured for each case. The density of Iba-1-positive microglial cells was assessed using FIJI Image J analysis software (https://imagej.net/ij/, accessed on 6 September 2023) as previously reported [43].

2.5. Statistical Analyses

Statistical analyses were performed using GraphPad Prism 8. Data are expressed as means ± SEM. One-way or two-way ANOVA followed by Tukey's post hoc analysis was used. A level of $p < 0.05$ was considered for statistical significance.

3. Results

3.1. Effect of Fenofibrate on Relapse-like Alcohol Consumption

Over a span of 60 days, voluntary alcohol consumption was tracked within the four experimental groups. As shown in Figure 1 (left side), alcohol consumption increased from 7.6 ± 1.5 g/kg/day on day 1 (10% ethanol solution) until stabilizing at 13.86 ± 1.19 g/kg/day between days 47 and 60 (10% plus 20% ethanol solutions). Of this cumulative ethanol consumption, an average of 65% was attributed to the 10% ethanol solution bottle, while the remaining 35% was derived from the 20% ethanol solution bottle (data not presented in Figure 1 left). No significant differences were detected among the four groups (ANOVA $F_{(3,1586)} = 1.023$; $p = 0.3816$).

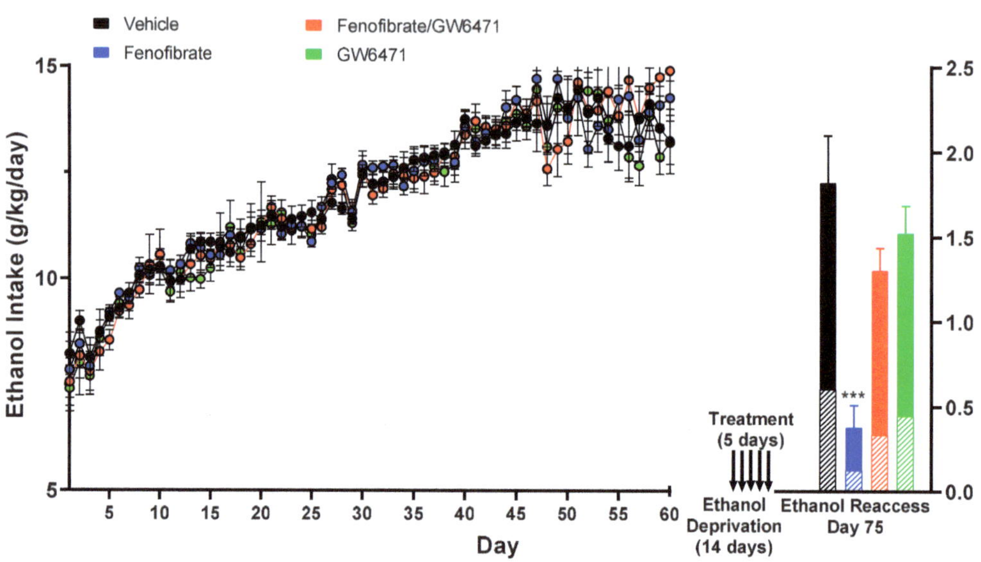

Figure 1. Fenofibrate administration during the ethanol withdrawal stage reduces alcohol relapse-like drinking. Following 46 days of chronic 10% (*v*/*v*) ethanol intake and then concurrent three-bottle choice access to 10% and 20% (*v*/*v*) ethanol solutions and water for 14 additional days, rats were deprived of ethanol for 14 days. In the last 5 days of deprivation, 4 groups were, respectively, treated with a daily dose of: (i) vehicle control (DMSO + water); (ii) fenofibrate 50 mg/kg/day + DMSO; (iii) fenofibrate 50 mg/kg/day plus GW6471 1 mg/kg/day; and (iv) GW6471 1 mg/kg/day + water. The day after finishing the 14 days of abstinence (which includes the 5 days of drug treatment), the animals were offered the 10% and 20% (*v*/*v*) ethanol solutions again for just 1 h, and the consumed volume was recorded. Hatched pattern in bars represents the amount of 20% ethanol ingested, while the filled pattern represents the amount of 10% ethanol; the total height of the bars and the standard errors shown correspond to the total amounts of alcohol consumed. Data are presented as mean ± SEM, n = 7–8 rats per experimental group. *** = $p < 0.001$ vs. Vehicle and Fenofibrate/GW6471 groups. ANOVA followed by Tukey's test for multiple comparisons.

Following this period, the animals were deprived of alcohol for 14 days. In the last five days of the withdrawal period, the four groups were, respectively, treated with: (i) vehicles (DMSO i.p. and water gavage); (ii) fenofibrate 50 mg/kg/day per gavage + DMSO,

(iii) fenofibrate 50 mg/kg/day by gavage plus GW6471 1 mg/kg/day i.p. (a PPAR-α specific antagonist); (iv) GW6471 1 mg/kg/day i.p. + water. GW6471 is a specific PPAR-α antagonist, with capacity to counteract the protective effects of PPAR-α on neuroinflammation [44]. As shown in Figure 1 (right side), the administration of fenofibrate in the withdrawal stage produced an 80% decrease in total voluntary alcohol consumption during the sole hour of re-access to alcohol. This reduction was markedly different from the control group, which solely received the vehicles (0.38 ± 0.13 g/kg/1 h vs. 1.82 ± 0.28 g/kg/1 h; ANOVA $F_{(3,26)} = 15.47$; $p < 0.001$). When fenofibrate was administered simultaneously with GW6471, the decrease in alcohol consumption during relapse was abrogated, since no significant differences were observed when compared to the control group that solely received the vehicles (1.31 ± 0.13 g/kg/1 h vs. 1.82 ± 0.28 g/kg/1 h; ANOVA $p = 0.2030$). Similarly, the administration of GW6471 alone also yielded no significant effect on alcohol consumption when compared to the control group (1.53 ± 0.16 g/kg/1 h vs. 1.82 ± 0.28 g/kg/1 h; ANOVA $p = 0.6788$). These findings provide strong evidence that the effect of fenofibrate on alcohol consumption during relapse is specifically due to PPAR-α activation.

Regardless of the total amount of alcohol ingested among the four groups, the ratio between 10% ethanol and 20% ethanol consumption during relapse remained statistically unchanged from that observed during the chronic consumption stage (vehicle group: 67%/33%; fenofibrate group: 66%/34%, fenofibrate+GW6471 group: 70%/30%, GW6471 group: 70%/30%), indicating that the different treatments did not change the preference for ethanol concentration in the ingested solution.

3.2. Effect of Fenofibrate on Ethanol-Induced Expression of Proinflammatory Cytokines and an Oxidative Stress Marker

In a previous study, we had reported the effectiveness of fenofibrate in reversing ethanol-induced increase in GFAP expression, and NF-κB activation in the hippocampus, prefrontal cortex, and the hypothalamus [37]. These findings align with fenofibrate´s propriety in inhibiting neuroinflammation in contexts beyond alcohol exposure, including aging, ischemia/reperfusion injury, and traumatic brain injury [32–36]. In our current investigation, we explored the ability of fenofibrate to counteract an increase in proinflammatory cytokines and a marker of oxidative stress, both induced by alcohol consumption. To quantify these effects, the expression of three well-known proinflammatory cytokines (*TNF-α*, *IL1-β*, and *IL-6*) as well as the expression of heme oxygenase-1 (*HO-1*), a gene that is induced in response to oxidative stress [44] was quantified by RT-qPCR in the hippocampus, nucleus accumbens (NAcc), and prefrontal cortex (PFC).

As shown in Figure 2a–c, fenofibrate was able to decrease mRNA *TNF-α* expression in the hippocampus (fenofibrate vs. vehicle: 26.7 ± 1.5% vs. 100.0 ± 17.8%; ANOVA $F_{(3,26)} = 28.0$; $p < 0.01$), NAcc (6.2 ± 0.9% vs. 100.0 ± 23.7%; ANOVA $F_{(3,26)} = 80.2$; $p < 0.01$), and PFC (16.3 ± 0.3% vs. 100.0 ± 23.7%; ANOVA $F_{(3,26)} = 12.0$; $p < 0.01$). Fenofibrate was also able to decrease the mRNA expression of *IL-β* in the hippocampus (fenofibrate vs. vehicle: 30.3 ± 3.5% vs. 100.0 ± 11.5%; ANOVA $F_{(3,26)} = 32.9$; $p < 0.01$), NAcc (7.8 ± 1.3% vs. 100.0 ± 25.1%; ANOVA $F_{(3,26)} = 53.0$; $p < 0.01$), and PFC (40.2 ± 7.8% vs. 100.0 ± 18.9%; ANOVA $F_{(3,26)} = 11.0$; $p < 0.05$). Similar results were observed regarding the mRNA expression of *IL-6*, since fenofibrate showed effects in the hippocampus (fenofibrate vs. vehicle: 16.8 ± 1.1% vs. 100.0 ± 13.2%; ANOVA $F_{(3,26)} = 46.2$; $p < 0.01$) and NAcc (0.0 ± 0.0% vs. 100.0 ± 33.1%; ANOVA $F_{(3,26)} = 12.0$; $p < 0.01$). In the case of PFC, although there was a decrease, it was, albeit marginally, not statistically significant (63.6 ± 15.0% vs. 100.0 ± 17.3%; ANOVA $F_{(3,26)} = 2.6$; $p = 0.07$). Regarding the oxidative stress marker gene *HO-1*, the administration of fenofibrate showed marked effects in reducing its expression in the hippocampus (fenofibrate vs. vehicle: 27.7 ± 1.3% vs. 100.0 ± 9.4%; ANOVA $F_{(3,26)} = 18.0$; $p < 0.01$), NAcc (11.0 ± 0.6% vs. 100.0 ± 19.2%; ANOVA $F_{(3,26)} = 43.7$; $p < 0.01$), and PFC (42.8 ± 3.2% vs. 100.0 ± 11.7%; ANOVA $F_{(3,26)} = 9.0$; $p < 0.05$), which would indicate its ability to reduce not only neuroinflammation but also ethanol-induced oxidative stress in these brain areas.

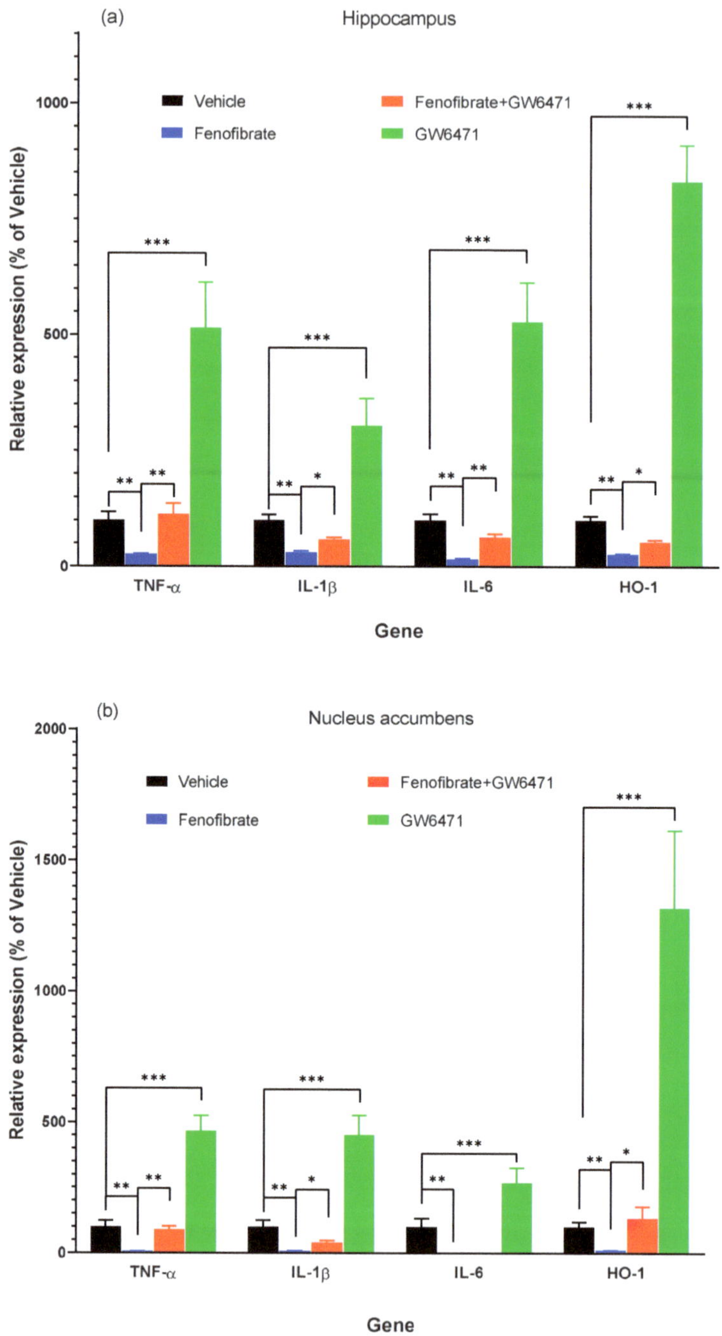

Figure 2. *Cont.*

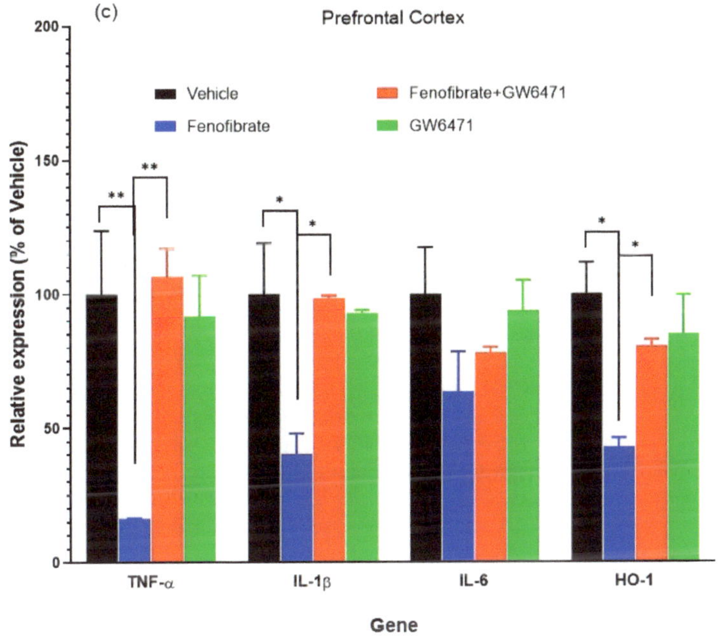

Figure 2. Fenofibrate administration during the ethanol withdrawal stage reduces the expression of proinflammatory cytokines and the marker of oxidative stress. Following the experiment shown in Figure 1, the expression *of TNF-α, IL-1β, IL-6*, and *HO-1* in the hippocampus (**a**), NAcc (**b**), and PFC (**c**) was analyzed by RT-qPCR in the 4 groups of animals, treated, respectively, with: (i) vehicle control (DMSO + water); (ii) fenofibrate 50 mg/kg/day + DMSO; (iii) fenofibrate 50 mg/kg/day plus GW6471 1 mg/kg/day; and (iv) GW6471 1 mg/kg/day + water. The graphs show the levels of gene expression as percentages of their vehicle-administered controls, normalized by the levels of expression of *ß-actin*. Data are presented as mean ± SEM, n = 7–8 rats per experimental group. * = $p < 0.05$, ** = $p < 0.01$, *** = $p < 0.001$ between the indicated groups. ANOVA followed by Tukey's test for multiple comparisons.

The co-administration of the antagonist GW6471 and fenofibrate abrogated the effect of the latter as in alcohol consumption during relapse (Figure 1), further reinforcing the notion that the reduced expression of proinflammatory cytokines and the oxidative stress marker is explicitly attributed to PPAR-α activation (Figure 2). [Fenofibrate + GW6471 vs. fenofibrate, *TNF-α*: hippocampus 113.0 ± 22.4% vs. 26.7 ± 1.5% ANOVA $F_{(3,26)} = 28.0$; $p < 0.01$; NAcc 88.7 ± 13.2% vs. 6.2 ± 0.9% ANOVA $F_{(3,26)} = 80.2$; $p < 0.01$; PFC 106.5 ± 10.5% vs. 16.3 ± 0.3% ANOVA $F_{(3,26)} = 12.0$; $p < 0.01$; *IL-1β*: hippocampus 58.5 ± 4.7% vs. 30.3 ± 3.5% ANOVA $F_{(3,26)} = 32.9$; $p < 0.05$; NAcc (40.9 ± 7.4% vs. 7.8 ± 1.3%; ANOVA $F_{(3,26)} = 53.0$; $p < 0.05$); PFC 98.3 ± 1.0% vs. 40.2 ± 7.8%; ANOVA $F_{(3,26)} = 11.0$; $p < 0.05$; *IL-6*: hippocampus 63.2 ± 7.2% vs. 16.8 ± 1.1% ANOVA $F_{(3,26)} = 46.2$; $p < 0.01$; NAcc 0.0 ± 0.0% vs. 0.0 ± 0.0% ANOVA $F_{(3,26)} = 12.0$, n.s.; PFC 77.9 ± 2.1% vs. 63.6 ± 15.0% ANOVA $F_{(3,26)} = 2.6$; $p = 0.07$; *HO-1*: hippocampus 53.7 ± 4.4% vs. 27.7 ± 1.3%; ANOVA $F_{(3,26)} = 18.0$; $p < 0.05$; NAcc (132.8 ± 45.9% vs. 11.0 ± 0.6%; ANOVA $F_{(3,26)} = 43.7$; $p < 0.05$; PFC 80.3 ± 2.5% vs. 42.8 ± 3.2% ANOVA $F_{(3,26)} = 9.0$; $p < 0.05$].

Notably, in the group administered only with the antagonist GW6471, a marked increase in all proinflammatory cytokines evaluated and in the marker of oxidative stress was detected, approximately between 3- and 13-fold in the hippocampus and NAcc (Figure 2) [GW6471 vs. vehicle, *TNF-α*: hippocampus 515.3 ± 98.1% vs. 100.0 ± 17.8% ANOVA $F_{(3,26)} = 28.0$; $p < 0.001$; NAcc 466.7 ± 57.9% vs. 100.0 ± 23.7% ANOVA $F_{(3,26)} = 80.1$;

$p < 0.001$; *IL-1β*: hippocampus 304.5 ± 58.6% vs. 100.0 ± 11.5%; ANOVA $F_{(3,26)} = 32.9$; $p < 0.001$; NAcc (452.9 ± 73.7% vs. 100.0 ± 25.1%; ANOVA $F_{(3,26)} = 53.0$; $p < 0.001$); *IL-6*: hippocampus 528.0 ± 85.5% vs. 100.0 ± 13.2%; ANOVA $F_{(3,26)} = 46.2$; $p < 0.01$; NAcc 267.2 ± 57.4% vs. 100.0 ± 33.1%; ANOVA $F_{(3,26)} = 12.0$; $p < 0.001$; *HO-1*: hippocampus 831.0 ± 79.3% vs. 100.0 ± 9.4%; ANOVA $F_{(3,26)} = 18.0$; $p < 0.001$; NAcc (1318.7 ± 296.5% vs. 100.0 ± 19.2%; ANOVA $F_{(3,26)} = 43.7$; $p < 0.001$)].

3.3. Effect of Fenofibrate on the Levels of the Antioxidant Glutathione

Figure 3 shows that the administration of fenofibrate decreased the oxidized glutathione/reduced the glutathione ratio (GSSG/GSH) in the hippocampus (fenofibrate vs. vehicle: 81.4 ± 2.5% vs. 100.0 ± 3.1%, ANOVA $F_{(3,26)} = 41.3$; $p < 0.01$). Like what was observed with the expression of proinflammatory cytokines and HO-1 (Figure 2), co-administration of the antagonist GW6471 decreased the protective effect of fenofibrate against oxidative stress (fenofibrate+GW6471 vs. fenofibrate: 130.6 ± 4.2% vs. 100.0 ± 3.1%, ANOVA $F_{(3,26)} = 41.3$; $p < 0.001$). Furthermore, administration of GW6471 alone produced per se an increase in oxidative stress in the hippocampus (GW6471 vs. vehicle: 118.4 ± 3.7% vs. 100.0 ± 3.1% ANOVA $F_{(3,26)} = 41.3$; $p < 0.01$). These findings indicate that fenofibrate can reduce not only neuroinflammation, but also oxidative stress levels in this brain area, which is consistent with the decrease in HO-1 expression levels (Figure 2a), together with the decrease in the relapse-like alcohol intake (Figure 1).

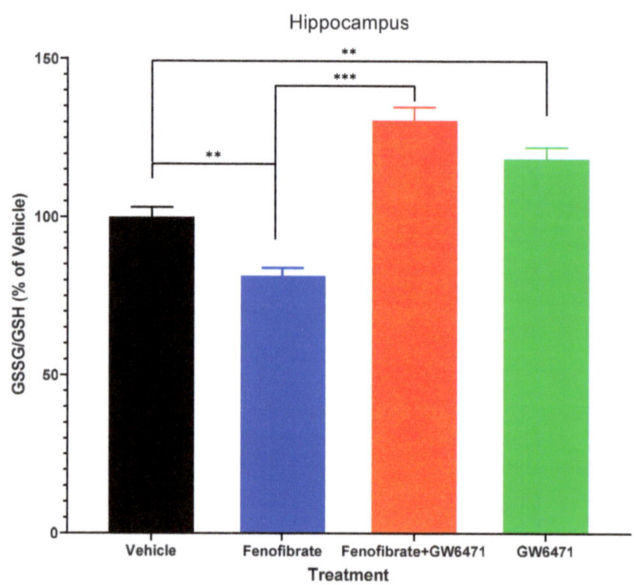

Figure 3. The administration of fenofibrate reduces the oxidative stress induced by ethanol. Oxidized glutathione (GSSG) and reduced glutathione (GSH) levels were determined in the hippocampus of the 4 groups of treated animals. The graph shows the GSSG/GSH ratios as percentages of their vehicle-administered controls. Data are presented as mean ± SEM, n = 7–8 rats per experimental group. ** = $p < 0.01$ and *** = $p < 0.001$ between the indicated groups. ANOVA followed by Tukey's test as a post hoc.

Similar to what was observed with the expression of proinflammatory cytokines and with the oxidative stress-induced gene (HO-1) in the hippocampus and NAcc (Figure 2), the administration of the antagonist GW6471, either in conjunction with fenofibrate or alone, produced a 30.6% and 18.4% increase in oxidative stress levels in the hippocampus, respectively (fenofibrate+GW6471 vs. vehicle: 130.6 ± 4.2% vs. 100.0 ± 3.1%, GW6471 vs. vehicle: 118.4 ± 3.6% vs. 100.0 ± 3.1%, ANOVA $F_{(3,26)} = 41.3$; $p < 0.01$).

3.4. Effect of Fenofibrate on Ethanol-Induced Microglial Reactivity

Regarding microglia reactivity, while not reaching statistical significance, fenofibrate showed a trend towards doubling the number of cells displaying an anti-inflammatory morphology (M2) in comparison to the vehicle group (Figure 4) (fenofibrate vs. vehicle: 1633 ± 375 vs. 823 ± 234, ANOVA $F_{(3,86)} = 1.76$; $p = 0.16$). In addition, the co-administration of the antagonist GW6471 together with fenofibrate attenuated the effect of the latter on the increase in the number of M2 microglial cells (fenofibrate+GW6471 vs. fenofibrate: 1266 ± 235 vs. 1633 ± 375, ANOVA $F_{(3,86)} = 1.76$; $p = 0.16$).

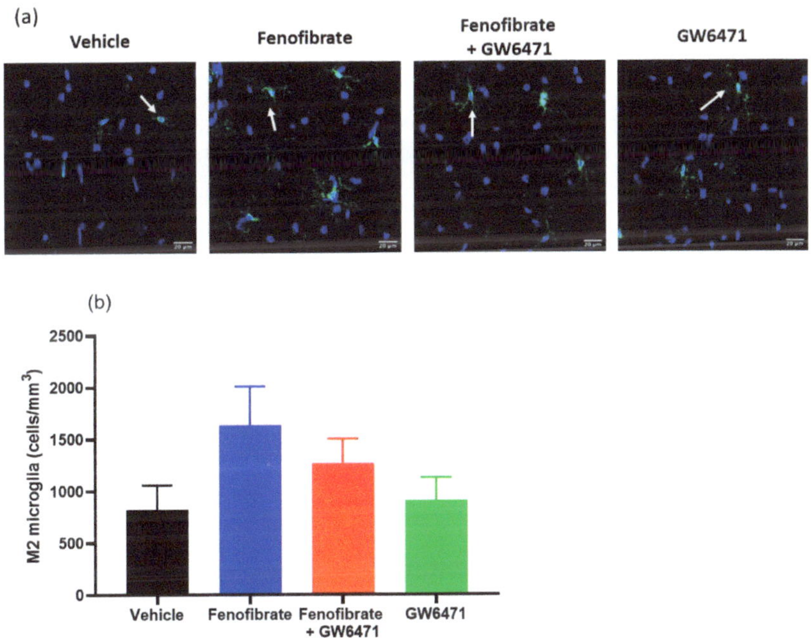

Figure 4. Effect of fenofibrate administration on microglial reactivity. The cells with the M2 (anti-inflammatory) phenotype were quantified in the hippocampus of the 4 groups of treated animals. (**a**) Representative microphotograph of microglia immunofluorescence (IBA-1 immunoreactivity: green; DAPI: blue) from the four experimental groups. Scale bar 20 µm. The arrows show an example of the morphology of the cells that were counted. (**b**) Densitometric analysis from Iba-1-positive cells/mm^3 present in 3 hippocampal slices from each animal. Data are presented as mean ± SEM, n = 7–8 rats per experimental group. ANOVA followed by Tukey's test as a post hoc.

We also quantified phagocytic microglial cells in the hippocampus. Phagocytic microglia are characterized by rounded macrophage-like morphology with no or few processes and are associated with maximal proinflammatory activation and oxidative-free radicals [11]. The group of animals treated with fenofibrate showed a decrease of 30.4% with respect to the control group, although the difference was statistically significant only when compared with the fenofibrate+GW6471 group (ANOVA $F_{(3,86)} = 2.66$ $p < 0.05$) (Figure 5).

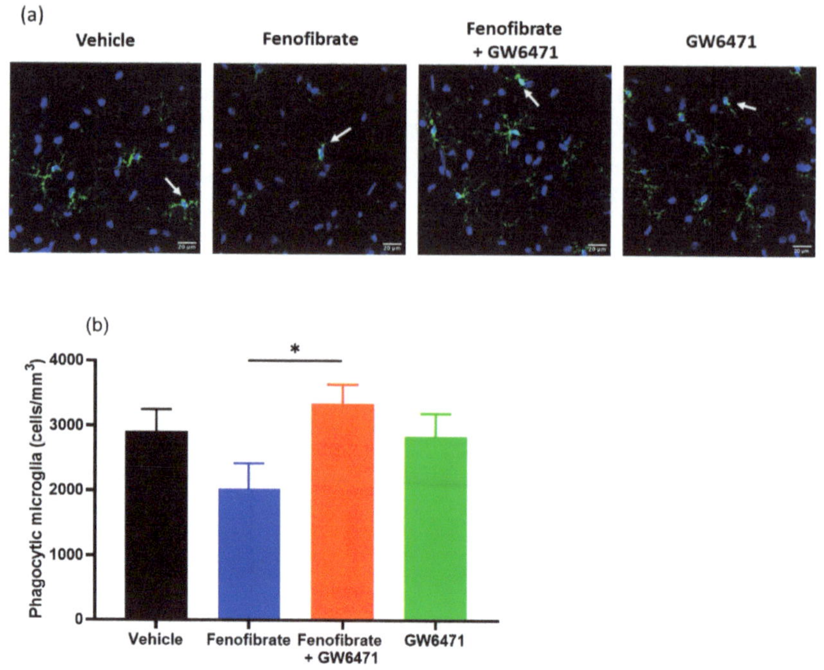

Figure 5. Effect of fenofibrate administration on microglial reactivity. The microglial cells with the phagocytic phenotype were quantified in the hippocampus of the 4 groups of treated animals. (**a**) Microglia immunofluorescence (IBA-1 immunoreactivity: green; DAPI: blue). Scale bar 20 μm. The arrows show examples of the morphology of the microglial cells in phagocytic process. (**b**) Densitometric analysis of microglial cells in the phagocytic process (phagocytic pouches). Data are presented as mean ± SEM, n = 7–8 rats per experimental group. * = $p < 0.05$, between the indicated groups. ANOVA followed by Tukey's test as a post hoc.

4. Discussion

Due to the marked effect of fenofibrate in decreasing post-withdrawal relapse-type alcohol consumption (Figure 1), it is reasonable to consider whether this could be attributed to effects that extend beyond the inhibition of the drinking reflex, e.g., producing memory and/or spatial orientation disorders, mobility impairment, sedation, anxiety, depression, etc. Although in this work we did not carry out behavioral or motor studies, there are studies by other authors that have shown that fenofibrate decreases the motivation of animals to obtain ethanol (operant self-administration) but not to self-administer sucrose [30], which could indicate that there are no effects on memory or spatial orientation, or other effects such as depression (depressed animals decrease their sucrose intake) or sedation. In another study, Blednov et al. [29] reported that fenofibrate did not alter preference for saccharin, nor motor response to novelty, reduced duration of ethanol-induced loss of righting reflex, and increased EtOH-induced conditioned taste aversion. That is, fenofibrate per se does not produce motor effects that could explain its effect in reducing alcohol consumption.

In addition to the decrease in relapse-type alcohol consumption, fenofibrate showed effects in a reduction in the expression of proinflammatory cytokines and of a gene that is induced by oxidative stress (*HO-1*), in the brain (Figure 2). Although the effects of PPAR-α in mitigating ethanol-induced brain oxidative stress had not previously been explored, its neuroprotective capabilities have been reported in distinct models [45], including traumatic brain injury [31], transient cerebral ischemia/reperfusion [34], and Alzheimer's disease [46]. Collino et al. [34] reported that the administration of a synthetic PPAR-α agonist (WY14643) reduced *HO-1* expression induced by ischemia/reperfusion brain injury. Additionally, oleoylethanolamide (a physiological PPAR-α agonist) administration in a model of chemically induced degeneration of substantia nigra dopamine neurons, led to a decrease in the number of HO-1 immunoreactive cells in the striatum when compared to untreated animals [47]. Nevertheless, there are also several reports where the expression of *HO-1* increases due to treatment with anti-inflammatory agents. It has been found that administration of the antioxidant N-acetylcysteine to UChB rats with chronic alcohol consumption was linked to increased *HO-1* expression [48]. The discordant findings concerning the decrease or increase in *HO-1* expression following particular anti-inflammatory treatments could be attributed to various factors, including the animal model employed, the nature of the anti-inflammatory drug, the route and duration of administration, and the timing of *HO-1* expression evaluation post-treatment. Specifically, in the UChB rat model, the distinction between the study conducted by Quintanilla et al. [48] and our findings lies in the fact that N-acetylcysteine was administered over a 14-day span during the period of alcohol consumption. In contrast, fenofibrate was administered for only 5 days during the withdrawal period. Nonetheless, since the induction of *HO-1* is mediated by ROS production [44] it remains unclear whether the reduction in *HO-1* expression is a direct outcome of fenofibrate or a consequence of diminished oxidative stress levels in rats to which fenofibrate was administered after chronic alcohol consumption.

Remarkably, the group that received only the antagonist GW6471 exhibited a considerable rise in all assessed proinflammatory cytokines, along with an elevated expression of *HO-1*, demonstrating an increase ranging from approximately 3- to 13-fold in the hippocampus and NAcc (Figure 2). The substantial increases induced by the administration of GW6471 alone (Figure 2) do not correlate with an increase in alcohol consumption of this group during the re-access stage (Figure 1). This discrepancy likely arises because the amount of alcohol that UChB rats can drink in 1 h has already reached its ceiling of ~1.5–1.8 g/kg/1 h, as evidenced in other reports from our group [43]. PPAR-α is not only activated by drugs of the fibrate family; it also responds to endogenous agonists such as palmitoylethanolamide [49] and oleoylethanolamide [50]. Furthermore, the anti-inflammatory and antioxidant actions of these agonists in the brain have been demonstrated [49,51,52]. In a similar way to fibrates, it has been reported that these endogenous agonists exert effects in reducing voluntary alcohol consumption in animals [53,54]. Thus, it is possible that the administration of GW6471 alone produced a degree of PPAR-α inhibition that prevented its activation by endogenous agonists, consequently inhibiting its anti-inflammatory and antioxidant functions against the deleterious effects of ethanol. The minor response within the PFC in comparison to the hippocampus and NAcc could be attributed to the fact that PFC has been observed to display a comparative lower response to alcohol administration [37]. In agreement, Kane et al. [55] reported that ethanol administration fails to increase the expression of pro-inflammatory cytokines like chemokine C-C motif ligand 2 (CCL2), IL-6, and TNF-α in the mouse cerebral cortex.

Additionally, fenofibrate showed effects in reducing oxidative stress in the hippocampus, measured as the GSSG/GSH ratio. In several reports, the reduction in relapse-like alcohol intake has been related to a decrease in cerebral oxidative stress through the administration of N-acetylcysteine (a precursor for the cellular synthesis of glutathione) [48], ibudilast (an anti-inflammatory phosphodiesterase inhibitor) [56], and mesenchymal stem cells and their secreted products known for potent anti-inflammatory actions [43]. In our study, it was found that the administration of fenofibrate produced approximately 20% reduction in the GSSG/GSH ratio. Quintanilla et al. reported a decrease of 70% in the GSSG/GSH ratio by N-acetylcysteine [48] and 45% due to the secretome derived from mesenchymal stem cells [43]. The significant reduction achieved by N-acetylcysteine may be attributed to its direct function as a precursor for the synthesis of glutathione, whereas the effect of fenofibrate seems to be indirect. Regarding the difference obtained for secretome derived from mesenchymal stem cells, this was administered in 3 doses spread over 18 days of treatment [43], a longer treatment compared to fenofibrate (5 days). Moreover, the mechanisms underlying these effects may not necessarily be identical. In agreement with our results, in a model of oxidative stress induced by valproic acid, fenofibrate decreased 33% the levels of GSH [57]. The administration of the GW6471 antagonist, whether in combination with fenofibrate or alone, resulted in a rise in oxidative stress levels in the hippocampus (Figure 3). As previously discussed in relation to Figure 2, it is plausible to consider that the administration of GW6471 resulted in a substantial inhibition of PPAR-α, preventing its activation by endogenous agonists and consequently inhibiting its antioxidant activity against the detrimental effects of ethanol.

We also studied the properties of fenofibrate on the reactivity of microglia in the hippocampus, as a parameter to evaluate its protective effects against ethanol-induced neuroinflammation. We observed a tendency to increase the number of microglia with anti-inflammatory phenotype (M2) (Figure 4) and to reduce phagocytic microglial cells (Figure 5). In contrast to the effects observed on alcohol consumption during relapse (Figure 1), the expression of proinflammatory cytokines (Figure 2) and oxidative stress levels (Figure 3), the relatively short treatment period of five days with fenofibrate might not have been sufficient to elicit a more pronounced impact on the microglial cell phenotype. In future experiments, we plan to evaluate the effect of longer treatments with fenofibrate. According to this idea, Quintanilla et al. [48] and Ezquer et al. [43] were able to observe differences in microglia cells in the hippocampus of UChB rats that had consumed alcohol and were then treated with N-acetylcysteine or mesenchymal stem cell-derived secretome, respectively, by employing extended treatment periods.

One of the considerations that we must make as caveat of the current study is that we did not include groups of rats that had not drunk alcohol. In several publications of our group (please see references [43,48], as examples), we have reported that this scheme of alcohol administration and subsequent withdrawal produces a neuroinflammation response and oxidative stress in the brain of UChB drinking rats compared to alcohol-naïve rats. The objective of the current work was to demonstrate that, if fenofibrate was administered during the withdrawal stage to rats that had previously voluntarily drunk alcohol chronically, it would decrease voluntary alcohol consumption when it was offered again to the animals (model of relapse-type drinking), and that these effects are mediated by PPAR-α activation. Thus, it would have been somewhat difficult to include in this scheme a group of alcohol-naïve animals, since we would not have been able to assess relapse in animals that had never drunk. Regarding the effect of fenofibrate on markers of neuroinflammation and oxidative stress when administered to naïve animals, there are some reports in the literature. For example, Barbiero et al. [58] reported that fenofibrate per se does not produce changes in oxidative stress (GSH levels, superoxide dismutase, lipid peroxidation) in the striatum of rats. In other studies, the administration of fenofibrate to control rats did not produce changes in the expression of TNF-α, IL-1β, or IL-6 in the spinal cords of mice [59] nor in inflammatory markers such as prostaglandin D2, thromboxane, arachidonic acid, cyclooxygenase-2, and TNF-α in rat primary astrocyte cultures [60]. For their part, Mirza and Sharma [57] reported that fenofibrate did not change the levels of GSH, IL-6, and TNF-α in the cerebellum, brainstem, and prefrontal cortex of rats. Based on this background, we believe that the administration of fenofibrate per se should not produce alterations in the proinflammatory markers when administered to alcohol-naïve animals.

Another limitation that we must consider is regarding the use of only female animals in this study. As we have stated above, UChB females, as well as females from other lines selected to drink alcohol, present a higher consumption compared to males. However, we cannot rule out that the response to fenofibrate could have different characteristics between male and female UChB rats. In this sense, Blednov et al. [61] reported that male and female mice that drank alcohol respond in different ways to treatment with PPAR agonists.

We are also aware that in these studies we do not include naive animals that have not been administered DMSO. In a recent systematic review, Dludla et al. [62] found that DMSO possesses essential antioxidant properties that are linked to its protective effect against oxidative damage. Interestingly, this antioxidant effect appears to be dose-dependent in vivo, since lower doses (1–3 g/kg) were the most effective in blocking oxidative stress-induced damage, mostly through a reduction in ROS, inhibition of lipid peroxidation, improving mitochondrial function, enhancement of intracellular antioxidants, and suppression of pro-inflammatory makers, in the different studies analyzed. On the other hand, high doses of DMSO have been shown to be toxic and induce oxidative stress and cellular damage [62]. In our studies, we administered a DMSO dose of 1 mL/kg—which is equivalent to 1.1 g/kg—that is, in the range of the lowest doses used in the literature. The possibility exists that the anti-inflammatory effects of DMSO might have overshadowed the anti-inflammatory and anti-oxidative stress effects of fenofibrate; however, we do not believe this would have been the case, as only animals receiving fenofibrate showed effects on downregulation of proinflammatory and oxidative stress cytokine genes, as well as a decreased GSSG/GSH ratio, while the four groups of rats indeed did receive DMSO.

5. Conclusions

Overall, our studies show that the administration of fenofibrate to rats during the withdrawal stage after chronic ethanol consumption decreases the severity of relapse when ethanol is offered again to the animals. This beneficial effect apparently is due to the capacity of fenofibrate to reduce ethanol-induced neuroinflammation. This was evident in the downregulation of proinflammatory cytokines and oxidative stress-induced genes in the brain, decreased oxidative stress (GSSG/GSH ratio), the decreased number of phagocytic microglial cells, and a trend to increase the number of anti-inflammatory microglia. These protective effects collectively could contribute to a reduction in ethanol craving, which is the main cause of relapse in patients undergoing detoxification. Our findings suggest that incorporating fenofibrate into the post-detoxification withdrawal phase could serve as a promising therapeutic approach to prevent or alleviate the severity of relapse. The ultimate goal of our research is to identify a novel and effective pharmacological treatment for AUD, based on a drug already approved for other clinical conditions. In this context, fenofibrate has been clinically used worldwide for decades, having received approval in Europe in the 1980s, and in USA since 1994 for the treating for dyslipidemia.

Author Contributions: Conceptualization, E.K. and P.M.; methodology, C.I., T.A. and D.P.-R.; writing—original draft preparation, E.K.; writing—review and editing, P.M. and M.E.Q.; supervision of animals, M.E.Q. All authors have read and agreed to the published version of the manuscript.

Funding: This research was funded by Agencia Nacional de Investigación y Desarrollo (ANID, Chile), grant number ANILLO ANID/ACT210012.

Institutional Review Board Statement: The animal study protocol was approved by the Bioethics Committee on Animal Research, Faculty of Medicine, Universidad de Chile (Protocol 22543-MED-UCH).

Data Availability Statement: We have no additional data available to share.

Conflicts of Interest: The authors declare no conflict of interest.

References

1. Maisel, N.C.; Blodgett, J.C.; Wilbourne, P.L.; Humphreys, K.; Finney, J.W. Meta-Analysis of Naltrexone and Acamprosate for Treating Alcohol Use Disorders: When Are These Medications Most Helpful? *Addiction* **2013**, *108*, 275–293. [CrossRef]
2. Plosker, G.L. Acamprosate: A Review of Its Use in Alcohol Dependence. *Drugs* **2015**, *75*, 1255–1268. [CrossRef]
3. Holleck, J.L.; Merchant, N.; Gunderson, C.G. Symptom-Triggered Therapy for Alcohol Withdrawal Syndrome: A Systematic Review and Meta-Analysis of Randomized Controlled Trials. *J. Gen. Intern. Med.* **2019**, *34*, 1018–1024. [CrossRef]
4. Coller, J.K.; Hutchinson, M.R. Implications of Central Immune Signaling Caused by Drugs of Abuse: Mechanisms, Mediators and New Therapeutic Approaches for Prediction and Treatment of Drug Dependence. *Pharmacol. Ther.* **2012**, *134*, 219–245. [CrossRef]
5. Flores-Bastías, O.; Karahanian, E. Neuroinflammation Produced by Heavy Alcohol Intake Is Due to Loops of Interactions between Toll-like 4 and TNF Receptors, Peroxisome Proliferator-Activated Receptors and the Central Melanocortin System: A Novel Hypothesis and New Therapeutic Avenues. *Neuropharmacology* **2018**, *128*, 401–407. [CrossRef]
6. Crews, F.T.; Zou, J.; Qin, L. Induction of Innate Immune Genes in Brain Create the Neurobiology of Addiction. *Brain. Behav. Immun.* **2011**, *25* (Suppl. S1), S4–S12. [CrossRef]

7. Cao, Q.; Mak, K.M.; Lieber, C.S. Cytochrome P4502E1 Primes Macrophages to Increase TNF-Alpha Production in Response to Lipopolysaccharide. *Am. J. Physiol.-Gastrointest. Liver Physiol.* **2005**, *289*, G95–G107. [CrossRef]
8. Chandel, N.S.; Trzyna, W.C.; McClintock, D.S.; Schumacker, P.T. Role of Oxidants in NF-Kappa B Activation and TNF-Alpha Gene Transcription Induced by Hypoxia and Endotoxin. *J. Immunol.* **2000**, *165*, 1013–1021. [CrossRef]
9. Qin, L.; He, J.; Hanes, R.N.; Pluzarev, O.; Hong, J.-S.; Crews, F.T. Increased Systemic and Brain Cytokine Production and Neuroinflammation by Endotoxin Following Ethanol Treatment. *J. Neuroinflamm.* **2008**, *5*, 10. [CrossRef]
10. Ferrier, L.; Bérard, F.; Debrauwer, L.; Chabo, C.; Langella, P.; Buéno, L.; Fioramonti, J. Impairment of the Intestinal Barrier by Ethanol Involves Enteric Microflora and Mast Cell Activation in Rodents. *Am. J. Pathol.* **2006**, *168*, 1148–1154. [CrossRef] [PubMed]
11. Crews, F.T.; Vetreno, R.P. Mechanisms of Neuroimmune Gene Induction in Alcoholism. *Psychopharmacology* **2016**, *233*, 1543–1557. [CrossRef] [PubMed]
12. Qin, L.; Wu, X.; Block, M.L.; Liu, Y.; Breese, G.R.; Hong, J.-S.; Knapp, D.J.; Crews, F.T. Systemic LPS Causes Chronic Neuroinflammation and Progressive Neurodegeneration. *Glia* **2007**, *55*, 453–462. [CrossRef] [PubMed]
13. Crews, F.T.; Sarkar, D.K.; Qin, L.; Zou, J.; Boyadjieva, N.; Vetreno, R.P. Neuroimmune Function and the Consequences of Alcohol Exposure. *Alcohol Res.* **2015**, *37*, 331–341,344–351.
14. Clapp, P.; Bhave, S.V.; Hoffman, P.L. How Adaptation of the Brain to Alcohol Leads to Dependence: A Pharmacological Perspective. *Alcohol Res. Health J. Natl. Inst. Alcohol Abus. Alcohol.* **2008**, *31*, 310–339.
15. Koob, G.F. A Role for GABA Mechanisms in the Motivational Effects of Alcohol. *Biochem. Pharmacol.* **2004**, *68*, 1515–1525. [CrossRef]
16. Hou, R.; Baldwin, D.S. A Neuroimmunological Perspective on Anxiety Disorders. *Hum. Psychopharmacol.* **2012**, *27*, 6–14. [CrossRef]
17. Simen, B.B.; Duman, C.H.; Simen, A.A.; Duman, R.S. TNFalpha Signaling in Depression and Anxiety: Behavioral Consequences of Individual Receptor Targeting. *Biol. Psychiatry* **2006**, *59*, 775–785. [CrossRef]
18. Koo, J.W.; Duman, R.S. Interleukin-1 Receptor Null Mutant Mice Show Decreased Anxiety-like Behavior and Enhanced Fear Memory. *Neurosci. Lett.* **2009**, *456*, 39–43. [CrossRef]
19. Reissner, K.J.; Kalivas, P.W. Using Glutamate Homeostasis as a Target for Treating Addictive Disorders. *Behav. Pharmacol.* **2010**, *21*, 514–522. [CrossRef]
20. Rao, P.S.S.; Bell, R.L.; Engleman, E.A.; Sari, Y. Targeting Glutamate Uptake to Treat Alcohol Use Disorders. *Front. Neurosci.* **2015**, *9*, 144. [CrossRef]
21. Dahchour, A.; De Witte, P. Taurine Blocks the Glutamate Increase in the Nucleus Accumbens Microdialysate of Ethanol-Dependent Rats. *Pharmacol. Biochem. Behav.* **2000**, *65*, 345–350. [CrossRef]
22. Scofield, M.D.; Heinsbroek, J.A.; Gipson, C.D.; Kupchik, Y.M.; Spencer, S.; Smith, A.C.W.; Roberts-Wolfe, D.; Kalivas, P.W. The Nucleus Accumbens: Mechanisms of Addiction across Drug Classes Reflect the Importance of Glutamate Homeostasis. *Pharmacol. Rev.* **2016**, *68*, 816–871. [CrossRef]
23. Berger, J.; Moller, D.E. The Mechanisms of Action of PPARs. *Annu. Rev. Med.* **2002**, *53*, 409–435. [CrossRef] [PubMed]
24. Schoonjans, K.; Staels, B.; Auwerx, J. Role of the Peroxisome Proliferator-Activated Receptor (PPAR) in Mediating the Effects of Fibrates and Fatty Acids on Gene Expression. *J. Lipid Res.* **1996**, *37*, 907–925. [CrossRef] [PubMed]
25. Cignarella, A.; Bellosta, S.; Corsini, A.; Bolego, C. Hypolipidemic Therapy for the Metabolic Syndrome. *Pharmacol. Res.* **2006**, *53*, 492–500. [CrossRef]
26. Karahanian, E.; Quintanilla, M.E.; Fernandez, K.; Israel, Y. Fenofibrate—A Lipid-Lowering Drug—Reduces Voluntary Alcohol Drinking in Rats. *Alcohol* **2014**, *48*, 665–670. [CrossRef]
27. Rivera-Meza, M.; Muñoz, D.; Jerez, E.; Quintanilla, M.E.; Salinas-Luypaert, C.; Fernandez, K.; Karahanian, E. Fenofibrate Administration Reduces Alcohol and Saccharin Intake in Rats: Possible Effects at Peripheral and Central Levels. *Front. Behav. Neurosci.* **2017**, *11*, 133. [CrossRef]
28. Blednov, Y.A.; Benavidez, J.M.; Black, M.; Ferguson, L.B.; Schoenhard, G.L.; Goate, A.M.; Edenberg, H.J.; Wetherill, L.; Hesselbrock, V.; Foroud, T.; et al. Peroxisome Proliferator-Activated Receptors α and γ Are Linked with Alcohol Consumption in Mice and Withdrawal and Dependence in Humans. *Alcohol. Clin. Exp. Res.* **2015**, *39*, 136–145. [CrossRef]
29. Blednov, Y.A.; Black, M.; Benavidez, J.M.; Stamatakis, E.E.; Harris, R.A. PPAR Agonists: II. Fenofibrate and Tesaglitazar Alter Behaviors Related to Voluntary Alcohol Consumption. *Alcohol. Clin. Exp. Res.* **2016**, *40*, 563–571. [CrossRef]
30. Haile, C.N.; Kosten, T.A. The Peroxisome Proliferator-Activated Receptor Alpha Agonist Fenofibrate Attenuates Alcohol Self-Administration in Rats. *Neuropharmacology* **2017**, *116*, 364–370. [CrossRef] [PubMed]
31. Barson, J.R.; Karatayev, O.; Chang, G.-Q.; Johnson, D.F.; Bocarsly, M.E.; Hoebel, B.G.; Leibowitz, S.F. Positive Relationship between Dietary Fat, Ethanol Intake, Triglycerides, and Hypothalamic Peptides: Counteraction by Lipid-Lowering Drugs. *Alcohol* **2009**, *43*, 433–441. [CrossRef]
32. Bordet, R.; Ouk, T.; Petrault, O.; Gelé, P.; Gautier, S.; Laprais, M.; Deplanque, D.; Duriez, P.; Staels, B.; Fruchart, J.C.; et al. PPAR: A New Pharmacological Target for Neuroprotection in Stroke and Neurodegenerative Diseases. *Biochem. Soc. Trans.* **2006**, *34 Pt 6*, 1341–1346. [CrossRef]

33. Chen, X.R.; Besson, V.C.; Palmier, B.; Garcia, Y.; Plotkine, M.; Marchand-Leroux, C. Neurological Recovery-Promoting, Anti-Inflammatory, and Anti-Oxidative Effects Afforded by Fenofibrate, a PPAR Alpha Agonist, in Traumatic Brain Injury. *J. Neurotrauma* **2007**, *24*, 1119–1131. [CrossRef]
34. Collino, M.; Aragno, M.; Mastrocola, R.; Benetti, E.; Gallicchio, M.; Dianzani, C.; Danni, O.; Thiemermann, C.; Fantozzi, R. Oxidative Stress and Inflammatory Response Evoked by Transient Cerebral Ischemia/Reperfusion: Effects of the PPAR-Alpha Agonist WY14643. *Free Radic. Biol. Med.* **2006**, *41*, 579–589. [CrossRef] [PubMed]
35. Poynter, M.E.; Daynes, R.A. Peroxisome Proliferator-Activated Receptor Alpha Activation Modulates Cellular Redox Status, Represses Nuclear Factor-KappaB Signaling, and Reduces Inflammatory Cytokine Production in Aging. *J. Biol. Chem.* **1998**, *273*, 32833–32841. [CrossRef] [PubMed]
36. Shehata, A.H.F.; Ahmed, A.-S.F.; Abdelrehim, A.B.; Heeba, G.H. The Impact of Single and Combined PPAR-α and PPAR-γ Activation on the Neurological Outcomes Following Cerebral Ischemia Reperfusion. *Life Sci.* **2020**, *252*, 117679. [CrossRef]
37. Villavicencio-Tejo, F.; Flores-Bastías, O.; Marambio-Ruiz, L.; Pérez-Reytor, D.; Karahanian, E. Fenofibrate (a PPAR-α Agonist) Administered during Ethanol Withdrawal Reverts Ethanol-Induced Astrogliosis and Restores the Levels of Glutamate Transporter in Ethanol-Administered Adolescent Rats. *Front. Pharmacol.* **2021**, *12*, 653175. [CrossRef] [PubMed]
38. Israel, Y.; Karahanian, E.; Ezquer, F.; Morales, P.; Ezquer, M.; Rivera-Meza, M.; Herrera-Marschitz, M.; Quintanilla, M.E. Acquisition, Maintenance and Relapse-Like Alcohol Drinking: Lessons from the UChB Rat Line. *Front. Behav. Neurosci.* **2017**, *11*, 57. [CrossRef]
39. Karahanian, E.; Rivera-Meza, M.; Tampier, L.; Quintanilla, M.E.; Herrera-Marschitz, M.; Israel, Y. Long-Term Inhibition of Ethanol Intake by the Administration of an Aldehyde Dehydrogenase-2 (ALDH2)-Coding Lentiviral Vector into the Ventral Tegmental Area of Rats. *Addict. Biol.* **2015**, *20*, 336–344. [CrossRef]
40. Dhaher, R.; McConnell, K.K.; Rodd, Z.A.; McBride, W.J.; Bell, R.L. Daily Patterns of Ethanol Drinking in Adolescent and Adult, Male and Female, High Alcohol Drinking (HAD) Replicate Lines of Rats. *Pharmacol. Biochem. Behav.* **2012**, *102*, 540–548. [CrossRef]
41. Loi, B.; Colombo, G.; Maccioni, P.; Carai, M.A.M.; Franconi, F.; Gessa, G.L. High Alcohol Intake in Female Sardinian Alcohol-Preferring Rats. *Alcohol* **2014**, *48*, 345–351. [CrossRef]
42. Gavzan, H.; Hashemi, F.; Babaei, J.; Sayyah, M. A Role for Peroxisome Proliferator-Activated Receptor α in Anticonvulsant Activity of Docosahexaenoic Acid against Seizures Induced by Pentylenetetrazole. *Neurosci. Lett.* **2018**, *681*, 83–86. [CrossRef] [PubMed]
43. Ezquer, F.; Quintanilla, M.E.; Morales, P.; Santapau, D.; Ezquer, M.; Kogan, M.J.; Salas-Huenuleo, E.; Herrera-Marschitz, M.; Israel, Y. Intranasal Delivery of Mesenchymal Stem Cell-Derived Exosomes Reduces Oxidative Stress and Markedly Inhibits Ethanol Consumption and Post-Deprivation Relapse Drinking. *Addict. Biol.* **2019**, *24*, 994–1007. [CrossRef] [PubMed]
44. Ryter, S.W.; Choi, A.M.K. Heme Oxygenase-1: Redox Regulation of a Stress Protein in Lung and Cell Culture Models. *Antioxid. Redox Signal.* **2005**, *7*, 80–91. [CrossRef] [PubMed]
45. Moreno, S.; Cerù, M.P. In Search for Novel Strategies towards Neuroprotection and Neuroregeneration: Is PPARα a Promising Therapeutic Target? *Neural Regen. Res.* **2015**, *10*, 1409–1412. [CrossRef]
46. Wójtowicz, S.; Strosznajder, A.K.; Jeżyna, M.; Strosznajder, J.B. The Novel Role of PPAR Alpha in the Brain: Promising Target in Therapy of Alzheimer's Disease and Other Neurodegenerative Disorders. *Neurochem. Res.* **2020**, *45*, 972–988. [CrossRef]
47. Galan-Rodriguez, B.; Suarez, J.; Gonzalez-Aparicio, R.; Bermudez-Silva, F.J.; Maldonado, R.; Robledo, P.; Rodriguez de Fonseca, F.; Fernandez-Espejo, E. Oleoylethanolamide Exerts Partial and Dose-Dependent Neuroprotection of Substantia Nigra Dopamine Neurons. *Neuropharmacology* **2009**, *56*, 653–664. [CrossRef]
48. Quintanilla, M.E.; Ezquer, F.; Morales, P.; Ezquer, M.; Olivares, B.; Santapau, D.; Herrera-Marschitz, M.; Israel, Y. N-Acetylcysteine and Acetylsalicylic Acid Inhibit Alcohol Consumption by Different Mechanisms: Combined Protection. *Front. Behav. Neurosci.* **2020**, *14*, 122. [CrossRef]
49. Lo Verme, J.; Fu, J.; Astarita, G.; La Rana, G.; Russo, R.; Calignano, A.; Piomelli, D. The Nuclear Receptor Peroxisome Proliferator-Activated Receptor-Alpha Mediates the Anti-Inflammatory Actions of Palmitoylethanolamide. *Mol. Pharmacol.* **2005**, *67*, 15–19. [CrossRef]
50. Fu, J.; Gaetani, S.; Oveisi, F.; Lo Verme, J.; Serrano, A.; Rodríguez De Fonseca, F.; Rosengarth, A.; Luecke, H.; Di Giacomo, B.; Tarzia, G.; et al. Oleylethanolamide Regulates Feeding and Body Weight through Activation of the Nuclear Receptor PPAR-Alpha. *Nature* **2003**, *425*, 90–93. [CrossRef]
51. Sayd, A.; Antón, M.; Alén, F.; Caso, J.R.; Pavón, J.; Leza, J.C.; Rodríguez de Fonseca, F.; García-Bueno, B.; Orio, L. Systemic Administration of Oleoylethanolamide Protects from Neuroinflammation and Anhedonia Induced by LPS in Rats. *Int. J. Neuropsychopharmacol.* **2014**, *18*, pyu111. [CrossRef] [PubMed]
52. Stahel, P.F.; Smith, W.R.; Bruchis, J.; Rabb, C.H. Peroxisome Proliferator-Activated Receptors: "Key" Regulators of Neuroinflammation after Traumatic Brain Injury. *PPAR Res.* **2008**, *2008*, 538141. [CrossRef] [PubMed]
53. Bilbao, A.; Serrano, A.; Cippitelli, A.; Pavón, F.J.; Giuffrida, A.; Suárez, J.; García-Marchena, N.; Baixeras, E.; Gómez de Heras, R.; Orio, L.; et al. Role of the Satiety Factor Oleoylethanolamide in Alcoholism. *Addict. Biol.* **2016**, *21*, 859–872. [CrossRef] [PubMed]
54. Orio, L.; Alen, F.; Pavón, F.J.; Serrano, A.; García-Bueno, B. Oleoylethanolamide, Neuroinflammation, and Alcohol Abuse. *Front. Mol. Neurosci.* **2019**, *11*, 490. [CrossRef]

55. Kane, C.J.M.; Phelan, K.D.; Douglas, J.C.; Wagoner, G.; Johnson, J.W.; Xu, J.; Phelan, P.S.; Drew, P.D. Effects of Ethanol on Immune Response in the Brain: Region-Specific Changes in Adolescent versus Adult Mice. *Alcohol. Clin. Exp. Res.* **2014**, *38*, 384–391. [CrossRef]
56. Bell, R.L.; Lopez, M.F.; Cui, C.; Egli, M.; Johnson, K.W.; Franklin, K.M.; Becker, H.C. Ibudilast Reduces Alcohol Drinking in Multiple Animal Models of Alcohol Dependence. *Addict. Biol.* **2015**, *20*, 38–42. [CrossRef]
57. Mirza, R.; Sharma, B. Benefits of Fenofibrate in Prenatal Valproic Acid-Induced Autism Spectrum Disorder Related Phenotype in Rats. *Brain Res. Bull.* **2019**, *147*, 36–46. [CrossRef]
58. Barbiero, J.K.; Santiago, R.; Tonin, F.S.; Boschen, S.; da Silva, L.M.; Werner, M.F.; da Cunha, C.; Lima, M.M.; Vital, M.A. PPAR-α agonist fenofibrate protects against the damaging effects of MPTP in a rat model of Parkinson's disease. *Prog. Neuropsychopharmacol. Biol. Psychiatry* **2014**, *53*, 35–44. [CrossRef]
59. Caillaud, M.; Patel, N.H.; White, A.; Wood, M.; Contreras, K.M.; Toma, W.; Alkhlaif, Y.; Roberts, J.L.; Tran, T.H.; Jackson, A.B.; et al. Targeting Peroxisome Proliferator-Activated Receptor-α (PPAR-α) to reduce paclitaxel-induced peripheral neuropathy. *Brain Behav. Immun.* **2021**, *93*, 172–185. [CrossRef]
60. Chistyakov, D.V.; Astakhova, A.A.; Goriainov, S.V.; Sergeeva, M.G. Comparison of PPAR Ligands as Modulators of Resolution of Inflammation, via Their Influence on Cytokines and Oxylipins Release in Astrocytes. *Int. J. Mol. Sci.* **2020**, *21*, 9577. [CrossRef]
61. Blednov, Y.A.; Black, M.; Benavidez, J.M.; Stamatakis, E.E.; Harris, R.A. PPAR Agonists: I. Role of Receptor Subunits in Alcohol Consumption in Male and Female Mice. *Alcohol. Clin. Exp. Res.* **2016**, *40*, 553–562. [CrossRef] [PubMed]
62. Dludla, P.V.; Nkambule, B.B.; Mazibuko-Mbeje, S.E.; Nyambuya, T.M.; Silvestri, S.; Orlando, P.; Mxinwa, V.; Louw, J.; Tiano, L. The impact of dimethyl sulfoxide on oxidative stress and cytotoxicity in various experimental models. In *Toxicology*; Patel, V.B., Preedy, V.R., Eds.; Academic Press: Cambridge, MA, USA, 2021; Chapter 25, pp. 243–261, ISBN 9780128190920. [CrossRef]

antioxidants

MDPI

Article

Synergistic Protective Effect of Fermented Schizandrae Fructus Pomace and Hoveniae Semen cum Fructus Extracts Mixture in the Ethanol-Induced Hepatotoxicity

Kyung-Hwan Jegal [1], Hye-Rim Park [2,3], Beom-Rak Choi [3], Jae-Kwang Kim [4,*] and Sae-Kwang Ku [2,*]

1 Department of Korean Medical Classics, College of Korean Medicine, Daegu Haany University, Gyeongsan 38610, Republic of Korea; jegalkh@dhu.ac.kr
2 Department of Anatomy and Histology, College of Korean Medicine, Daegu Haany University, Gyeongsan 38610, Republic of Korea; hrpark@nutracore.co.kr
3 Nutracore Co., Ltd., Suwon 16514, Republic of Korea; brchoi@nutracore.co.kr
4 Department of Physiology, College of Korean Medicine, Daegu Haany University, Gyeongsan 38610, Republic of Korea
* Correspondence: kim-jk@dhu.ac.kr (J.-K.K.); gucci200@hanmail.net (S.-K.K.)

Abstract: Schizandrae Fructus (SF), fruits of *Schisandra chinensis* (Turcz.) Baill. and Hoveniae Semen cum Fructus (HSCF), the dried peduncle of *Hovenia dulcis* Thunb., have long been used for alcohol detoxification in the traditional medicine of Korea and China. In the current study, we aimed to evaluate the potential synergistic hepatoprotective effect of a combination mixture (MSH) comprising fermented SF pomace (fSFP) and HSCF hot water extracts at a 1:1 (*w:w*) ratio against ethanol-induced liver toxicity. Subacute ethanol-mediated hepatotoxicity was induced by the oral administration of ethanol (5 g/kg) in C57BL/6J mice once daily for 14 consecutive days. One hour after each ethanol administration, MSH (50, 100, and 200 mg/kg) was also orally administered daily. MSH administration significantly reduced the serum activities of alanine aminotransferase, aspartate aminotransferase, alkaline phosphatase, and γ-glutamyl transpeptidase. Histological observation indicated that MSH administration synergistically and significantly decreased the fatty changed region of hepatic parenchyma and the formation of lipid droplet in hepatocytes. Moreover, MSH significantly attenuated the hepatic triglyceride accumulation through reducing lipogenesis genes expression and increasing fatty acid oxidation genes expression. In addition, MSH significantly inhibited protein nitrosylation and lipid peroxidation by lowering cytochrome P450 2E1 enzyme activity and restoring the glutathione level, superoxide dismutase and catalase activity in liver. Furthermore, MSH synergistically decreased the mRNA level of tumor necrosis factor-α in the hepatic tissue. These findings indicate that MSH has potential for preventing alcoholic liver disease through inhibiting hepatic steatosis, oxidative stress, and inflammation.

Keywords: alcoholic liver disease; ethanol; Schizandrae Fructus; Hoveniae Semen cum Fructus; antioxidant; anti-steatosis

check for updates

Citation: Jegal, K.-H.; Park, H.-R.; Choi, B.-R.; Kim, J.-K.; Ku, S.-K. Synergistic Protective Effect of Fermented Schizandrae Fructus Pomace and Hoveniae Semen cum Fructus Extracts Mixture in the Ethanol-Induced Hepatotoxicity. *Antioxidants* **2023**, *12*, 1602. https://doi.org/10.3390/antiox12081602

Academic Editor: Marco Fiore

Received: 22 July 2023
Revised: 4 August 2023
Accepted: 10 August 2023
Published: 11 August 2023

1. Introduction

Alcohol-related death was estimated to reach 3.3 million annually, accounting for 3–5% of total global deaths [1,2]. Among the various health issues associated with alcohol consumption, liver disease is the most significantly linked to heavy alcohol intake. A public health study evaluating the impact of alcohol on liver-related mortality in Europe estimated that 60–80% of deaths caused by liver disease are associated with alcohol consumption [3]. Primarily, alcohol disrupts the metabolism of liver cells, causing the accumulation of fat and acting as a hepatotoxic substance that induces oxidative stress, ultimately resulting in hepatocyte death. Alcoholic liver disease (ALD) progresses through various stages, starting with fatty liver and steatohepatitis, and it can ultimately lead to fibrosis, cirrhosis, and hepatocellular carcinoma [4]. However, the absence of reliable treatment options for severe stages of ALD has highlighted the significance of preventive measures for alcoholism and early intervention in ALD [5].

The initial stage of ALD is hepatic steatosis, which is characterized by the accumulation of fat, mainly triglycerides (TG), in hepatocytes. This pathological change is a result of the alcohol-mediated direct regulation of transcription factors associated with lipid metabolism. Acetaldehyde, a metabolite of ethanol, directly stimulates the expression of sterol regulatory element-binding protein (SREBP)-1c, which is an essential transcriptional factor that regulates the expression of genes involved in the synthesis of fatty acid [6]. On the other hand, ethanol inhibits fatty acid oxidation by suppressing the DNA binding and transcriptional activity of the peroxisome proliferator-activated receptor (PPAR) α, which governs various genes related to fatty acid transport and oxidation [7]. These imbalances between fatty acid synthesis and oxidation serves as the fundamental mechanism behind the development of fatty liver.

Oxidative stress is also a major contributing factor to the pathogenesis of ALD [4]. Most of the absorbed alcohol is metabolized by alcohol dehydrogenase to acetaldehyde, and it is further oxidized to acetic acid by aldehyde dehydrogenase in the liver. However, excessive alcohol consumption activates the 2E1 isoform of cytochrome P450 (CYP2E1) in the hepatic endoplasmic reticulum as an alternative pathway [8]. Ethanol metabolism by the CYP2E1 causes the generation of reactive oxygen species (ROS) such as hydrogen peroxide, superoxide anion, and hydroxyl radical. An improper increase in ROS in hepatocytes by ethanol induces oxidative stress and lipid peroxidation, thereby leading to the cellular damage of hepatocytes and the impairment of liver function [9]. Research on alcohol-induced liver injury using experimental animals has revealed that prolonged ethanol consumption decreases the activities of superoxide dismutase (SOD) and catalase (CAT) and the content glutathione (GSH) within liver tissue, which act as endogenous antioxidants [10–12]. Therefore, reducing oxidative stress through antioxidant therapies may be a potential approach to mitigate the progress of ALD.

Diverse medicinal herbs and natural products with anti-steatosis and antioxidant properties have been suggested as potent prophylactic agents for ALD [13,14]. Schizandrae Fructus (SF), fruits of *Schisandra chinensis* (Turcz.) Baill. and Hoveniae Semen cum Fructus (HSCF), dried peduncle of *Hovenia dulcis* Thunb., have been used as a remedy for alcohol-related symptoms in the traditional medicine of Korea and China [15,16]. Based on traditional usage, both herbs have been researched for the hepatoprotective effect against ethanol-induced liver injury. SF prevented the liver injury and hepatic steatosis induced by 5 weeks of ethanol administration [17]. And the identified active constituents such as lignans, triterpenoids, polysaccharides of SF have been shown to exert the anti-oxidative stress, anti-steatosis, and anti-inflammatory effects against ethanol-induced liver injury [10,18–20]. As for HSCF, the hepatoprotective effect against ethanol consumption has been reported by reducing oxidative stress and fatty changes of liver as well as enhancing endogenous antioxidant activity [12]. However, research about the hepatoprotective effect of SF and HSCF combination mixtures has not been conducted.

In previous research, we reported that various ratios of fermented SF pomace (fSFP) and HSCF combination mixtures (fSFP:HSCF *(w:w)*; 1:1, 1:2, 1:4, 1:6, 2:1, 4:1, 6:1, and 8:1) prevented the carbon tetrachloride (CCl_4)-induced acute liver injury [21]. Moreover, we found that a mixture of 1:1 (*w:w*) fSFP and HSCF (MSH) exhibited the most potent hepatoprotective effect against CCl_4 through enhancing the expression and the enzymatic activity of endogenous antioxidants in liver. In the present study, we further evaluated the additional hepatoprotective effect of MSH against ethanol-induced hepatotoxicity. And we also investigated the mechanisms of MSH involved in fatty acid metabolism, oxidative stress, and inflammation caused by ethanol.

2. Materials and Methods

2.1. Preparation of Silymarin, fSFP, HSCF, and MSH

Silymarin was purchased from Sigma-Aldrich (St. Louis, MO, USA). fSFP, HSCF, and MSH (1:1) extracts were supplied from Nutracore (Suwon, Republic of Korea). Briefly, 100 kg of each raw material (fSFP and HSCF) was extracted with hot water and then filtered. Finally, the resulting extracts were concentrated using an evaporator and dried using a spray drier (Figure S1). Mixed formulas (MSH) were prepared by dissolving the same amount of fSFP and HSCF in the distilled water. High-performance liquid chromatographic (HPLC) analysis determined that the MSH contained 0.6 mg/g of schizandrin (a marker compound of fSFP) and 0.17 mg/g of myricetin (a marker compound of HSCF) in MSH (Figure S2). The comprehensive procedures for the preparation of fSFP and HSCF extracts, along with the HPLC analysis of marker compounds in each extract, have been documented in the Supplementary Materials. Silymarin (–20 °C) and all test herb extracts (–4 °C) were stored in a refrigerator until usage.

2.2. Animal Husbandry and Experiment

A total of 80 male C57BL/6J mice (6 weeks old) were supplied from Daehan Bio Link (DBL, Eumseong, Republic of Korea), and they were acclimatized for 7 days before experiments with a 12/12 h light-cycle, 50–55% humidity, and 20–25 °C temperature-controlled conditions. Mice were allocated in eight groups (ten mice per group): vehicle, ethanol, silymarin (ethanol + 250 mg/kg of silymarin), fSFP (ethanol + 200 mg/kg of fSFP), HSCF (ethanol + 200 mg/kg of HSCF), MSH50 (ethanol + 50 mg/kg of MSH), MSH100 (ethanol + 100 mg/kg of MSH), and MSH200 (ethanol + 200 mg/kg of MSH). Ethanol (5 g/kg) was orally administered once daily for 14 consecutive days. Then, 1 h after each ethanol administration, silymarin, fSFP, HSCF, and three doses of MSH dissolved in distilled water were orally administered. The same volume of isocaloric maltose solution (as a vehicle for ethanol) or distilled water (as a vehicle for each herb extracts) was orally administered as vehicle treatment. Afterwards, 24 h after the last administration, all mice were sacrificed, and their blood and livers were collected for the further evaluation.

2.3. Measurements of Body and Liver Weight

Body weight was measured at 1 day before the first oral gavage, the day of first administration (day 0), and at 1, 7, 10, 13, 14 days after the first administration (total 7 times) using an electronic balance (XB320M, Precisa Instrument, Dietikon, Switzerland). Body weight gains were calculated as subtracting the body weight on the last day of administration (day 14) from the body weight on the first day of administration (day 0). Liver weight was measured immediately at sacrifice and represented as a ratio to the body weight of the individual (g/g).

2.4. Serum Biochemistry

First, 1 mL of venous blood was collected from the vena cava at sacrifice and centrifuged (12,500 rpm, 10 min, 4 °C) to separate serum. Activities of alanine aminotransferase (ALT), aspartate aminotransferase (AST), alkaline phosphatase (ALP), γ-glutamyl transpeptidase (GGT), and level of TG in serum were measured using an automated blood analyzer (Dri-Chem NX500i; Fuji Medical System, Tokyo, Japan).

2.5. Histopathological Analysis and Immunohistochemistry

After the sacrifice, the left lateral lobes of the harvested liver were fixed in 10% neutral-buffered formalin. Equal regions of fixed left lateral lobes were crossly trimmed and divided into two parts. One was embedded in paraffin and serially sectioned (3–4 μm thickness, three sections per single liver tissue) for general histopathological observation using hematoxylin and eosin or immunohistochemistry. The other part of the liver tissues was dehydrated in 30% sucrose solution, sectioned by cryostat, and stained with oil red. The representative histological profiles of individual samples were observed using a light microscope (Model Eclipse 80*i*, Nikon, Tokyo, Japan). The numbers of hepatocytes occupied with over 20% of lipid droplets in cytoplasm (cells/1000 hepatocytes), the percentages of fatty changed regions (%/mm^2), and the mean hepatocytes diameter (μm/hepatocytes) in ten fields of single slide were calculated using an automated image analyzer (*i*Solution FL ver 9.1, IMT *i*-solution Inc., Burnaby, BC, Canada), according to the previously established method [11,12]. The average values of ten mice were represented as the results.

For the immunohistochemistry, deparaffinized sections were heated (95–100 °C) in 10 mM citrate buffer for antigen retrievals. Furthermore, sections were incubated in methanol and 0.3% hydrogen peroxide for 30 min to inhibit endogenous peroxidase activity and blocked with normal horse serum to prevent the non-specific binding of antibodies. After incubation with primary and biotinylated secondary antibodies, immunoreactive cells were visualized with an avidin–biotin complex kit (Vector Labs, Burlingame, CA, USA). The numbers of cells with more than 20% immunoreactivity, regarded as positive, in the restricted view area of the hepatic parenchyma was measured using an automated image analyzer (IMT *i*-solution Inc.), as previously described [11,12].

2.6. Enzyme-Linked Immunosorbent Assay (ELISA)

To measure hepatic TG contents, the right lobes of liver were homogenized with phosphate-buffered saline (PBS) using a bead homogenizer (taco™Pre, GeneReach Biotechnology, Taichung, Taiwan) and ultrasonic disruptor (KS-750, Madell Technology, Ontario, CA, USA). The TG contents in liver homogenates were measured using a commercial ELISA kit (Mybiosource, San Diego, CA, USA), as previously reported [12]. Finally, hepatic TG contents were normalized with a total protein content.

To measure the level of tumor necrosis factor (TNF)-α, liver tissues with ice-cold radioimmunoprecipitation assay buffer were centrifuged at 20,000× *g* for 15 min at 4 °C temperature. The TNF-α concentration in the resulting supernatant was quantified with a commercial ELISA kit (Mybiosource). Each of the measured values was normalized with the total protein content of the samples.

Contents of total protein in each sample were measured by the Lowry method using bovine serum albumin as the standard [22].

2.7. Quantitative Reverse Transcription Polymerase Chain Reaction (RT-qPCR)

The total RNA was isolated using TRIzol reagent (Invitrogen, Carlsbad, CA, USA), and cDNA was synthesized using the High-Capacity cDNA Reverse Transcription kit (Thermo Fisher Scientific, Rockford, IL, USA), according to the manufacturer's instructions. Real-time PCR was performed using CFX96™ (Bio-Rad, Hercules, CA, USA) for the quantification of mRNA expression. Primer sequences used for the amplification of each target mRNA are listed in Table 1. For the quantitative measure of each mRNA expression, the expression in hepatic tissue of the vehicle-treated group was used as the control, and the expression of mRNA was normalized with the expression of β-actin mRNA. The relative mRNA expression of each gene was calculated using the $2^{-\Delta\Delta Ct}$ method [23].

Table 1. Information of oligonucleotides used in RT-qPCR.

Target Gene	Primer Sequence (Forward, Reverse)	Gene ID	Product Size
SREBP-1c	5′-GATGTGCGAACTGGACACAG-3′, 5′-CATAGGGGGCGTCAAACAG-3′	20787	104 bp
SCD1	5′-CCCCTGCGGATCTTCCTTAT-3′, 5′-AGGGTCGGCGTGTGTTTCT-3′	20249	114 bp
ACC1	5′-GCCATTGGTATTGGGGCTTAC-3′, 5′-CCCGACCAAGGACTTTGTTG-3′	107476	112 bp
FAS	5′-GCTGCGGAAACTTCAGGAAAT-3′, 5′-AGAGACGTGTCACTCCTGGACTT-3′	14104	84 bp
PPARγ	5′-AGTGGAGACCGCCCAGG-3′, 5′ CCACCAGCTTGTCTTGGATGT-3′	19016	64 bp

Table 1. *Cont.*

Target Gene	Primer Sequence (Forward, Reverse)	Gene ID	Product Size
DGAT2	5′-AGTGGCAATGCTATCATCATCGT-3′, 5′-AAGGAATAAGTGGGAACCAGATCA-3′	67800	149 bp
PPARα	5′-ATGCCAGTACTGCCGTTTTC-3′, 5′-GGCCTTGACCTTGTTCATGT-3′	19013	220 bp
ACO	5′-GCCCAACTGTGACTTCCATT-3′, 5′-GGCATGTAACCCGTAGCACT-3′	74121	113 bp
CPT1	5′-GCACTGCAGCTCGCACATTACAA-3′, 5′-CTCAGACAGTACCTCCTTCAGGAAA-3′	12894	324 bp
Nrf2	5′-CGAGATATACGCAGGAGAGGTAAGA-3′, 5′-GCTCGACAATGTTCTCCAGCTT-3′	18024	79 bp
β-actin	5′-CTGTCGAGTCGCGTCCA CCCGCGAG-3′, 5′-CTCGCGGGTGGACGCGACTCGACAG-3′	11461	516 bp

RT-qPCR, Quantitative reverse transcription polymerase chain reaction; SREBP-1c, Sterol regulatory element-binding protein-1c; SCD1, Stearoyl-CoA desaturase 1; ACC1, Acetyl-CoA carboxylase 1; FAS, Fatty acid synthase; PPAR, Peroxisome proliferator-activated receptor; DGAT2, Diacylglycerol acyltransferase 2; ACO, Acyl-CoA oxidase; CPT1, Carnitine palmitoyltransferase 1; Nrf2, Nuclear factor erythroid 2-related factor-2.

2.8. Measurement of Liver Lipid Peroxidation

Liver lipid peroxidation was estimated in terms of malondialdehyde (MDA) concentration in hepatic homogenates. Hepatic tissues were homogenized in the ice-cold 0.01 M Tris-HCl (pH 7.4) solutions and centrifuged at 12,000× *g* for 15 min. After reaction with thiobarbituric acid (TBA), the MDA–TBA adduct was quantified by detecting absorbance at 525 nm. MDA concentration was normalized with the total protein content of each sample.

2.9. Measurement of Antioxidant Capacities in the Liver

Liver tissues with PBS were homogenized using a bead homogenizer (GeneReach Biotechnology) and ultrasonic disruptor (Madell Technology). After centrifugation, the resulting supernatant was collected as liver homogenates. The GSH level, SOD activity, and CAT activity were measured in hepatic homogenates as previously reported [11]. Briefly, to measure the GSH level, liver homogenates were mixed with 100 μL of 25% trichloroacetic acid and centrifuged at 4200 rpm for 40 min. After reaction with 2-nitrobenzoic acid, absorbance at 412 nm was detected. Here, 1 unit of CAT activity was defined as the amount of enzyme needed to catalyze the decomposition of 1 nM of hydrogen peroxide per minute at 25 °C and pH 7.8 conditions. And CAT activity was calculated by assessing the breakdown of hydrogen peroxide by CAT within the liver homogenate using absorbance readings at 240 nm. Superoxide radicals react with nitrotetrazolium blue (NBT) to produce formazan dye. SOD activity was measured by assessing the extent of inhibition of this reaction. One unit of SOD activity was defined as the amount of enzyme needed to inhibit NBT reduction by 50% within 1 min. Superoxide radicals were generated by xanthine and xanthine oxidase, and the degree of NBT reduction inhibition by SOD within liver homogenates was measured through absorbance readings at 560 nm. Measurement results of SOD and CAT activity were normalized with the protein content of each sample and expressed as U/mg protein.

2.10. Determination of Liver CYP2E1 Activity

CYP2E1 catalyzes the hydroxylation of *p*-nitrophenol to 4-nitrocatechol. Therefore, the concentration of the created 4-nitrocatechol by reaction in liver homogenates was determined by colorimetrical analysis [12,24]. Briefly, liver homogenates were centrifuged at 10,000× *g* for 15 min, and supernatants were further centrifuged (105,000× *g*, 60 min) to obtain microsomes. Absorbance at 535 nm was measured to detect the concentration of 4-nitrocatechol. The CYP2E1 activities were expressed as micromoles of 4-nitrocatechol formed per minute per mg of microsomal protein (μM/min/mg protein).

2.11. Statistical Analysis

All values were expressed as mean ± standard deviation. According to the homogeneity of variance, a one-way analysis of variance (ANOVA) test or Kruskal–Wallis H test was conducted. If significant difference was observed among the group, Tukey's Honest Significant Difference test or a Mann–Whitney U test was conducted to determine the statistically significant difference in the specific pairs of groups. All statistical analyses were conducted using SPSSS 14.0 (IBM SPSS Inc., Armonk, NY, USA).

3. Results

3.1. MSH Synergistically Ameliorated Ethanol-Induced Liver Injury in Mice

To explore the hepatoprotective effects of MSH, mice were orally administered with ethanol (5 g/kg) once daily for 14 consecutive days. One hour after each ethanol administration, Silymarin (250 mg/kg), fSFP (200 mg/kg), HSCF (200 mg/kg), and three doses of MSH (50, 100, and 200 mg/kg) were administrated for 14 days. Silymarin consistently demonstrated a protective effect against animal experimental models of liver disease using CCl_4, acetaminophen, or thioacetamide [25–27]. Additionally, in the experimental model of ALD, it proved to be effective via reducing ethanol-induced lipid peroxidation, TG accumulation, and TNF-α secretion in the liver [24]. In previous reports, we also confirmed its effectiveness against 14 days of daily ethanol consumption (5 g/kg) at a concentration of 250 mg/kg [11,12]. Due to these consistent and reliable effects, silymarin was selected as the positive control drug for this study.

Gains of body weight between the first (day 0) and the last day (day 14) of administration showed no statistical significance among all treatment groups (Figure 1a). As previously reported [11,12], the relative liver weight, representing absolute liver weight divided by the body weight, was significantly increased after 14 days of ethanol administration compared to the vehicle control group. However, silymarin, fSFP, HSCF, and three doses of MSH (50, 100, and 200 mg/kg) significantly reduced the elevation of relative liver weight by ethanol administration. Especially, the reduction in relative liver weight achieved through 100 and 200 mg/kg of MSH administration was significantly greater than those observed in fSFP or HSCF single administration (Figure 1b).

Next, we assessed the effect of MSH on the serum biomarkers of liver function. We found that 14 days of ethanol administration significantly increased the serum activities of ALT, AST, ALT and GGT, indicating hepatic injury. However, silymarin, fSFP, HSCF, and MSH (50–200 mg/kg) administration significantly reduced the activities of ALT, AST, ALP, and GGT in the serum. Especially, 100 and 200 mg/kg of MSH administration showed a significantly greater reductive effect on the ALT, AST, ALT, and GGT activities than the single administration of fSFP or HSCF (Figure 1c).

3.2. MSH Synergistically Alleviated Ethanol-Induced Hepatic Steatosis in Mice

Hepatic steatosis is a major pathological feature of alcohol mediated-liver disease [4]. Therefore, we assessed fatty changes of hepatic tissues by histological observation using hematoxylin–eosin and oil red O stain (Figure 2a). Histomorphological analysis revealed that the percentage of fatty changed regions, the number of fatty changed hepatocytes, and the mean hepatocyte diameters were significantly increased after 14 days of ethanol administration, indicating fatty changes of the liver. However, silymarin, fSFP, HSCF, and MSH (50–200 mg/kg) administration significantly reduced hepatocytes with lipid droplets, fatty regions of hepatic parenchyma, and means of hepatocytes diameter. Especially, the reductions of hepatic fatty changes achieved through 100 and 200 mg/kg of MSH administration were significantly greater than those observed after the single administration of either fSFP or MSH (Figure 2b).

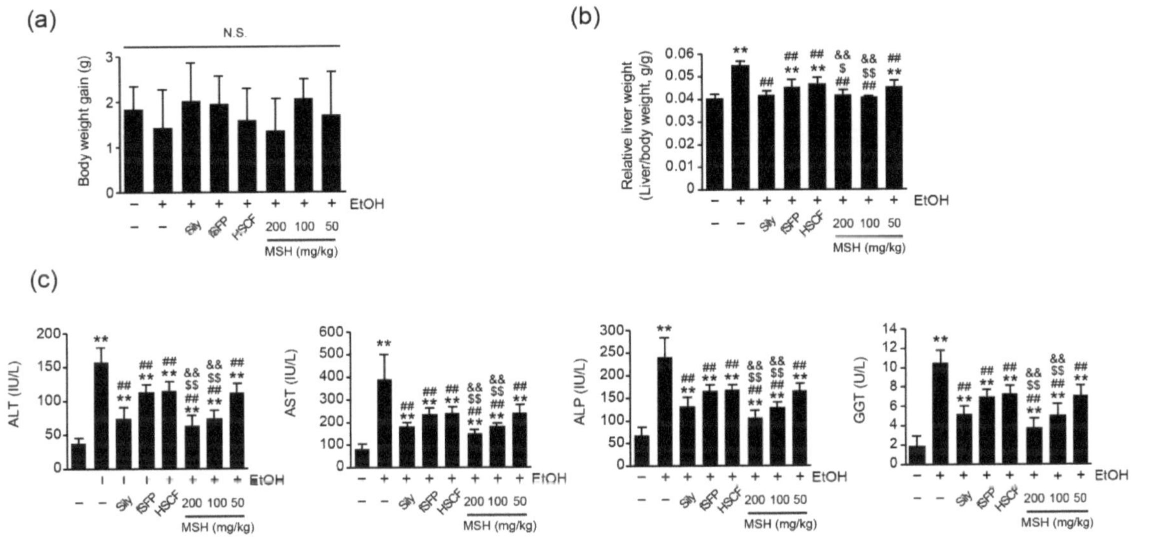

Figure 1. Effect of fermented Schizandrae Fructus Pomace (fSFP) and Hoveniae Semen cum Fructus (HSCF) combination mixture (MSH) on body weight gain, liver weight, and serum activities of liver enzymes in ethanol intoxicated mice. (**a**) Body weight gain was calculated by subtracting the body weight of day 0 from those of day 14. (**b**) Relative liver weight. Absolute liver weight was divided by the body weight of each individual. (**c**) Enzymatic activity of liver enzymes in serum. Activities of alanine aminotransferase (ALT), aspartate aminotransferase (AST), alkaline phosphatase (ALP), and γ-glutamyl transpeptidase (GGT) were measured in serum. All values were expressed as the mean ± standard deviation of 10 mice. Significant versus vehicle group, ** $p < 0.01$; versus EtOH group, ## $p < 0.01$; versus fSFP-treated group, $ $p < 0.05$, $$ $p < 0.01$; versus HSCF-treated group, && $p < 0.01$. N.S., not significant; EtOH, ethanol; Sily, silymarin.

3.3. MSH Alleviated Ethanol-Induced Hepatic TG Accumulation via Modulating Fatty Acid Metabolism

Next, we investigated whether MSH ameliorates ethanol-induced lipid accumulation in hepatic tissue through modulating fatty acid metabolism. First, hepatic and serum TG levels were assessed to elucidate the effect of MSH on the TG synthesis. We found that 14 days of ethanol administration significantly increased the levels of hepatic and serum TG. However, silymarin, fSFP, HSCF, and three doses of MSH (50–200 mg/kg) significantly decreased the levels of hepatic and serum TG (Figure 3a). Furthermore, the mRNA levels of genes related to fatty acid metabolism were assessed using RT-qPCR to elucidate the modulative effect of MSH on the fatty acid metabolism. As is known [11,28], 14 days of ethanol administration significantly increased the relative mRNA expressions of genes including SREBP-1c, PPARγ, stearoyl-CoA desaturase 1 (SCD1), acetyl-CoA carboxylase 1 (ACC1), fatty acid synthase (FAS), and diglyceride acyltransferase 2 (DGAT2), which are involved in fatty acid and TG synthesis. Conversely, mRNA expressions of PPARα, acyl-CoA oxidase (ACO), and carnitine palmitoyltransferase 1 (CPT1), related to fatty acid oxidation, were significantly inhibited by ethanol administration. However, silymarin, fSFP, HSCF, and MSH (50–200 mg/kg) significantly restored the ethanol-induced changes in the mRNA expression of genes responsible for fatty acid metabolism (Figure 3b,c). Moreover, the restorative effect of MSH (100 and 200 mg/kg) on the elevation of lipogenic genes expression and the reduction in lipolytic genes expression by ethanol was significantly greater than those after the single administration of fSFP or HSCF (Figure 3b,c).

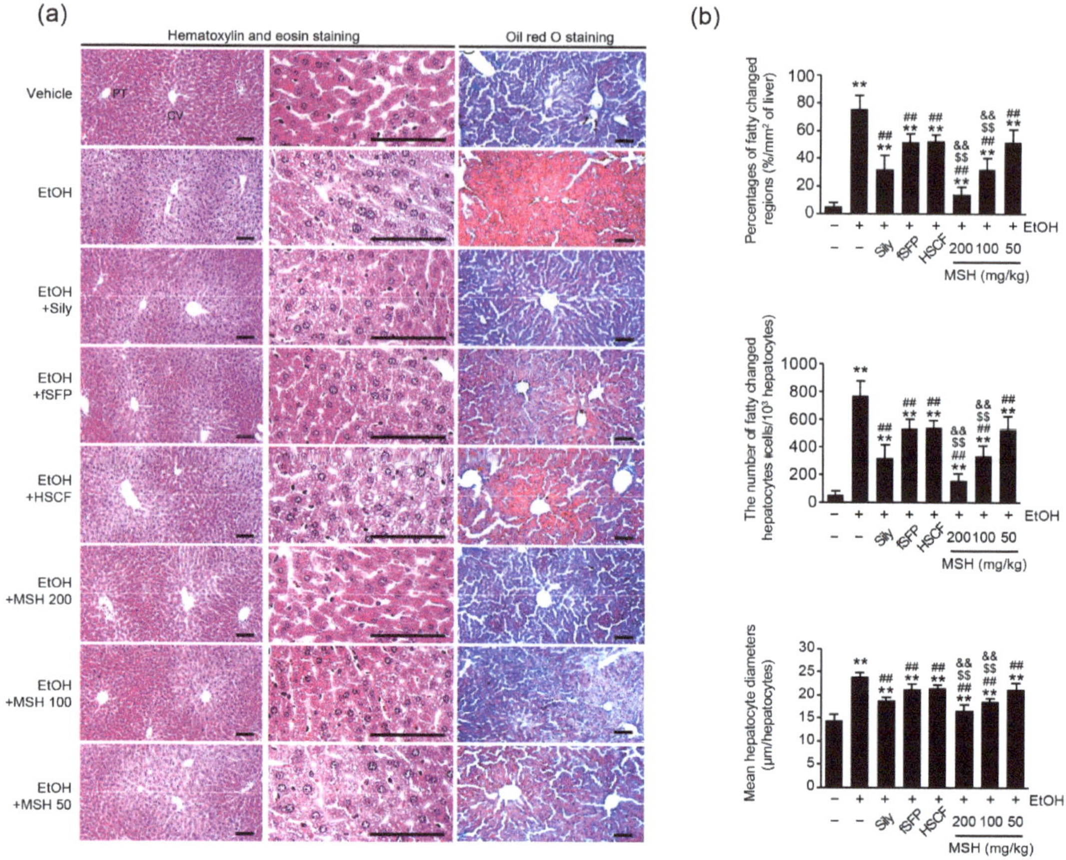

Figure 2. Effect of MSH on the hepatic steatosis in ethanol-intoxicated mice. (**a**) Representative profile images of hematoxylin and eosin-stained (**left**) or oil red O-stained (**right**) liver tissues for histopathological observation. In the oil red O staining, nuclei were counterstained with hematoxylin solution. Scale bars indicate 200 μm. (**b**) Histopathological analysis. Percentage of fatty changed regions (**upper**), the number of fatty changed hepatocytes (**middle**), and mean hepatocyte diameter (**lower**) were observed using an automated image analyzer. Significant versus vehicle group, ** $p < 0.01$; versus EtOH group, ## $p < 0.01$; versus fSFP-treated group, $$ $p < 0.01$; versus HSCF-treated group, && $p < 0.01$. CT, central vein; PT, portal triad; EtOH, ethanol; Sily, silymarin.

3.4. MSH Inhibited Ethanol-Mediated Oxidative Stress and Inflammation in Mice

Along with steatosis, oxidative stress and inflammation are the main etiologies of ethanol-induced hepatotoxicity [4]. To elucidate the effect of MSH on the ethanol-induced oxidative stress in liver, we measured the expressions of nitrotyrosine (NT; a nitrosative stress marker) and 4-hydroxynonenal (4-HNE; a lipid peroxidation marker) using immunohistochemistry (Figure 4a). Results showed that ethanol administration significantly increased the numbers of NT- and 4-HNE immunopositive cells in the hepatic parenchyma. However, administration of silymarin, fSFP, HSCF, and MSH (50–200 mg/kg) significantly decreased the numbers of NT- and 4-HNE positive cells (Figure 4b). Moreover, the content of MDA, another product of lipid peroxidation, in the liver homogenates was elevated by ethanol administration. However, it was significantly decreased by silymarin, fSFP, HSCF and MSH (50–200 mg/kg) administration (Figure 4c). In addition, the ethanol-mediated elevation of TNF-α, which known as a pro-inflammatory cytokine, in hepatic homogenates was significantly inhibited by silymarin, fSFP, HSCF, and MSH (50–200 mg/kg) administrations (Figure 4d). Especially, the reduction in these oxidative stress (NT, 4-HNE, and MDA) and inflammation (TNF-α) markers achieved through 100 and 200 mg/kg of MSH administration was significantly greater than those observed in either fSFP or HSCF single administration.

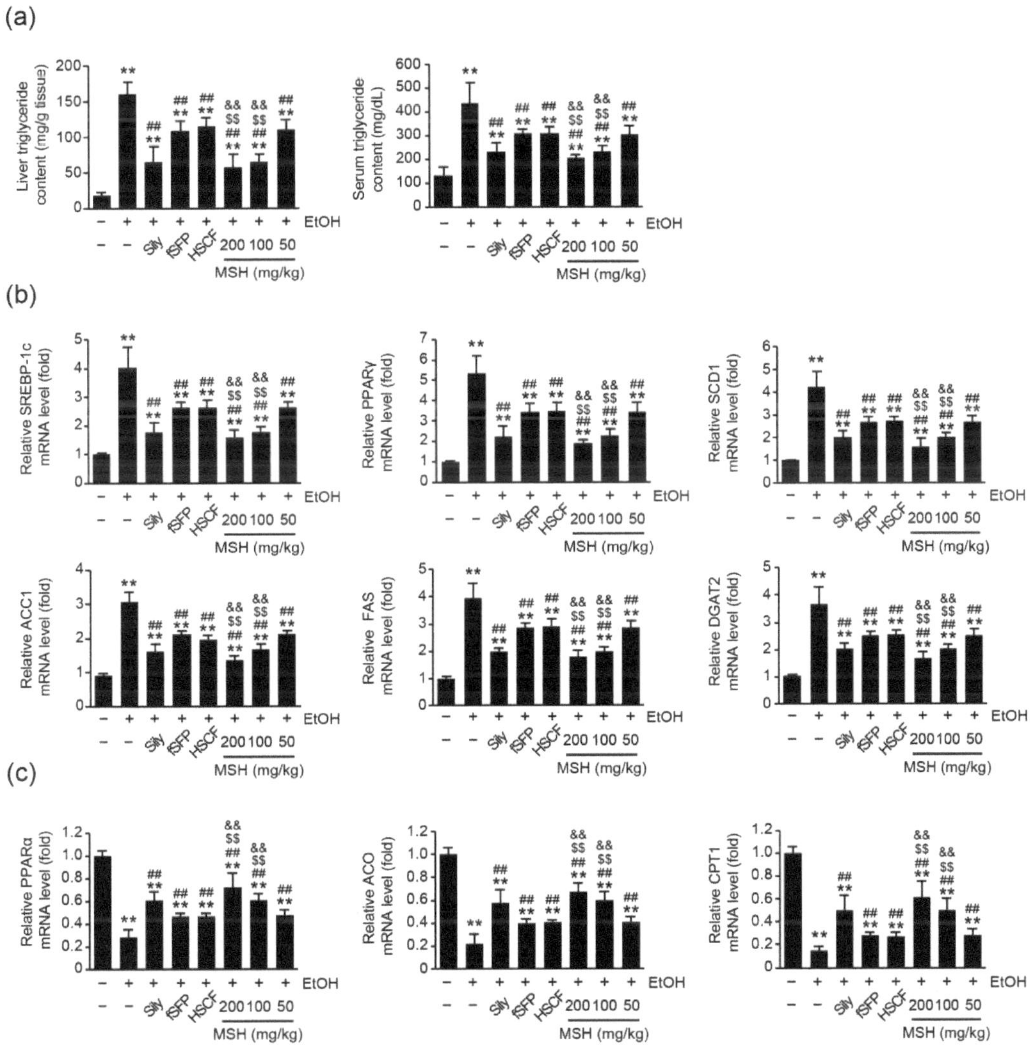

Figure 3. Effect of MSH on triglycerides synthesis and fatty acid metabolism in the ethanol-intoxicated mice liver. (**a**) Triglyceride level in hepatic tissue (**left**) and serum (**right**). (**b**) Expression of genes related to fatty acid and triglyceride synthesis. Relative mRNA levels of SREBP-1c, PPARγ, SCD1, ACC1, FAS, and DGAT2 were measured. (**c**) Expression of genes related to fatty acid oxidation. Relative mRNA levels of PPARα, ACO and CPT1 were measured. mRNA level of each gene was measured using RT-qPCR and normalized with β-actin gene expression. Significant versus vehicle group, ** $p < 0.01$; versus EtOH group, ## $p < 0.01$; versus fSFP-treated group, $$ $p < 0.01$; versus HSCF-treated group, && $p < 0.01$. EtOH, ethanol; Sily, silymarin.

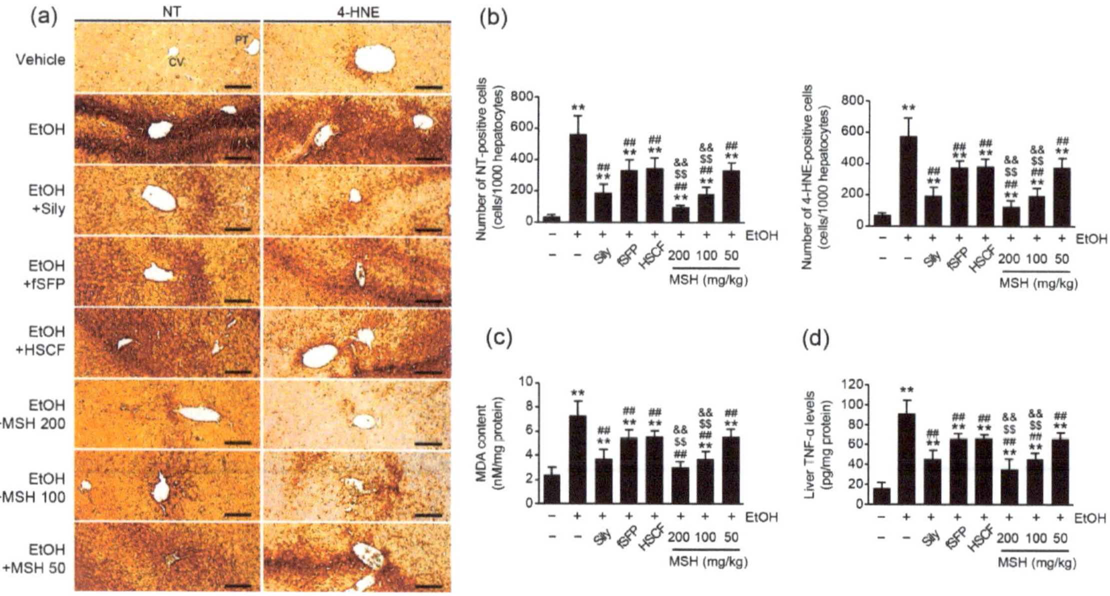

Figure 4. Effect of MSH on the ethanol-induced oxidative stress and inflammation. (**a**) Representative profiles of immunohistochemistry using anti-NT and 4-HNE antibodies. Scale bars indicate 200 µm. (**b**) The numbers of NT- (**left**) and 4-HNE- (**right**) positive cells. Cells showing more than 20% of immunoreactivity were counted. (**c**) MDA content and (**d**) tumor necrosis factor (TNF)-α in the hepatic tissue were quantified using liver homogenates. Significant versus vehicle group, ** $p < 0.01$; versus EtOH group, ## $p < 0.01$; versus fSFP-treated group, $$ $p < 0.01$; versus HSCF-treated group, && $p < 0.01$. CT, central vein; PT, portal triad; EtOH, ethanol; Sily, silymarin; NT, nitrotyrosine; 4-HNE, 4-hydroxynonenal; MDA, malondialdehyde.

3.5. MSH Enhanced the Endogenous Antioxidant Capacity and Lowered the CYP2E1 Activity in Liver

Numerous researchers have reported that ethanol consumption in the experimental animal model induced the depletion of endogenous antioxidant capacities [11,12,24,29]. Therefore, we measured the level of GSH and activities of SOD and CAT in the hepatic tissue to investigate whether the hepatoprotective effect of MSH on ethanol-induced toxicity is exerted through restoring endogenous antioxidants. Ethanol administration significantly reduced the level of GSH as well as activities of SOD and CAT in liver. However, silymarin, fSFP, HSCF, and MSH (50–200 mg/kg) administration significantly restored the level of GSH as well as activities of SOD and CAT. Moreover, the extent of elevations in the GSH level, SOD and CAT activities by MSH (100 and 200 mg/kg) was significantly greater than those by fSFP or HSCF single administration (Figure 5a). In addition, MSH (50–200 mg/kg) administration significantly and dose-dependently increased the mRNA level of nuclear factor erythroid 2-related factor-2 (Nrf2), which is known as a master transcriptional regulator of antioxidants genes. And its extent of increase was significantly greater than those observed in fSFP or HSCF single administration (Figure 5b).

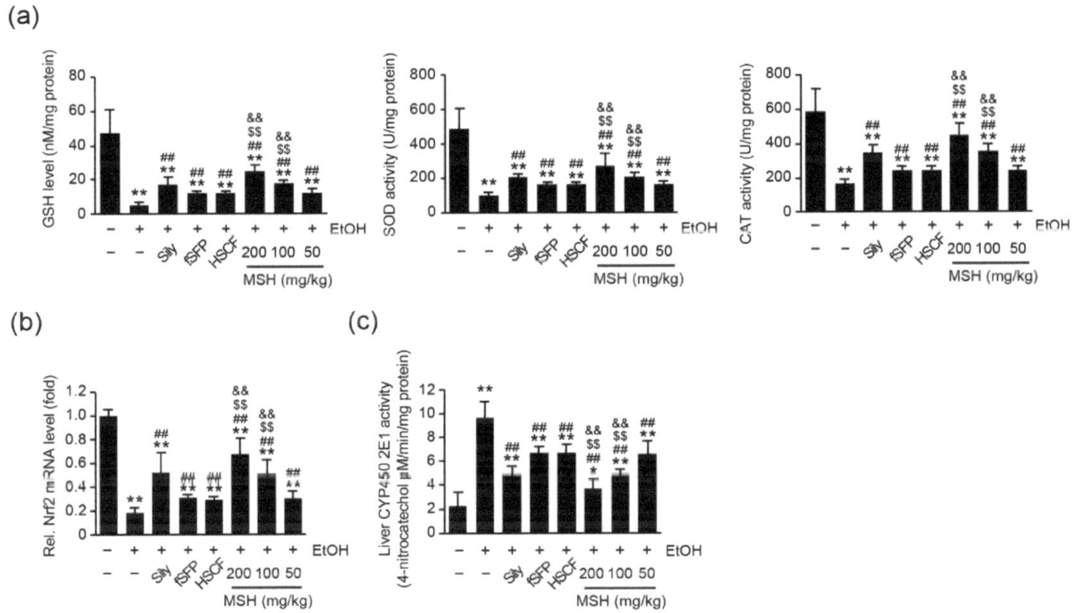

Figure 5. Effect of MSH on the antioxidant system and CYP2E1 activity in ethanol-intoxicated mice liver. (**a**) Endogenous antioxidant capacities. The glutathione (GSH) level (**left**), superoxide dismutase (SOD) (**middle**), and catalase (CAT) (**right**) activities were measured in liver homogenates. (**b**) Relative level of Nrf2 mRNA. Nrf2 mRNA level was measured using RT-qPCR and normalized with β-actin mRNA expression. (**c**) CYP450 2E1 activity. Significant versus vehicle group, * $p < 0.05$, ** $p < 0.01$; versus EtOH group, ## $p < 0.01$; versus fSFP-treated group, $$ $p < 0.01$; versus HSCF-treated group, && $p < 0.01$. EtOH, ethanol; Sily, silymarin.

CYP2E1 is considered as one of the major contributors on ethanol-mediated ROS generation in liver [4,30,31]. We assessed the activity of CYP2E1 in a liver homogenate to investigate whether MSH contributes to the reduction in ethanol-induced ROS via the regulating activity of CYP2E1. As a result, MSH (50–200 mg/kg) administration significantly reduced the activity of CYP2E1 (Figure 5c). These results suggested that MSH lowered the oxidative stress in the hepatic tissue via enhancing antioxidant capacities and lowering CYP2E1 activities.

4. Discussion

Despite the profound impact of alcohol abuse on public health, few therapeutic approaches have been approved for the treatment of patients with ALD [5]. Therefore, numerous studies were conducted to find the medicinal herbs and natural compounds possessing hepatoprotective effects for the prevention of ALD [13,14]. Especially, *S. chinensis* and *H. dulcis* have been used as a key ingredient herb for treating alcohol detoxification in Korean and Chinese traditional medicine [15,16]. The ethnopharmacological evidence has prompted research about the hepatoprotective effect of SF and HSCF against ALD [12,17,32]. In previous study, we evaluated the hepatoprotective effect fSFP and HSCF (*w:w*; 1:1, 1:2, 1:4, 1:6, 2:1, 4:1, 6:1 and 8:1) mixtures at different ratios to develop a novel functional food for preventing liver disease [21]. Among the various ratios of mixtures, we concluded that the 1:1 mixture of fSFP and HSCF (MSH) showed the most potent hepatoprotective effect in CCl_4-induced liver injury mice via lowering oxidative stress and inflammatory response. And this mixture showed a greater hepatoprotective effect than fSFP or HSCF single administration. Therefore, in the current study, we aimed to investigate the synergistic protective effect of a fSFP and HSCF 1:1 mixture (MSH) on ethanol-induced liver toxicity.

One of the early pathological responses to prolonged alcohol consumption is the fat accumulation in hepatocytes, which is a condition called hepatic steatosis [28]. With fatty changes of liver, ethanol-induced oxidative stress and inflammatory response intensify the damage of liver [4]. In the present study, histomorphometrical observation showed the significant fatty changes of hepatic parenchyma and hepatocytes after 14 days of ethanol feeding. Moreover, ALT, AST, ALP, and GGT activities, as serum markers of liver damage, were elevated, as previous research showed [11]. However, ethanol-induced hepatic steatosis was significantly and dose-dependently reduced by MSH administration (50–200 mg/kg), showing decreases in the percentage of fatty changed regions, the number of fatty changed hepatocytes, and the mean of hepatocyte diameters (Figure 2). And activities of ALT, AST, ALP, and GGT in the serum were also significantly reduced by MSH administration (50–200 mg/kg) (Figure 1c). In addition, a significantly greater restorative effect of MSH on fatty changes of liver and serum markers, compared to fSFP or HSCF single administration, suggested the synergistic protective effect of MSH against ethanol-induced steatosis and liver damage.

An imbalance between lipid synthesis and fatty acid oxidation results in the excessive accumulation of fat in the liver, playing a crucial role in the progression of ALD [28]. In the current study, we found that ethanol increased the hepatic TG accumulation and mRNA expression of SREBP-1c, PPARγ, and its target genes (Figure 3a,b). SREBP-1c, the predominant isoform of SREBP in liver, acts as a transcriptional factor regulating the expression of genes related to fatty acid synthesis, such as SCD1, ACC1, and FAS. It has been reported that ethanol treatment promoted fatty acid synthesis via enhancing the transcriptional activity of SREBP-1 [6]. PPARγ, a type II nuclear receptor, also controls the expression of lipogenic genes such as DGAT2, which is involved in TG synthesis. The hepatic knockdown of PPARγ-attenuated chronic alcohol feeding induced hepatic steatosis and injury, inhibiting the expression of SREBP-1c [33]. In addition to elevated lipogenesis, reduced fatty acid oxidation also contributes to the development of fatty liver by ethanol consumption. In our experiment, 14 days of ethanol administration decreased the mRNA expression of PPARα, CPT1, and ACO, which is involved in fatty acid oxidation. As a subtype of the PPAR superfamily, PPARα regulates the expression of genes involved in the transport and mitochondrial oxidation of fatty acid [34]. Research showed that ethanol feeding impaired the expression of PPARα response genes [35]. Conversely, agonists of PPARα showed a therapeutic effect against the ethanol-induced liver injury via modulating lipid metabolism [35–37]. In our results, MSH administration restored the gene expression of not only fatty acid synthesis but also fatty acid oxidation in hepatic tissue (Figure 3). These results suggested that MSH inhibited ethanol-induced TG accumulation in the liver by modulating lipid metabolism.

Ethanol is mainly oxidized by alcohol dehydrogenase to acetaldehyde in hepatocytes. However, when alcohol dehydrogenase becomes saturated by excessive ethanol, CYP2E1-mediated-alcohol metabolism is activated as an alternative pathway [8]. The chronic consumption of alcohol increases the expression and activity of CYP2E1, thereby promoting the generation of ROS during the metabolization of ethanol [28,38]. Numerous research studies have reported that CYP2E1 expressions and activity are responsible for the severity of ethanol-mediated liver injury. Studies using transgenic mice showed that CYP2E1 overexpression aggravates ethanol-induced hepatic injury [30,31]. On the other hand, genetic deletion and the pharmacological inhibition of CYP2E1 attenuated alcohol-induced hepatic damage [39,40]. Taken together, the evidence provides support for CYP2E1 being an important factor in the progress of ALD. Therefore, the development of drugs targeting CYP2E1 has been attempted for the treatment of early stages of ALD [38]. The inhibitory effects of *S. chinensis* and *H. dulcis* on ethanol-induced CYP2E1 activation have also been studied. The total lignans of *S. chinensis* significantly reduced the protein and mRNA level of CYP2E1 in alcohol-induced liver injury rats [32]. In addition, we reported that HSCF administration significantly inhibited CYP2E1 enzymatic activities in ethanol-intoxicated mice [12]. In the present study, the elevated CYP2E1 activity by ethanol administration was significantly reduced by MSH administration. Especially, 100 and 200 mg/kg of MSH administration exhibited the greater inhibitory effect on CYP2E1 activity than either fSFP or HSCF single administration (Figure 5c). Eventually, the generation of ROS by chronic ethanol exposure causes the lipid peroxidation. The reaction between generated radical and hepatic lipid led to the formation of reactive aldehyde substances like 4-HNE and MDA. Because of their high reactivity, 4-HNE and MDA can affect the structure and function of major biomolecules such as proteins and DNA [41]. One study reported that patients with ALD showed increases in the MDA and 4-HNE level in the serum [42,43]. In the present study, MSH administration significantly reduced the elevated number of 4-HNE-positive cells in hepatic parenchyma and content of MDA in hepatic homogenates by ethanol administration (Figure 4). These results suggested that MSH attenuated ethanol-induced oxidative stress in liver by inhibiting CYP2E1-mediated ROS generation and lipid peroxidation.

The abnormal expression of pro-inflammatory cytokines is another major feature of ALD [44]. Especially, chronic alcohol consumption increased TNF-α production in the hepatic tissue, contributing to liver injury [45,46]. Monocytes and Kupffer cells become more sensitive to lipopolysaccharide after continuous alcohol consumption, leading to further TNF-α production [47,48]. Elevated TNF-α secretion is considered a major determinant of susceptibility to ethanol-induced liver toxicity. TNF-α type I receptor-deficient mice fed long-term ethanol did not develop liver injury [49]. Moreover, the neutralization of TNF-α by specific antibodies attenuated hepatic inflammation and necrosis in the ethanol-fed rats [46]. In the current study, subacute ethanol administration elevated the hepatic TNF-α production in hepatic tissue (Figure 4d). However, MSH (50–200 mg/kg) administration significantly reduced the hepatic TNF-α expression induced by ethanol, suggesting that MSH attenuated the severity of ethanol-induced liver injury via modulating the inflammatory response.

Normally, SOD and CAT, as enzymatic antioxidants, contribute to the redox homeostasis of liver tissue by scavenging superoxide and hydrogen peroxide respectively. As a non-enzymatic antioxidant, GSH is involved in the detoxification of xenobiotics and neutralizing harmful free radicals [50]. One study using transgenic mice showed that SOD1 knockout promoted the ethanol-induced oxidative stress and liver injury [51]. In addition, the depletion of mitochondrial GSH increased the susceptibility of hepatocytes to TNF-α-mediated cell death in ethanol-fed rat hepatocytes [52]. Although hepatocytes are equipped with these antioxidant systems to defend against oxidative stress, studies using laboratory animals revealed that chronic alcohol exposure reduces the activity of SOD, CAT, and depleted GSH content, thereby causing oxidative stress-mediated liver injury [12,29]. Consistent with these facts, we found that ethanol administration significantly reduced the SOD, CAT activities, and GSH level in the hepatic homogenates. However, MSH administration significantly restored the SOD, CAT activities and GSH level in the hepatic tissue. These results suggested that MSH attenuated ethanol-induced oxidative stress via restoring the compromised capacity of the liver antioxidant system.

Nrf2 is a transcriptional factor that regulates the expression of antioxidant and phase II detoxification genes. As with other liver diseases mediated by oxidative stress, the protective role of Nrf2 in ALD has also been highlighted [53,54]. The hepatocyte-specific knockout of Kelch-like ECH-associated protein 1 (Keap1) in mice, which induces the activation of Nrf2, offered protection against liver injury by ethanol [55]. On the other hand, Nrf2-deficient mice showed susceptibility to ethanol exposure, exhibiting GSH depletion and increased mortality [56]. Research has reported that the total extract or lignans of *S. chinensis* exerted a hepatoprotective effect on the ethanol-induced liver injury via increasing the stability and transcriptional activity of Nrf2 [32,57]. Moreover, HSCF treatment increased the transcriptional activity of Nrf2 and the expression of its target genes in HepG2 cells [58]. And HSCF administration also increased the relative expression of Nrf2 mRNA levels in ethanol-intoxicated mice [12]. In the current study, MSH administration increased the relative mRNA expression of Nrf2, suggesting that the hepatoprotective effects of MSH against ethanol-induced oxidative stress are mediated by the modulation of the antioxidant system via Nrf2 activation.

Research has been conducted about the composition and preventive effect of SF and HSCF. The main active chemical constituents of *S. chinensis* include lignans and polysaccharides [59]. Gomisin N, one of the lignans in SF, alleviated ethanol-induced hepatic steatosis, lipogenesis, and inflammation in mice [19]. The polysaccharide fraction of *S. chinensis* exerted the protective effect on ethanol-induced oxidative stress via reducing CYP2E1 expression and elevating SOD activity [10]. As for *H. dulcis*, several flavonoids such as myricetin, dihydromyricetin, quercetin, gallocatechin and polysaccharides have been identified as active components in its seeds, fruits, and peduncles [15]. Specifically, inhibitory effects of myricetin on the fatty acid biosynthesis in ethanol-fed mice have been reported [60]. And the hepatoprotective effect of polysaccharides isolated from peduncles of *H. dulcis* have also been reported [61].

The chemical composition of herbs and their extracts can be influenced by various factors, including the timing of harvest and the method of extraction. Furthermore, the specific parts or species of plants used may vary depending on traditional medicine practices or regional customs. Such variations could affect the biological effectiveness of herbs, highlighting the importance of carefully controlling these factors to achieve reproducible experimental outcomes. In this research, the fruit of *S. chinensis* and the peduncle of *H. dulcis* were utilized to manufacture each hot water extract. The manufacturing processes of fSFP, HSCF, and MSH were documented for standardization and quality control. Additionally, the indicative constituents of both herbs, schizandrin and myricetin, in MSH were identified using HPLC analysis (Figures S1 and S2). We expect that these findings will improve the reliability of the efficacy and reproducibility of future studies. Nevertheless, the current study did not provide a specific identification of the constituent components within MSH that exhibited synergistic protective effects. Therefore, additional studies on the pharmacological interactions, bioavailability, and content of active components that are responsible for MSH's synergistic effect are needed.

Over the years, preclinical experimental models of ALD have been continuously developed for the comprehensive understanding of its pathogenesis. The binge-feeding method, in which ethanol is administered acutely through oral gavage, is widely used due to its simplicity in replicating liver damage caused by excessive alcohol consumption. The chronic feeding method, providing animals with free access to ethanol through drinking water for an extended period of times, mimics chronic alcohol consumption [62]. Moreover, chronic plus binge feeding has been developed for reflecting a drinking pattern of alcoholic hepatitis patients [63]. In the present study, we adopted subacute ethanol-mediated liver injury using 14 days of binge feeding with simultaneous MSH administration to investigate its protective effects. However, the current model is known to allow the observation of early phenomena such as liver cell death and steatosis in the pathogenesis of ALD, while severe progression such as fibrosis or mortality is not observed. Therefore, to validate the efficacy of MSH on a broad spectrum of ALD, it would be necessary to conduct additional research using chronic and severe experimental models such as chronic plus binge feeding. Furthermore, exploring the therapeutic effects of post-treatment with MSH after the occurrence of hepatic damage and steatosis by ethanol administration will provide a more profound understanding of MSH's effect in treating ALD.

5. Conclusions

The current investigation found that 50, 100, and 200 mg/kg of HSCF administration significantly alleviated the ethanol-induced liver toxicity. MSH lowered lipid accumulation in liver via modulating the expression of genes involved in fatty acid synthesis and oxidation. Moreover, MSH exerted an anti-oxidative property through enhancing the antioxidant defense system capacity by Nrf2 activation. Furthermore, the significantly greater effects of MSH in anti-steatosis, anti-oxidative stress, and anti-inflammation, compared to fSFP or HSCF single administration, indicate the synergistic hepatoprotective effect against ethanol-induced liver toxicity.

Supplementary Materials: The following supporting information can be downloaded at: https://www.mdpi.com/article/10.3390/antiox12081602/s1, Figure S1: Flow chart for manufacturing hot water extracts. Figure S2: Identification of MSH using high-performance liquid chromatography (HPLC) analysis. Table S1: Instrument condition of HPLC analysis.

Author Contributions: Conceptualization, B.-R.C. and S.-K.K.; methodology, B.-R.C. and S.-K.K.; formal analysis, K.-H.J., H.-R.P. and S.-K.K.; investigation, K.-H.J., J.-K.K. and S.-K.K.; writing—original draft preparation, K.-H.J. and J.-K.K.; writing—review and editing, J.-K.K. and S.-K.K.; visualization, K.-H.J. and J.-K.K. All authors have read and agreed to the published version of the manuscript.

Funding: This work was supported by the Technology Development Program (S2940975) funded by the Ministry of SMEs and Startups (MSS, Republic of Korea).

Institutional Review Board Statement: All animal experiments in the current study were conducted according to the international regulations of the usage and welfare of laboratory animals and approved by the Institutional Animal Care and Use Committee in Daegu Haany University (Approval No. DHU-2021-022, 22 March 2021).

Informed Consent Statement: Not applicable.

Data Availability Statement: Not applicable.

Conflicts of Interest: B.-R.C. and H.-R.P. are employed by Nutracore; however, in this research, they were only involved in the preparation and analysis of the raw materials to a limited extent. K.-H.J., J.-K.K., and S.-K.K. declare no conflict of interest.

References

1. Rehm, J.; Mathers, C.; Popova, S.; Thavorncharoensap, M.; Teerawattananon, Y.; Patra, J. Global burden of disease and injury and economic cost attributable to alcohol use and alcohol-use disorders. *Lancet* **2009**, *373*, 2223–2233. [CrossRef] [PubMed]
2. Thursz, M.; Gual, A.; Lackner, C.; Mathurin, P.; Moreno, C.; Spahr, L.; Sterneck, M.; Cortez-Pinto, H. EASL Clinical Practice Guidelines: Management of alcohol-related liver disease. *J. Hepatol.* **2018**, *69*, 154–181. [CrossRef] [PubMed]
3. Sheron, N. Alcohol and liver disease in Europe--Simple measures have the potential to prevent tens of thousands of premature deaths. *J. Hepatol.* **2016**, *64*, 957–967. [CrossRef] [PubMed]
4. Gao, B.; Bataller, R. Alcoholic Liver Disease: Pathogenesis and New Therapeutic Targets. *Gastroenterology* **2011**, *141*, 1572–1585. [CrossRef] [PubMed]
5. Subramaniyan, V.; Chakravarthi, S.; Jegasothy, R.; Seng, W.Y.; Fuloria, N.K.; Fuloria, S.; Hazarika, I.; Das, A. Alcohol-associated liver disease: A review on its pathophysiology, diagnosis and drug therapy. *Toxicol. Rep.* **2021**, *8*, 376–385. [CrossRef] [PubMed]
6. You, M.; Fischer, M.; Deeg, M.A.; Crabb, D.W. Ethanol Induces Fatty Acid Synthesis Pathways by Activation of Sterol Regulatory Element-binding Protein (SREBP). *J. Biol. Chem.* **2002**, *277*, 29342–29347. [CrossRef] [PubMed]
7. Galli, A.; Pinaire, J.; Fischer, M.; Dorris, R.; Crabb, D.W. The Transcriptional and DNA Binding Activity of Peroxisome Proliferator-activated Receptor α Is Inhibited by Ethanol Metabolism. *J. Biol. Chem.* **2001**, *276*, 68–75. [CrossRef]
8. Cederbaum, A.I. Alcohol metabolism. *Clin. Liver. Dis.* **2012**, *16*, 667–685. [CrossRef]
9. Cederbaum, A.I.; Lu, Y.; Wu, D. Role of oxidative stress in alcohol-induced liver injury. *Arch. Toxicol.* **2009**, *83*, 519–548. [CrossRef]
10. Yuan, R.; Tao, X.; Liang, S.; Pan, Y.; He, L.; Sun, J.; Wenbo, J.; Li, X.; Chen, J.; Wang, C. Protective effect of acidic polysaccharide from Schisandra chinensis on acute ethanol-induced liver injury through reducing CYP2E1-dependent oxidative stress. *Biomed. Pharmacother.* **2018**, *99*, 537–542. [CrossRef]
11. Choi, B.R.; Cho, I.J.; Jung, S.J.; Kim, J.K.; Park, S.M.; Lee, D.G.; Ku, S.K.; Park, K.M. Lemon balm and dandelion leaf extract synergistically alleviate ethanol-induced hepatotoxicity by enhancing antioxidant and anti-inflammatory activity. *J. Food Biochem.* **2020**, *44*, e13232. [CrossRef] [PubMed]
12. Cho, I.; Kim, J.; Jung, J.; Sung, S.; Kim, J.; Lee, N.; Ku, S. Hepatoprotective effects of hoveniae semen cum fructus extracts in ethanol intoxicated mice. *J. Exerc. Nutrition. Biochem.* **2016**, *20*, 49–64. [CrossRef]
13. Ding, R.B.; Tian, K.; Huang, L.L.; He, C.W.; Jiang, Y.; Wang, Y.T.; Wan, J.B. Herbal medicines for the prevention of alcoholic liver disease: A review. *J. Ethnopharmacol.* **2012**, *144*, 457–465. [CrossRef] [PubMed]
14. Yan, J.; Nie, Y.; Luo, M.; Chen, Z.; He, B. Natural Compounds: A Potential Treatment for Alcoholic Liver Disease? *Front. Pharmacol.* **2021**, *12*, 694475. [CrossRef] [PubMed]

15. Sferrazza, G.; Brusotti, G.; Zonfrillo, M.; Temporini, C.; Tengattini, S.; Bononi, M.; Tateo, F.; Calleri, E.; Pierimarchi, P. Hovenia dulcis Thumberg: Phytochemistry, Pharmacology, Toxicology and Regulatory Framework for Its Use in the European Union. *Molecules* **2021**, *26*, 903. [CrossRef]
16. Yang, K.; Qiu, J.; Huang, Z.; Yu, Z.; Wang, W.; Hu, H.; You, Y. A comprehensive review of ethnopharmacology, phytochemistry, pharmacology, and pharmacokinetics of Schisandra chinensis (Turcz.) Baill. and Schisandra sphenanthera Rehd. et Wils. *J. Ethnopharmacol.* **2022**, *284*, 114759. [CrossRef]
17. Park, H.J.; Lee, S.J.; Song, Y.; Jang, S.H.; Ko, Y.G.; Kang, S.N.; Chung, B.Y.; Kim, H.D.; Kim, G.S.; Cho, J.H. Schisandra chinensis prevents alcohol-induced fatty liver disease in rats. *J. Med. Food* **2014**, *17*, 103–110. [CrossRef] [PubMed]
18. Zhu, L.; Li, B.; Liu, X.; Meng, X. Hepatoprotective Effects of Triterpenoid Isolated from Schizandra chinensis against Acute Alcohol-Induced Liver Injury in Mice. *Food Sci. Technol. Res.* **2013**, *19*, 1003–1009. [CrossRef]
19. Nagappan, A.; Jung, D.Y.; Kim, J.H.; Lee, H.; Jung, M.H. Gomisin N Alleviates Ethanol-Induced Liver Injury through Ameliorating Lipid Metabolism and Oxidative Stress. *Int. J. Mol. Sci.* **2018**, *19*, 2601. [CrossRef]
20. Li, W.; Qu, X.-N.; Han, Y.; Zheng, S.-W.; Wang, J.; Wang, Y.-P. Ameliorative Effects of 5-Hydroxymethyl-2-furfural (5-HMF) from Schisandra chinensis on Alcoholic Liver Oxidative Injury in Mice. *Int. J. Mol. Sci.* **2015**, *16*, 2446–2457. [CrossRef]
21. Park, H.; Jegal, K.H.; Choi, B.; Kim, J.K.; Ku, S.K. Hepatoprotective effect of fermented Schizandrae Fructus Pomace extract and Hoveniae Semen Cum Fructus extract combination mixtures against carbon tetrachloride-induced acute liver injured mice. *Herb. Formula Sci.* **2023**, *31*, 53–65. [CrossRef]
22. Lowry, O.; Rosebrough, N.; Farr, A.L.; Randall, R. Protein Measurement with the Folin Phenol Reagent. *J. Biol. Chem.* **1951**, *193*, 265–275. [CrossRef] [PubMed]
23. Livak, K.J.; Schmittgen, T.D. Analysis of relative gene expression data using real-time quantitative PCR and the 2(-Delta Delta C(T)) Method. *Methods* **2001**, *25*, 402–408. [CrossRef] [PubMed]
24. Song, Z.; Deaciuc, I.; Song, M.; Lee, D.Y.; Liu, Y.; Ji, X.; McClain, C. Silymarin protects against acute ethanol-induced hepatotoxicity in mice. *Alcohol. Clin. Exp. Res.* **2006**, *30*, 407–413. [CrossRef] [PubMed]
25. Tsai, J.H.; Liu, J.Y.; Wu, T.T.; Ho, P.C.; Huang, C.Y.; Shyu, J.C.; Hsieh, Y.S.; Tsai, C.C.; Liu, Y.C. Effects of silymarin on the resolution of liver fibrosis induced by carbon tetrachloride in rats. *J. Viral Hepat.* **2008**, *15*, 508–514. [CrossRef] [PubMed]
26. Chen, I.S.; Chen, Y.-C.; Chou, C.-H.; Chuang, R.-F.; Sheen, L.-Y.; Chiu, C.-H. Hepatoprotection of silymarin against thioacetamide-induced chronic liver fibrosis. *J. Sci. Food Agric.* **2012**, *92*, 1441–1447. [CrossRef] [PubMed]
27. Papackova, Z.; Heczkova, M.; Dankova, H.; Sticova, E.; Lodererova, A.; Bartonova, L.; Poruba, M.; Cahova, M. Silymarin prevents acetaminophen-induced hepatotoxicity in mice. *PLoS ONE* **2018**, *13*, e0191353. [CrossRef] [PubMed]
28. Seitz, H.K.; Bataller, R.; Cortez-Pinto, H.; Gao, B.; Gual, A.; Lackner, C.; Mathurin, P.; Mueller, S.; Szabo, G.; Tsukamoto, H. Alcoholic liver disease. *Nat. Rev. Dis. Primers.* **2018**, *4*, 16. [CrossRef]
29. Chen, L.H.; Xi, S.; Cohen, D.A. Liver antioxidant defenses in mice fed ethanol and the AIN-76A diet. *Alcohol* **1995**, *12*, 453–457. [CrossRef]
30. Butura, A.; Nilsson, K.; Morgan, K.; Morgan, T.R.; French, S.W.; Johansson, I.; Schuppe-Koistinen, I.; Ingelman-Sundberg, M. The impact of CYP2E1 on the development of alcoholic liver disease as studied in a transgenic mouse model. *J. Hepatol.* **2009**, *50*, 572–583. [CrossRef]
31. Morgan, K.; French, S.W.; Morgan, T.R. Production of a cytochrome P450 2E1 transgenic mouse and initial evaluation of alcoholic liver damage. *Hepatology* **2002**, *36*, 122–134. [CrossRef] [PubMed]
32. Su, L.; Li, P.; Lu, T.; Mao, C.; Ji, D.; Hao, M.; Huang, Z. Protective effect of Schisandra chinensis total lignans on acute alcoholic-induced liver injury related to inhibiting CYP2E1 activation and activating the Nrf2/ARE signaling pathway. *Rev. Bras. Farmacogn.* **2019**, *29*, 198–205. [CrossRef]
33. Zhang, W.; Sun, Q.; Zhong, W.; Sun, X.; Zhou, Z. Hepatic Peroxisome Proliferator-Activated Receptor Gamma Signaling Contributes to Alcohol-Induced Hepatic Steatosis and Inflammation in Mice. *Alcohol. Clin. Exp. Res.* **2016**, *40*, 988–999. [CrossRef] [PubMed]
34. Nan, Y.-M. Peroxisome proliferator-activated receptor α, a potential therapeutic target for alcoholic liver disease. *World J. Gastroenterol.* **2014**, *20*, 8055–8060. [CrossRef] [PubMed]
35. Nanji, A.A.; Dannenberg, A.J.; Jokelainen, K.; Bass, N.M. Alcoholic liver injury in the rat is associated with reduced expression of peroxisome proliferator-alpha (PPARalpha)-regulated genes and is ameliorated by PPARalpha activation. *J. Pharmacol. Exp. Ther.* **2004**, *310*, 417–424. [CrossRef] [PubMed]
36. Kong, L.; Ren, W.; Li, W.; Zhao, S.; Mi, H.; Wang, R.; Zhang, Y.; Wu, W.; Nan, Y.; Yu, J. Activation of peroxisome proliferator activated receptor alpha ameliorates ethanol induced steatohepatitis in mice. *Lipids Health Dis.* **2011**, *10*, 246. [CrossRef] [PubMed]
37. Fischer, M.; You, M.; Matsumoto, M.; Crabb, D.W. Peroxisome Proliferator-activated Receptor α (PPARα) Agonist Treatment Reverses PPARα Dysfunction and Abnormalities in Hepatic Lipid Metabolism in Ethanol-fed Mice. *J. Biol. Chem.* **2003**, *278*, 27997–28004. [CrossRef] [PubMed]
38. Diesinger, T.; Buko, V.; Lautwein, A.; Dvorsky, R.; Belonovskaya, E.; Lukivskaya, O.; Naruta, E.; Kirko, S.; Andreev, V.; Buckert, D.; et al. Drug targeting CYP2E1 for the treatment of early-stage alcoholic steatohepatitis. *PLoS ONE* **2020**, *15*, e0235990. [CrossRef]
39. Lu, Y.; Wu, D.; Wang, X.; Ward, S.C.; Cederbaum, A.I. Chronic alcohol-induced liver injury and oxidant stress are decreased in cytochrome P4502E1 knockout mice and restored in humanized cytochrome P4502E1 knock-in mice. *Free. Radic. Biol. Med.* **2010**, *49*, 1406–1416. [CrossRef]

40. Shi, Y.; Liu, Y.; Wang, S.; Huang, J.; Luo, Z.; Jiang, M.; Lu, Y.; Lin, Q.; Liu, H.; Cheng, N.; et al. Endoplasmic reticulum-targeted inhibition of CYP2E1 with vitamin E nanoemulsions alleviates hepatocyte oxidative stress and reverses alcoholic liver disease. *Biomaterials* **2022**, *288*, 121720. [CrossRef]
41. Hartley, D.P.; Kolaja, K.L.; Reichard, J.; Petersen, D.R. 4-Hydroxynonenal and malondialdehyde hepatic protein adducts in rats treated with carbon tetrachloride: Immunochemical detection and lobular localization. *Toxicol. Appl. Pharmacol.* **1999**, *161*, 23–33. [CrossRef] [PubMed]
42. Aleynik, S.I.; Leo, M.A.; Aleynik, M.K.; Lieber, C.S. Increased Circulating Products of Lipid Peroxidation in Patients with Alcoholic Liver Disease. *Alcohol. Clin. Exp. Res.* **1998**, *22*, 192–196. [CrossRef] [PubMed]
43. Gupta, S.; Pandey, R.; Katyal, R.; Aggarwal, H.K.; Aggarwal, R.P.; Aggarwal, S.K. Lipid peroxide levels and antioxidant status in alcoholic liver disease. *Indian J. Clin. Biochem.* **2005**, *20*, 67–71. [CrossRef] [PubMed]
44. Kawaratani, H.; Tsujimoto, T.; Douhara, A.; Takaya, H.; Moriya, K.; Namisaki, T.; Noguchi, R.; Yoshiji, H.; Fujimoto, M.; Fukui, H. The effect of inflammatory cytokines in alcoholic liver disease. *Mediat. Inflamm* **2013**, *2013*, 495156. [CrossRef] [PubMed]
45. Kitazawa, T.; Nakatani, Y.; Fujimoto, M.; Tamura, N.; Uemura, M.; Fukui, H. The Production of Tumor Necrosis Factor-α by Macrophages in Rats With Acute Alcohol Loading. *Alcohol. Clin. Exp. Res.* **2003**, *27*, 72S–75S. [CrossRef]
46. Iimuro, Y.; Gallucci, R.M.; Luster, M.I.; Kono, H.; Thurman, R.G. Antibodies to tumor necrosis factor alfa attenuate hepatic necrosis and inflammation caused by chronic exposure to ethanol in the rat. *Hepatology* **1997**, *26*, 1530–1537. [CrossRef] [PubMed]
47. Aldred, A.; Nagy, L.E. Ethanol dissociates hormone-stimulated cAMP production from inhibition of TNF-alpha production in rat Kupffer cells. *Am. J. Physiol.* **1999**, *276*, G98–G106. [CrossRef] [PubMed]
48. McClain, C.J.; Cohen, D.A. Increased tumor necrosis factor production by monocytes in alcoholic hepatitis. *Hepatology* **1989**, *9*, 349–351. [CrossRef]
49. Yin, M.; Wheeler, M.D.; Kono, H.; Bradford, B.U.; Gallucci, R.M.; Luster, M.I.; Thurman, R.G. Essential role of tumor necrosis factor α in alcohol-induced liver injury in mice. *Gastroenterology* **1999**, *117*, 942–952. [CrossRef]
50. Vairetti, M.; Di Pasqua, L.G.; Cagna, M.; Richelmi, P.; Ferrigno, A.; Berardo, C. Changes in Glutathione Content in Liver Diseases: An Update. *Antioxidants* **2021**, *10*, 364. [CrossRef]
51. Kessova, I. Alcohol-induced liver injury in mice lacking Cu, Zn-superoxide dismutase. *Hepatology* **2003**, *38*, 1136–1145. [CrossRef] [PubMed]
52. Colell, A.; Garcia-Ruiz, C.; Miranda, M.; Ardite, E.; Mari, M.; Morales, A.; Corrales, F.; Kaplowitz, N.; Fernandez-Checa, J.C. Selective glutathione depletion of mitochondria by ethanol sensitizes hepatocytes to tumor necrosis factor. *Gastroenterology* **1998**, *115*, 1541–1551. [CrossRef] [PubMed]
53. Sun, J.; Fu, J.; Li, L.; Chen, C.; Wang, H.; Hou, Y.; Xu, Y.; Pi, J. Nrf2 in alcoholic liver disease. *Toxicol. Appl. Pharmacol.* **2018**, *357*, 62–69. [CrossRef] [PubMed]
54. Liu, J.; Wu, K.C.; Lu, Y.F.; Ekuase, E.; Klaassen, C.D. Nrf2 protection against liver injury produced by various hepatotoxicants. *Oxid. Med. Cell Longev.* **2013**, *2013*, 305861. [CrossRef] [PubMed]
55. Okawa, H.; Motohashi, H.; Kobayashi, A.; Aburatani, H.; Kensler, T.W.; Yamamoto, M. Hepatocyte-specific deletion of the keap1 gene activates Nrf2 and confers potent resistance against acute drug toxicity. *Biochem. Biophys Res. Commun.* **2006**, *339*, 79–88. [CrossRef]
56. Lamle, J.; Marhenke, S.; Borlak, J.; von Wasielewski, R.; Eriksson, C.J.; Geffers, R.; Manns, M.P.; Yamamoto, M.; Vogel, A. Nuclear factor-eythroid 2-related factor 2 prevents alcohol-induced fulminant liver injury. *Gastroenterology* **2008**, *134*, 1159–1168. [CrossRef]
57. He, J.L.; Zhou, Z.W.; Yin, J.J.; He, C.Q.; Zhou, S.F.; Yu, Y. Schisandra chinensis regulates drug metabolizing enzymes and drug transporters via activation of Nrf2-mediated signaling pathway. *Drug. Des. Devel. Ther.* **2015**, *9*, 127–146. [CrossRef] [PubMed]
58. Cho, I.J.; Kim, J.W.; Jung, J.J.; Sung, S.H.; Ku, S.K.; Choi, J.-S. In vitro protective effects of Hoveniae Semen cum Fructus extracts against oxidative stress. *Toxicol. Environ. Health Sci.* **2016**, *8*, 19–27. [CrossRef]
59. Szopa, A.; Ekiert, R.; Ekiert, H. Current knowledge of Schisandra chinensis (Turcz.) Baill. (Chinese magnolia vine) as a medicinal plant species: A review on the bioactive components, pharmacological properties, analytical and biotechnological studies. *Phytochem. Rev.* **2017**, *16*, 195–218. [CrossRef]
60. Guo, C.; Xue, G.; Pan, B.; Zhao, M.; Chen, S.; Gao, J.; Chen, T.; Qiu, L. Myricetin Ameliorates Ethanol-Induced Lipid Accumulation in Liver Cells by Reducing Fatty Acid Biosynthesis. *Mol. Nutr. Food Res.* **2019**, *63*, 1801393. [CrossRef]
61. Wang, M.; Zhu, P.; Jiang, C.; Ma, L.; Zhang, Z.; Zeng, X. Preliminary characterization, antioxidant activity in vitro and hepatoprotective effect on acute alcohol-induced liver injury in mice of polysaccharides from the peduncles of Hovenia dulcis. *Food Chem. Toxicol.* **2012**, *50*, 2964–2970. [CrossRef]
62. Buyco, D.G.; Martin, J.; Jeon, S.; Hooks, R.; Lin, C.; Carr, R. Experimental models of metabolic and alcoholic fatty liver disease. *World J. Gastroenterol.* **2021**, *27*, 1–18. [CrossRef]
63. Bertola, A.; Mathews, S.; Ki, S.H.; Wang, H.; Gao, B. Mouse model of chronic and binge ethanol feeding (the NIAAA model). *Nat. Protoc.* **2013**, *8*, 627–637. [CrossRef]

antioxidants

MDPI

Case Report

Oxidative Stress in a Mother Consuming Alcohol during Pregnancy and in Her Newborn: A Case Report

Martina Derme [1], Maria Grazia Piccioni [1], Roberto Brunelli [1], Alba Crognale [1], Marika Denotti [1], Paola Ciolli [1], Debora Scomparin [1], Luigi Tarani [1], Roberto Paparella [1], Gianluca Terrin [1], Maria Di Chiara [1], Alessandro Mattia [2], Simona Nicotera [2,3], Alberto Salomone [3], Mauro Ceccanti [4], Marisa Patrizia Messina [1], Nunzia La Maida [5], Giampiero Ferraguti [6], Carla Petrella [7] and Marco Fiore [7,*]

1 Department of Maternal Infantile and Urological Sciences, Sapienza University of Rome, 00185 Roma, Italy
2 Dipartimento Della Pubblica Sicurezza, Direzione Centrale di Sanità, Centro di Ricerche e Laboratorio di Tossicologia Forense, Ministero dell'Interno, 00185 Roma, Italy
3 Department of Chemistry, University of Turin, 10125 Turin, Italy
4 SITAC, Società Italiana per il Trattamento Dell'alcolismo e le sue Complicanze, 00185 Rome, Italy
5 National Centre on Addiction and Doping, Istituto Superiore di Sanità, 00161 Rome, Italy
6 Department of Experimental Medicine, Sapienza University of Rome, 00185 Rome, Italy
7 Institute of Biochemistry and Cell Biology (IBBC-CNR), Department of Sensory Organs, Sapienza University of Rome, 00185 Roma, Italy
* Correspondence: marco.fiore@cnr.it

Abstract: Fetal alcohol spectrum disorder (FASD) is a set of conditions resulting from prenatal alcohol exposure (PAE). FASD is estimated to affect between 2% and 5% of people in the United States and Western Europe. The exact teratogenic mechanism of alcohol on fetal development is still unclear. Ethanol (EtOH) contributes to the malfunctioning of the neurological system in children exposed in utero by decreasing glutathione peroxidase action, with an increase in the production of reactive oxygen species (ROS), which causes oxidative stress. We report a case of a mother with declared alcohol abuse and cigarette smoking during pregnancy. By analyzing the ethyl glucuronide (EtG, a metabolite of alcohol) and the nicotine/cotinine in the mother's hair and meconium, we confirmed the alcohol and smoking abuse magnitude. We also found that the mother during pregnancy was a cocaine abuser. As a result, her newborn was diagnosed with fetal alcohol syndrome (FAS). At the time of the delivery, the mother, but not the newborn, had an elevation in oxidative stress. However, the infant, a few days later, displayed marked potentiation in oxidative stress. The clinical complexity of the events involving the infant was presented and discussed, underlining also the importance that for cases of FASD, it is crucial to have more intensive hospital monitoring and controls during the initial days.

Keywords: fetal alcohol spectrum disorder (FASD); prenatal alcohol exposure (PAE); oxidative stress; ethyl glucuronide

Citation: Derme, M.; Piccioni, M.G.; Brunelli, R.; Crognale, A.; Denotti, M.; Ciolli, P.; Scomparin, D.; Tarani, L.; Paparella, R.; Terrin, G.; et al. Oxidative Stress in a Mother Consuming Alcohol during Pregnancy and in Her Newborn: A Case Report. *Antioxidants* **2023**, *12*, 1216. https://doi.org/10.3390/antiox12061216

Academic Editor: Gwonhwa Song

Received: 13 May 2023
Revised: 1 June 2023
Accepted: 2 June 2023
Published: 4 June 2023

1. Introduction

Prenatal alcohol exposure (PAE) is the leading preventable cause of developmental disabilities and congenital abnormalities [1,2]. Fetal alcohol spectrum disorder (FASD) is an umbrella term describing the plethora of conditions resulting from PAE [3]. FASD comprises alcohol-related neurodevelopmental disorder (ARND), alcohol-related birth defects (ARBD), fetal alcohol syndrome (FAS), and partial fetal alcohol syndrome (pFAS) [4]. FASD is thought to affect between 2% and 5% of people in Western Europe and the United States [5]. Over the years, different diagnostic guidelines for FASD have been developed; among the most recent are the guidelines by Hoyme [6], used in our case report.

Both animal models and clinical investigations have demonstrated that ethanol (EtOH) distributes through the placenta and diffuses quickly into the fetus [7]. Fetal blood alcohol levels approach maternal levels within two hours of maternal intake [8]. Once alcohol has reached the fetal circulation, the fetus attempts to metabolize EtOH through pathways similar to those of the adult [7,9]. Although the metabolic ability of the fetal liver is limited, the available findings demonstrate that the key enzyme involved in EtOH oxidation—alcohol dehydrogenases (ADH)—is present in the fetus at 2 months of gestation [9,10]. However, the metabolic capability of the fetus is severely reduced, working at a rate of about 5 to 10% of adult metabolic activity. Upon exposure, the fetus removes the EtOH through two different mechanisms: renal and pulmonary. The fetus's reduced capability to metabolize EtOH potentiates the duration of exposure. Despite the fetus's capability, although reduced, to remove EtOH as pulmonary and renal excretions, EtOH remains present in the amniotic fluid, leading to reabsorption, prolonging too the exposure time. Fetal swallowing has begun by 11 weeks of gestation and represents the main reabsorption component providing a path for EtOH reentry into the fetal circulation. The amniotic fluid amount swallowed by the fetus (500 to 1000 mL per day) does not account for the quantity of pulmonary secretions and urine (970 to 1370 mL per day) entering the amniotic fluid [11].

There are many different proposed mechanisms of alcohol teratogenicity [12]. EtOH can compromise endogenous antioxidant capacity, for example, by decreasing glutathione peroxidase levels or generating free radicals. Free radicals and reactive oxygen species (ROS), such as superoxide (O_2^-) and hydroxide (HO^-) ions, are generally considered to be responsible for fetal brain damage by inducing uncontrolled apoptosis [13–15]. FAS facial morphology is probably linked to the apoptotic effects of alcohol on cranial neural crest cells [16].

The neuropsychiatric effects of FASD may be explained by EtOH, favoring the apoptosis of serotonergic neurons as shown in a mouse model [17]. FASD features appear differently in every child, but all FASD children possess behavioral and/or intellectual damage [18,19]. These children often have disrupted cognitive functioning leading to problems following directions, learning disabilities, poor memory skills, hyperactive behavior, speech and language delays, inattentiveness, and difficulties in understanding the consequences of their actions [4,20].

Many FASD individuals have also anomalous facial features such as a smooth ridge between the nose and upper lip, a flat nasal bridge, short palpebral fissures, extra crease in the outer ears, a thin vermillion border of the upper lip, an upturned nose, and a curved pinky finger [21]. Many children born with FASD display a smaller head size, shorter than average height, and low body weight [22]. In addition to these physical features, there are numerous medical matters that seem to be caused by fetal alcohol exposure.

There is a strong association between alcohol use during pregnancy and congenital heart defects [23,24]. Significant associations were reported with atrial and ventricular septal defects. Furthermore, mothers abusing alcohol during gestation have a 1.64-fold times elevated risk of having an infant affected by conotruncal defect subtypes such as Great Arteries transposition. These data suggest that both binge drinking and prenatal heavy drinking are intensely associated with an overall elevated risk of having babies with congenital heart defects [25]. There is no a direct FASD cure, but early intervention and life-long support may help those born with FASD to cope with the complications that come with it [26].

It should be noted that recent studies refer to FASD and the fetal programming theory as a new concept [27,28]. This transformation expands FASD from being solely a neural disorder to a "whole body disorder" that affects multiple organs and systems [27,28]. This, in turn, increases the potential risk for developing chronic conditions such as cardiovascular disease (CVD) or diabetes later in life. According to these considerations, it becomes even more crucial to prioritize intensive hospital monitoring and controls during the initial days of infants affected by FASD.

In this study, we reported a case of a mother, with alcohol abuse during pregnancy, and her newborn diagnosed with FAS, presenting both increased serum ROS and oxidative stress. A significant novelty of the present case report is the disclosure, throughout 9 months of gestation, of long-lasting substance abuse (alcohol, smoking, and cocaine) by subtle hair analyses.

2. Case Report

A 29-year-old Caucasian woman (gravida 1, para 0) was referred and admitted to our hospital for the management of fetal growth restriction (FGR) and oligohydramnios at 33 weeks and 6 days of gestation. The patient had iatrogenic hypothyroidism, treated with Levotiroxine 75 mcg, due to hemithyroidectomy for a benign thyroid nodule in 2016. She started Enoxaparin 4000 UI two weeks before hospitalization.

The patient reported smoking five cigarettes per day and consuming four glasses of wine per day during all the pregnancy. She also reported using cocaine during the first two months of pregnancy. The first ultrasound at 12 weeks of gestation described a crown-rump length (CRL) corresponding to the anamnestic gestational age. The combined test in the first trimester resulted in low risk for the main aneuploidies, and a second trimester screening at 20 weeks of gestation described normal fetal anatomy.

Upon admission to the hospital, we performed an ultrasound exam with evidence of reduced amniotic fluid (amniotic fluid index—AFI: 73 mm) and an estimated fetal weight of 1559 g, equal to the first percentile (z-score −3.303) for gestational age [29–31]. Fetal Doppler showed an umbilical artery pulsatility index (PI) of 1.23, equal to the 96th percentile (z-score 1.697), middle cerebral artery PI of 1.62, equal to the 16th percentile (z-score −0.995), with the cerebroplacental ratio of 1.32, equal to the 3rd percentile (z-score −2.004) for gestational age. Antenatal corticosteroids (betamethasone 12 mg intramuscular, two doses 24 h apart) were administered in order to prevent neonatal respiratory distress syndrome [32].

According to the Truffle Study [33], we monitored the patient every three days with an ultrasound fetal Doppler and cardiotocographic examination twice a day. We also practiced a urine screen test for amphetamine, benzodiazepines, cocaine, methadone, opiates, and cannabinoids, with all tests resulting negative. A patient hair sample (10 cm from the root), with her consent, was collected in order to evaluate the use of alcohol, smoking, and cocaine during pregnancy. In addition, a urine sample was used in order to test alcohol abuse.

A rapid test for the diagnosis of premature rupture of membranes resulted negative. Cardiotocographic examinations always showed normal baseline heart rate, presence of accelerations, absence of decelerations, and no uterine contractions, and the short-term variability was always >4.5 ms [33]. An ultrasound exam after seven days of hospitalization (34 weeks and 6 days of gestational age) confirmed reduced amniotic fluid (AFI: 70 mm) and showed umbilical artery PI of 1.27, equal to the 98th percentile (z-score 2.005), middle cerebral artery PI of 1.21, equal to the 1st percentile (z-score −2.813), with cerebroplacental ratio of 0.95, equal to the 1st percentile (z-score −3.507) for gestational age (Figures 1 and 2).

The day after (35 weeks of gestational age), according to our local protocols, we decided for the induction of labor with vaginal Dinoprostone. Three hours after starting induction, the cardiotocography showed repetitive variable decelerations and the absence of uterine activity. The vaginal dispositive was removed after 27 min of monitoring. Intravenous fluids were administered and cardiotocography was continued. After sixty minutes of repetitive variable decelerations, loss of major accelerations, reduced baseline variability, and fetal tachycardia (mean fetal heart rate of 170 bpm), we decided to perform an emergency cesarean section.

The day of delivery, maternal and neonatal blood samples were collected to evaluate the oxidative stress status. The newborn was a female, APGAR scores were 8 at the 1st minute and 10 at the 5th minute, pH of umbilical cord blood was 7.31, the birth weight was 1515 g (below the third percentile), the length was 39 cm (below the third percentile), and the head circumference was 29.3 cm (3° centile). The newborn was admitted to the neonatal intensive care unit (NICU) due to low birth weight (LBW), although in good general clinical condition. The urine toxicology screen came out negative.

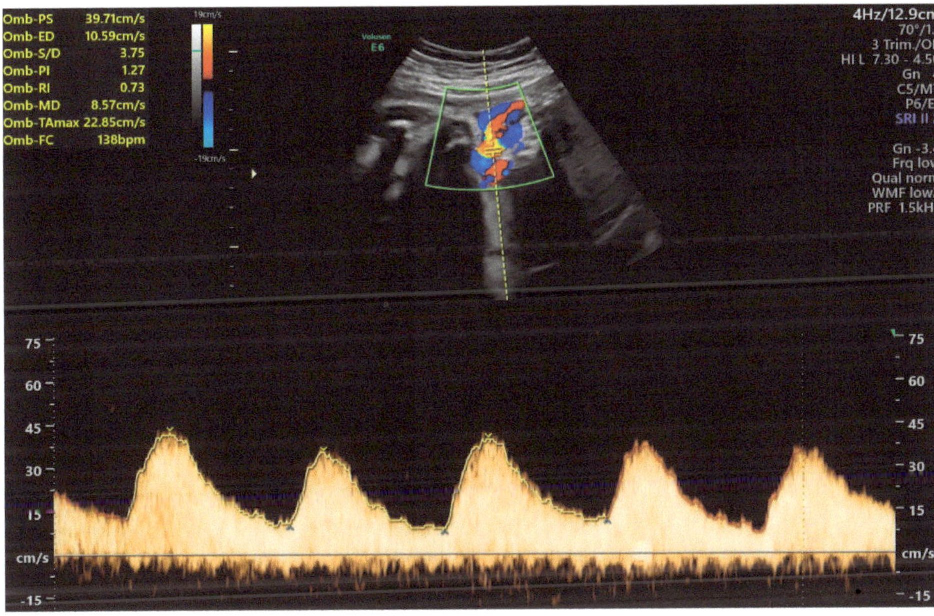

Figure 1. Ultrasound Doppler of the umbilical artery pulsatility index at 34 weeks and 6 days.

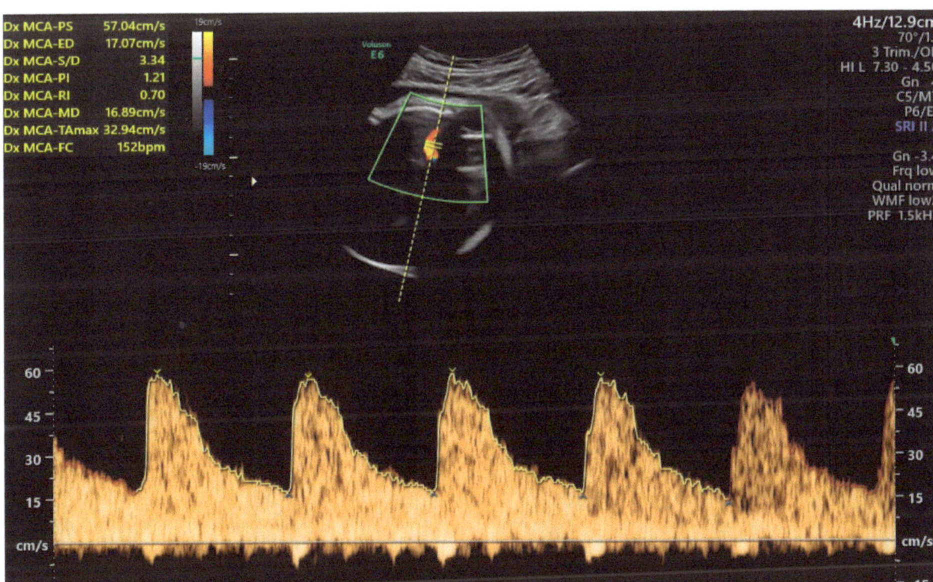

Figure 2. Ultrasound Doppler of the middle cerebral artery pulsatility index at 34 weeks and 6 days.

According also to the indications of the juvenile magistrate, to avoid the risk of contamination with maternal milk, that could contain EtOH, breastfeeding was not allowed by the mother. Bottle feeding containing the standard nutritional elements of the hospital department of maternal infantile was indeed the best choice for the baby.

A physical examination performed by the neonatologists and the pediatric geneticist was significant for smooth philtrum and thin upper vermilion border, which raised suspicion for FASD along with FGR and microcephaly. FASD diagnosis was corroborated by the detection of ethyl glucuronide (EtG) in newborn meconium. Screening for further malformations was also performed. Otoacoustic emissions, abdominal ultrasound, and whole-body X-ray were unremarkable. Echocardiography displayed a foramen ovale aneurysm with a hemodynamically insignificant left-to-right shunt. Cerebral ultrasound and ophthalmic examination showed delayed gyral and sulcal development and incomplete vascularization of the peripheral retina, respectively, the latter of which resolved with the newborn's growth.

Complete blood count, liver and renal function tests, and electrolytes were all within the normal limits. Fifty-two days after birth, the neurological examination resulted negative. Eighty-one days after birth, a neonatal blood sample was used in order to evaluate the oxidative stress status and the result was positive, although the day of the delivery was in normal range.

3. Experimental Methods and Results

3.1. FORT (Free Oxygen Radicals Test) Investigation

By using a special kit (Callegari, Parma, Italy) for the analysis of ROS [34,35], we measured maternal and neonatal serum oxidative stress status on the day of delivery in cord blood serum [36]. FORT is a colorimetric assay based on the ability of transition metals such as iron to catalyze, in the presence of hydroperoxides (ROOH), the formation of free radicals (reaction 1–2), which are then trapped by an amine derivative, $CrNH_2$ [14,35]. The amine reacts with free radicals, forming a colored, fairly long-lived radical cation, detectable at 505 nm (reaction 3). The intensity of the color correlates directly to the number of radical compounds and the hydroperoxides concentration and, consequently, to the oxidative status of the sample according to the Lambert–Beer law [14]. According to the instructions provided by the manufacturer, FORT values below 300 units (U) indicate an optimal condition of oxidative stress, for values between 300 and 330 U, a condition of latent oxidative stress, while for values superior to 330 U, a condition of oxidative stress in progress. It should be noted that there are no established reference values for women during delivery and for newborns. On the day of delivery, the maternal results described a high oxidative stress status (475 Fort U/3.61 mmol/L H_2O_2), while the neonatal results were normal (<160 Fort U/<1.22 mmol/L H_2O_2). Eighty-one days after birth, the neonatal results were positive (471 Fort U/3.58 mmol/L H_2O_2).

3.2. Maternal Hair EtG, Nicotine/Cotinine, and Cocaine Analyses

The EtG analysis followed the protocol described by Mattia et al. [37]. Briefly, 50 mg of hair samples was decontaminated by two consecutive washes in 5 mL of methanol and 5 mL of dichloromethane, manually shredded, and incubated overnight at 60 °C in water with EtG-D5 as the internal standard. The extract was purified with solid phase extraction polymeric cartridges, air-dried, and reconstituted with acetonitrile (CAN) and N-Methyl-N-(trimethylsilyl)trifluoroacetamide (MSTFA) as a derivatizing agent for injection in an Agilent Technologies (Santa Clara, CA, USA) 7890B gas chromatograph coupled to a 7000C tandem mass selective detector operating in EI ionization mode. Hair nicotine and cotinine analyses were performed on an aliquot of 20 mg of hair samples that were decontaminated by a sonication for 15 min in 5 mL of dichloromethane, manually shredded, incubated overnight at 60 °C in NaOH 1 M with Nicotine-D4 and Cotinine-D3 as internal standard, and purified with liquid extraction in 1 mL of dichloromethane and 1 mL of 25% K_2CO_3. The organic phase was recovered and added with 100 μL of methanol before being concentrated under nitrogen flow to about 100 μL for GC-MS/MS analysis performed by an Agilent Technologies 7890B gas chromatograph coupled to a 7000C tandem mass selective detector operating in EI ionization mode. For the analysis of cocaine and its metabolite, benzoylecgonine, an aliquot of 40 mg of hair was decontaminated with three washes, respectively, in methanol, dichloromethane, and methanol. The last wash was analyzed to verify the absence of external cocaine contamination. The hair was dried and finely minced before being incubated overnight at 40 °C in pH6 phosphate buffer, cocaine-D3, and benzoylecgonine-D3 as internal standards. The extract was purified with anion exchange mixed-bed solid-phase extraction (SPE) cartridges. The eluate was dried under nitrogen flow and reconstituted in 50 µL of N,O-Bis(trimethylsilyl)trifluoroacetamide (BSTFA) before the analysis. The analysis was performed with an Agilent 7890B gas chromatograph coupled to a 5977B mass spectrometer detector. In our case, the EtG results were 107 pg/mg in the first half of pregnancy; 151 pg/mg in the second half of pregnancy (cut off < 30 pg/mg). The nicotine result was 9.7 ng/mg (cut-off: 0.16 ng/mg) and the cotinine result was 0.3 ng/mg (cut-off: 0.07 ng/mg). Cocaine and benzoylecgonine analyses were positive with a value of 6.25 ng/mg (cut-off: 0.15 ng/mg) and 24 ng/mg (cut-off: 0.12 ng/mg), respectively.

3.3. EtG Determination in Maternal Urine Sample

The DRI Ethyl Glucuronide Assay (W1510011723, Instrumentation Laboratory SpA, V.le Monza, 338, 20128 Milan, Italy, www.werfen.com; accessed on 20 May 2023) used in this study is intended for the qualitative and semiquantitative determination of EtG in human urine at a cut-off of 100 ng/mL as a direct value of alcohol consumption, as described by the manufacturer [38]. Briefly, this immunoenzymatic assay uses specific antibodies which can detect EtG without generating significant cross-reactivity with other glucuronic compounds. The test is based on the competition between the drug conjugate with the glucose-6-phosphate dehydrogenase (G6PDH) and the free drug in the urine sample for a fixed number of specific binding sites for the antibody [39]. In the absence of free drug in the sample, the antibody binds to the drug conjugate with G6PDH, causing a decrease in enzyme activity. This phenomenon creates a direct relationship between the concentration of the drug in urine and enzyme activity. Active enzyme converts the NAD to NADH, producing an alteration in absorbance, which can be measured with a spectrophotometric examination at 340 nm [38]. The cut-off set at 100 ng/mL was used to discriminate exposure to alcohol from non-beverage sources, or incidental exposure, which can lead to false positives [39]. The sources of possible exposure in the environment may include alcohol in mouthwash, foods, over-the-counter medications containing ethanol, and even the inhalation of alcohol from topical use [39]. In our case, the EtG in the urine sample was 124 ng/mL.

3.4. EtG Determination in Neonatal Meconium

The EtG in neonatal meconium was determined according to the methods previously described [40]. Briefly, 1 mL of acetonitrile and 2 µL EtG-D5 internal standard solution were added to 100 mg of meconium and ultrasonicated at 40 °C for ten minutes. Solid-phase extraction was performed using Bond Elut-NH2 cartridges, and EtG was eluted with 1 mL HCl methanolic solution (2% *v*/*v*). The eluate was dried under nitrogen stream and derivatized with N,O-Bis(trimethylsilyl)trifluoroacetamide solution (BSTFA + 1%TMS) at 70 °C for 30 min. The extract was injected into a 7890B GC chromatographic system coupled to a 7000C triple quadruple mass spectrometer operating in multiple reaction monitoring mode acquisition [40]. EtG quantification was performed using six-point calibration curves linear in the range from 10 to 200 ng/g with a determination coefficient greater than 0.99. The bias and precision values were within the ranges acceptable in forensic toxicology (±20% bias; <20% Coefficient of Variation) [41,42]. In our case, the detection of EtG in newborn's meconium was 37 ng/g (cut off as limit of detection, 10 ng/g).

4. Discussion

In this case report, we show that in a mother smoking cigarettes, drinking alcohol, and abusing cocaine during pregnancy, serum oxidative stress was strongly potentiated during delivery. We also disclosed that her newborn displayed FASD facies with in-range serum oxidative stress during delivery. However, a marked potentiation in oxidative stress measured as serum FORT was clearly evidenced in the newborn a few days after delivery. This difference is truly interesting, considering that childbirth is a physiologically stressful event in human life, and stress is usually associated with an increase in oxidative stress. The observed oxidative stress elevation in the mother during delivery could be both explained by a direct effect of the discovered gestational abuse in smoking, drinking, and using cocaine, or in combination with the stressful event of the labor and delivery. The increase in oxidative stress in the newborn disclosed a few days after birth may indicate a further long-lasting effect of the mother's substance abuse. However, the stress induced in the newborn by hospitalization in the pediatric department after delivery for early treatments should be considered.

It should be noted that no breastfeeding was allowed to avoid contaminations with the maternal milk that could contain EtOH. Indeed, in experimental preclinical studies, it has been described that when mothers consume ethanol during the lactation period, the effects on the offspring can be even more detrimental compared to if the mother consumed ethanol during gestation. Therefore, despite the fact that there are less studies analyzing the effects of EtOH consumption during breastfeeding and the consequent damage caused to children, it has been found in rats that the pups exposed to EtOH only during weaning presented a lower body weight at the end of breastfeeding than those who were exposed to EtOH only during gestation [43]. Furthermore, as shown in numerous human and animal model studies, malnutrition can be accompanied with alcohol and cocaine abuse during pregnancy [44–48]. Indeed, deficiency in nutrients in utero may also play a role in the developmental disorders related to oxidative stress and prenatal alcohol exposure.

As for oxidative stress in FASD, oxidative stress is a condition determined by an imbalance between antioxidant and oxidative factors [13]. Oxidative stress may elicit both necrosis and cell apoptosis, determining increased carcinogenesis, cellular aging, increased autoimmune diseases, vascular stiffening, and muscle decay. In the framework of pediatric syndromes, including FASD, oxidative stress may play a subtle role in the first order [13]. Preclinical studies have indeed shown that gestational alcohol exposure disrupts the ability of specific potassium channels to dilate cerebral arterioles, which appears to be mediated also by increases in oxidative stress [49]. Furthermore, neonatal ethanol exposure induces deficits in depressive-like behavior and in context-dependent fear learning in a rat model that were associated with elevated oxidative stress findings in the prefrontal cortex and hippocampus [35].

FASD is a severe neonatal condition due to PAE [50–52]. Damage deriving from PAE can be permanent and have no cure, so proper identification and early treatment can help to prevent and mitigate neurological consequences that can affect the individual later in life. The severity of FASD depends on the quantity of alcohol consumed and the frequency of drinking, as well as the gestational age at which the alcohol was consumed by the mother [53–56]. Intervention services for the prevention and sensibilization of mothers may reduce the incidence of FASD. FASD is entirely preventable by avoiding alcohol. Since there is not a specific pharmacologic cure to treat alcohol-induced damage, early primary and secondary interventions are the only tools which can be used to care for children with FASD.

Diagnosing FASD in patients requires a thorough assessment, and proof of alcohol abuse during pregnancy may help clinicians in the differential diagnosis with other syndromic conditions. In the neonatal phase, FASD may be disclosed by the presence of a small bay for the gestational age (SGA), with microcephaly and typical dysmorphic characters (short palpebral fissures, middle-facial hypoplasia, thin upper lip, telecanthus, jaw hypoplasia, elongated and flat nasolabial filter, and anomalies of the ears) [57–60]. In the neonatal or early pediatric period, heart defects can be detected. Central nervous system (CNS) anomalies, neurobehavioral conditions, and cognitive impairment represent some of the worst consequences which can be detected later in the stages of the development of the infant/adolescent.

The exact teratogenic mechanism of alcohol on fetal development is still unclear. EtOH contributes to the malfunctioning of the neurological system in children exposed in utero by decreasing glutathione peroxidase action, with an increase in the production of ROS [30,31,61]. It causes oxidative stress and the oxidation of polyunsaturated fatty acids with the formation of aldehyde toxicants [14]. Acute and chronic alcohol exposure during prenatal development also results in impaired mitochondrial morphology and function, another important cause of cell oxidative stress. Most recent studies point to mitochondria as the potential link between prenatal alcohol exposure and brain damage [62].

In a mouse model, extended PAE generates an increased fraction of immature mitochondria in the fetal brain [63]. Depressed mitochondrial functioning is detected in the early postnatal phase in brain and liver tissues, including the cerebellar neurons of rats exposed prenatally. Mitochondria dysfunctions play a subtle role in the progressive decline detected during aging, in the development of neurocognitive disorders (such as Alzheimer's disorder), and substance abuse [64]. Moreover, EtOH can directly alter the function of neurons in the CNS by disrupting the synthesis and release of neurotrophic factors, such as brain-derived neurotrophic factor (BDNF) and nerve growth factor (NGF) [65,66]. These neurotrophic factors play important roles in the growth, nutrition, survival, and development of neuronal and non-neuronal cells.

Indeed, during fetal life, disruptions in BDNF and NGF may affect the development of the limbic system, with long-lasting changes in neuronal connections and predisposition to neuropsychiatric disorders [67,68]. To the best of our knowledge, few studies have been conducted on human beings, and most of the available data are based on studies conducted on animals. The aim of our research was to provide a starting point for future studies on a larger scale in order to clarify other physiopathological mechanisms leading to FASD and to understand the real role of EtOH in oxidative stress processes and subsequent neurocognitive impairment in children with FASD.

The main limitation of this study is the fact that there are no established FORT reference values for women during delivery and for newborns.

5. Conclusions

In conclusion, (i) these data do underline once again the importance, on the part of pregnant women, of carrying on the pregnancy in the healthiest way possible, avoiding any type of unnecessary stress, which includes the consumption of alcohol, smoking, and, of course, any type of drugs, as well as (ii) the importance of postnatal intensive hospital monitoring and controls in children when FASD is suspected.

Author Contributions: Conceptualization, M.F., M.D. (Martina Derme), M.G.P., A.C., M.C., L.T., G.T., C.P., P.C. and A.M., methodology, A.C., M.D.C., M.D. (Marika Denotti), S.N., M.P.M., N.L.M., R.P., D.S. and G.F., validation, M.F., L.T., M.G.P., G.T. and R.B.; formal analysis, M.F., C.P. and M.C.; investigation, M.F., M.D. (Martina Derme), A.S., D.S., M.G.P., A.C., M.P.M., M.C., L.T., G.T., C.P., P.C. and A.M.; data curation, M.F., A.S., A.M. and M.C.; writing—original draft preparation, M.D. (Martina Derme) and M.F., writing—review and editing, M.D. (Martina Derme) and M.F., visualization, M.D. (Martina Derme) and M.F. All authors have read and agreed to the published version of the manuscript.

Funding: This research received no external funding.

Institutional Review Board Statement: The study was approved with the protocol number 0382/2023.

Informed Consent Statement: Informed consent was obtained from the patient involved in the study.

Data Availability Statement: Data are available on request.

Acknowledgments: The authors thank the Sapienza university of Rome, IBBC-CNR, and SITAC.

Conflicts of Interest: The authors declare no conflict of interest.

References

1. Chudley, A.E.; Conry, J.; Cook, J.L.; Loock, C.; Rosales, T.; LeBlanc, N. Fetal alcohol spectrum disorder: Canadian guidelines for diagnosis. *CMAJ* **2005**, *172*, S1–S21. [CrossRef] [PubMed]
2. Manning, M.A.; Eugene Hoyme, H. Fetal alcohol spectrum disorders: A practical clinical approach to diagnosis. *Neurosci. Biobehav. Rev.* **2007**, *31*, 230–238. [CrossRef] [PubMed]
3. Walker, D.S.; Edwards, W.E.R.; Herrington, C. Fetal alcohol spectrum disorders: Prevention, identification, and intervention. *Nurse Pract.* **2016**, *41*, 28–34. [CrossRef] [PubMed]
4. Brems, C.; Johnson, M.E.; Metzger, J.S.; Dewane, S.L. College students' knowledge about Fetal Alcohol Spectrum Disorder. *J. Popul. Ther. Clin. Pharmacol.* **2014**, *21*, e159–e166.
5. Lange, S.; Probst, C.; Gmel, G.; Rehm, J.; Burd, L.; Popova, S. Global prevalence of fetal alcohol spectrum disorder among children and youth: A systematic review and meta-analysis. *JAMA Pediatr.* **2017**, *171*, 948–956. [CrossRef]
6. Hoyme, H.E.; Kalberg, W.O.; Elliott, A.J.; Blankenship, J.; Buckley, D.; Marais, A.S.; Manning, M.A.; Robinson, L.K.; Adam, M.P.; Abdul-Rahman, O.; et al. Updated clinical guidelines for diagnosing fetal alcohol spectrum disorders. *Pediatrics* **2016**, *138*, e20154256. [CrossRef]
7. Gupta, K.K.; Gupta, V.K.; Shirasaka, T. An Update on Fetal Alcohol Syndrome—Pathogenesis, Risks, and Treatment. *Alcohol. Clin. Exp. Res.* **2016**, *40*, 1594–1602. [CrossRef]
8. Burd, L.; Blair, J.; Dropps, K. Prenatal alcohol exposure, blood alcohol concentrations and alcohol elimination rates for the mother, fetus and newborn. *J. Perinatol.* **2012**, *32*, 652–659. [CrossRef]
9. Guerri, C.; Sanchis, R. Acetaldehyde and alcohol levels in pregnant rats and their fetuses. *Alcohol* **1985**, *2*, 267–270. [CrossRef]
10. Tanaka, H.; Suzuki, N.; Arima, M. Experimental studies on the influence of male alcoholism on fetal development. *Brain Dev.* **1982**, *4*, 1–6. [CrossRef]
11. Pikkarainen, P.H. Metabolism of ethanol and acetaldehyde in perfused human fetal liver. *Life Sci.* **1971**, *10*, 1359–1364. [CrossRef]
12. Kaufman, M.H. The teratogenic effects of alcohol following exposure during pregnancy, and its influence on the chromosome constitution of the pre-ovulatory egg. *Alcohol Alcohol.* **1997**, *32*, 113–128. [CrossRef] [PubMed]
13. Micangeli, G.; Menghi, M.; Profeta, G.; Tarani, F.; Mariani, A.; Petrella, C.; Barbato, C.; Ferraguti, G.; Ceccanti, M.; Tarani, L.; et al. The Impact of Oxidative Stress on Pediatrics Syndromes. *Antioxidants* **2022**, *11*, 1983. [CrossRef] [PubMed]
14. Fiore, M.; Petrella, C.; Coriale, G.; Rosso, P.; Fico, E.; Ralli, M.; Greco, A.; De Vincentiis, M.; Minni, A.; Polimeni, A.; et al. Markers of Neuroinflammation in the Serum of Prepubertal Children with Fetal Alcohol Spectrum Disorders. *CNS Neurol. Disord.-Drug Targets* **2022**, *21*, 854–868. [CrossRef] [PubMed]
15. Rahman, I.; Biswas, S.K.; Kode, A. Oxidant and antioxidant balance in the airways and airway diseases. *Eur. J. Pharmacol.* **2006**, *533*, 222–239. [CrossRef] [PubMed]
16. Cartwright, M.M.; Smith, S.M. Stage-Dependent Effects of Ethanol on Cranial Neural Crest Cell Development: Partial Basis for the Phenotypic Variations Observed in Fetal Alcohol Syndrome. *Alcohol. Clin. Exp. Res.* **1995**, *19*, 1454–1462. [CrossRef] [PubMed]
17. Sari, Y.; Zhou, F.C. Prenatal alcohol exposure causes long-term serotonin neuron deficit in mice. *Alcohol. Clin. Exp. Res.* **2004**, *28*, 941–948. [CrossRef]
18. May, P.A.; Baete, A.; Russo, J.; Elliott, A.J.; Blankenship, J.; Kalberg, W.O.; Buckley, D.; Brooks, M.; Hasken, J.; Abdul-Rahman, O.; et al. Prevalence and Characteristics of Fetal Alcohol Spectrum Disorders. *Pediatrics* **2014**, *134*, 855–866. [CrossRef]
19. Rasmussen, C.; Horne, K.; Witol, A. Neurobehavioral functioning in children with fetal alcohol spectrum disorder. *Child Neuropsychol.* **2006**, *12*, 453–468. [CrossRef]
20. Coriale, G.; Fiorentino, D.; Lauro, F.D.I.; Marchitelli, R.; Scalese, B.; Fiore, M.; Maviglia, M.; Ceccanti, M. Fetal Alcohol Spectrum Disorder (FASD): Neurobehavioral profile, indications for diagnosis and treatment. *Rivista di Psichiatria* **2013**, *48*, 359–369. [CrossRef]
21. Williams, J.F.; Smith, V.C. Fetal Alcohol Spectrum Disorders. *Pediatrics* **2015**, *136*, e1395–e1406. [CrossRef] [PubMed]
22. Murawski, N.J.; Moore, E.M.; Thomas, J.D.; Riley, E.P. Advances in diagnosis and treatment of fetal alcohol spectrum disorders: From animal models to human studies. *Alcohol Res. Curr. Rev.* **2015**, *37*, 97–108.
23. Zhang, S.; Wang, L.; Yang, T.; Chen, L.; Zhao, L.; Wang, T.; Chen, L.; Ye, Z.; Zheng, Z.; Qin, J. Parental alcohol consumption and the risk of congenital heart diseases in offspring: An updated systematic review and meta-analysis. *Eur. J. Prev. Cardiol.* **2020**, *27*, 410–421. [CrossRef] [PubMed]
24. Terracina, S.; Ferraguti, G.; Tarani, L.; Messina, M.P.; Lucarelli, M.; Vitali, M.; De Persis, S.; Greco, A.; Minni, A.; Polimeni, A.; et al. Transgenerational Abnormalities Induced by Paternal Preconceptual Alcohol Drinking. Findings from Humans and Animal Models. *Curr. Neuropharmacol.* **2021**, *19*, 1158–1173. [CrossRef] [PubMed]
25. Yang, J.; Qiu, H.; Qu, P.; Zhang, R.; Zeng, L.; Yan, H. Prenatal alcohol exposure and congenital heart defects: A meta-analysis. *PLoS ONE* **2015**, *10*, e0130681. [CrossRef] [PubMed]
26. Wilhoit, L.F.; Scott, D.A.; Simecka, B.A. Fetal Alcohol Spectrum Disorders: Characteristics, Complications, and Treatment. *Community Ment. Health J.* **2017**, *53*, 711–718. [CrossRef] [PubMed]
27. Ojeda, M.L.; Nogales, F.; Romero-Herrera, I.; Carreras, O. Fetal programming is deeply related to maternal selenium status and oxidative balance; experimental offspring health repercussions. *Nutrients* **2021**, *13*, 2085. [CrossRef]

28. Himmelreich, M.; Lutke, C.J.; Hargrove, E.T. The lay of the land: Fetal alcohol spectrum disorder (FASD) as a whole-body diagnosis. In *The Routledge Handbook of Social Work and Addictive Behaviors*; Routledge: Oxfordshire, UK, 2020; pp. 191–215; ISBN 9780429203121.
29. Panda, S.; Jayalakshmi, M.; Shashi Kumari, G.; Mahalakshmi, G.; Srujan, Y.; Anusha, V. Oligoamnios and Perinatal Outcome. *J. Obstet. Gynecol. India* **2017**, *67*, 104–108. [CrossRef]
30. Yang, M.; Zhou, X.; Tan, X.; Huang, X.; Yuan, L.; Zhang, Z.; Yang, Y.; Xu, M.; Wan, Y.; Li, Z. The Status of Oxidative Stress in Patients with Alcohol Dependence: A Meta-Analysis. *Antioxidants* **2022**, *11*, 1919. [CrossRef]
31. McDonough, K.H. The role of alcohol in the oxidant antioxidant balance in heart. *Front. Biosci.* **1999**, *4*, d601. [CrossRef]
32. Roberts, D.; Dalziel, S. Antenatal corticosteroids for accelerating fetal lung maturation for women at risk of preterm birth. *Cochrane Database Syst. Rev.* **2006**, *3*, CD004454. [CrossRef]
33. Lees, C.; Marlow, N.; Arabin, B.; Bilardo, C.M.; Brezinka, C.; Derks, J.B.; Duvekot, J.; Frusca, T.; Diemert, A.; Ferrazzi, E.; et al. Perinatal morbidity and mortality in early-onset fetal growth restriction: Cohort outcomes of the trial of randomized umbilical and fetal flow in Europe (TRUFFLE). *Ultrasound Obstet. Gynecol.* **2013**, *42*, 400–408. [CrossRef] [PubMed]
34. Palmieri, B.; Sblendorio, V. Oxidative stress tests: Overview on reliability and use. Part II. *Eur. Rev. Med. Pharmacol. Sci.* **2007**, *11*, 383–399. [PubMed]
35. Lopatynska-mazurek, M.; Komsta, L.; Gibula-tarlowska, E.; Kotlinska, J.H. Aversive learning deficits and depressive-like behaviors are accompanied by an increase in oxidative stress in a rat model of fetal alcohol spectrum disorders: The protective effect of rapamycin. *Int. J. Mol. Sci.* **2021**, *22*, 7083. [CrossRef] [PubMed]
36. Petrella, C.; Carito, V.; Carere, C.; Ferraguti, G.; Ciafrè, S.; Natella, F.; Bello, C.; Greco, A.; Ralli, M.; Mancinelli, R.; et al. Oxidative stress inhibition by resveratrol in alcohol-dependent mice. *Nutrition* **2020**, *79–80*, 110783. [CrossRef]
37. Mattia, A.; Moschella, C.; David, M.C.; Fiore, M.; Gariglio, S.; Salomone, A.; Vincenti, M. Development and Validation of a GC-EI-MS/MS Method for Ethyl Glucuronide Quantification in Human Hair. *Front. Chem.* **2022**, *10*, 858205. [CrossRef]
38. Ferraguti, G.; Merlino, L.; Battagliese, G.; Piccioni, M.G.; Barbaro, G.; Carito, V.; Messina, M.P.; Scalese, B.; Coriale, G.; Fiore, M.; et al. Fetus morphology changes by second-trimester ultrasound in pregnant women drinking alcohol. *Addict. Biol.* **2020**, *25*, e12724. [CrossRef]
39. Ceci, F.M.; Fiore, M.; Agostinelli, E.; Tahara, T.; Greco, A.; Ralli, M.; Polimeni, A.; Lucarelli, M.; Colletti, R.; Angeloni, A.; et al. Urinary Ethyl Glucuronide for the Assessment of Alcohol Consumption During Pregnancy: Comparison between Biochemical Data and Screening Questionnaires. *Curr. Med. Chem.* **2021**, *29*, 3125–3141. [CrossRef]
40. La Maida, N.; Di Trana, A.; Mannocchi, G.; Zaami, S.; Busardò, F.P. Sensitive and reliable gas chromatography tandem mass spectrometry assay for ethyl glucuronide in neonatal meconium. *J. Pharm. Biomed. Anal.* **2019**, *175*, 112743. [CrossRef]
41. American Academy of Forensic Science. Standard Practices for Method Validation in Forensic Toxicology. Available online: https://www.aafs.org/asb-standard/standard-practices-method-validation-forensic-toxicology (accessed on 23 May 2023).
42. La Maida, N.; Di Giorgi, A.; Pellegrini, M.; Ceccanti, M.; Caruso, S.; Ricci, G.; Neri, I.; Lana, S.; Minutillo, A.; Berretta, P.; et al. Reduced prevalence of fetal exposure to alcohol in Italy: A nationwide survey. *Am. J. Obstet. Gynecol. MFM* **2023**, *5*, 100944. [CrossRef]
43. Murillo-Fuentes, M.L.; Artillo, R.; Ojeda, M.L.; Delgado, M.J.; Murillo, M.L.; Carreras, O. Effects of prenatal or postnatal ethanol consumption on zinc intestinal absorption and excretion in rats. *Alcohol Alcohol.* **2007**, *42*, 3–10. [CrossRef] [PubMed]
44. Zaichkin, J.; Houston, R.F. The drug-exposed mother and infant: A regional center experience. *Neonatal Netw.* **1993**, *12*, 41–49. [PubMed]
45. Jacobson, J.L.; Jacobson, S.W.; Sokol, R.J.; Martier, S.S.; Ager, J.W.; Shankaran, S. Effects of alcohol use, smoking, and illicit drug use on fetal growth in black infants. *J. Pediatr.* **1994**, *124*, 757–764. [CrossRef] [PubMed]
46. Knight, E.M.; James, H.; Edwards, C.H.; Spurlock, B.G.; Oyemade, U.J.; Johnson, A.A.; West, W.L.; Cole, O.J.; Westney, L.S.; Westney, O.E.; et al. Relationships of serum illicit drug concentrations during pregnancy to maternal nutritional status. *J. Nutr.* **1994**, *124*, 973S–980S. [CrossRef] [PubMed]
47. Harsham, J.; Keller, J.H.; Disbrow, D. Growth patterns of infants exposed to cocaine and other drugs in utero. *J. Am. Diet. Assoc.* **1994**, *94*, 999–1007. [CrossRef] [PubMed]
48. Mahony, D.L.; Murphy, J.M. Neonatal drug exposure: Assessing a specific population and services provided by visiting nurses. *Pediatr. Nurs.* **1999**, *25*, 27–34, 108.
49. Saha, P.S.; Knecht, T.M.; Arrick, D.M.; Watt, M.J.; Scholl, J.L.; Mayhan, W.G. Prenatal exposure to alcohol impairs responses of cerebral arterioles to activation of potassium channels: Role of oxidative stress. *Alcohol. Clin. Exp. Res.* **2023**, *47*, 87–94. [CrossRef]
50. Chang, G. Reducing Prenatal Alcohol Exposure and the Incidence of FASD: Is the Past Prologue? *Alcohol Res.* **2023**, *43*, 2. [CrossRef]
51. Glass, L.; Moore, E.M.; Mattson, S.N. Current considerations for fetal alcohol spectrum disorders: Identification to intervention. *Curr. Opin. Psychiatry* **2023**, *36*, 249–256. [CrossRef]
52. Basavarajappa, B.S. Epigenetics in fetal alcohol spectrum disorder. *Prog. Mol. Biol. Transl. Sci.* **2023**, *197*, 211–239. [CrossRef]
53. Popova, S.; Dozet, D.; Shield, K.; Rehm, J.; Burd, L. Alcohol's impact on the fetus. *Nutrients* **2021**, *13*, 3452. [CrossRef] [PubMed]
54. Nutt, D.; Hayes, A.; Fonville, L.; Zafar, R.; Palmer, E.O.C.; Paterson, L.; Lingford-Hughes, A. Alcohol and the brain. *Nutrients* **2021**, *13*, 3938. [CrossRef] [PubMed]

55. McCormack, J.C.; Chu, J.T.W.; Marsh, S.; Bullen, C. Knowledge, attitudes, and practices of fetal alcohol spectrum disorder in health, justice, and education professionals: A systematic review. *Res. Dev. Disabil.* **2022**, *131*, 104354. [CrossRef] [PubMed]
56. Popova, S.; Charness, M.E.; Burd, L.; Crawford, A.; Hoyme, H.E.; Mukherjee, R.A.S.; Riley, E.P.; Elliott, E.J. Fetal alcohol spectrum disorders. *Nat. Rev. Dis. Prim.* **2023**, *9*, 11. [CrossRef]
57. Bastons-Compta, A.; Astals, M.; Andreu-Fernandez, V.; Navarro-Tapia, E.; Garcia-Algar, O. Postnatal nutritional treatment of neurocognitive deficits in fetal alcohol spectrum disorder. *Biochem. Cell Biol.* **2018**, *96*, 213–221. [CrossRef]
58. Petrelli, B.; Bendelac, L.; Hicks, G.G.; Fainsod, A. Insights into retinoic acid deficiency and the induction of craniofacial malformations and microcephaly in fetal alcohol spectrum disorder. *Genesis* **2019**, *57*, e23278. [CrossRef]
59. Siqueira, M.; Stipursky, J. Blood brain barrier as an interface for alcohol induced neurotoxicity during development. *Neurotoxicology* **2022**, *90*, 145–157. [CrossRef]
60. del Campo, M.; Jones, K.L. A review of the physical features of the fetal alcohol spectrum disorders. *Eur. J. Med. Genet.* **2017**, *60*, 55–64. [CrossRef]
61. Kurose, I.; Higuchi, H.; Kato, S.; Miura, S.; Ishii, H. Ethanol-induced oxidative stress in the liver. *Alcohol. Clin. Exp. Res.* **1996**, *20*, 77A–85A. [CrossRef]
62. Bukiya, A.N. Fetal cerebral artery mitochondrion as target of prenatal alcohol exposure. *Int. J. Environ. Res. Public Health* **2019**, *16*, 1586. [CrossRef]
63. Xi, Q.; Cheranov, S.Y.; Jaggar, J.H. Mitochondria-derived reactive oxygen species dilate cerebral arteries by activating Ca^{2+} sparks. *Circ. Res.* **2005**, *97*, 354–362. [CrossRef] [PubMed]
64. Carvalho, C.; Moreira, P.I. Oxidative stress: A major player in cerebrovascular alterations associated to neurodegenerative events. *Front. Physiol.* **2018**, *9*, 806. [CrossRef]
65. Ciafrè, S.; Ferraguti, G.; Greco, A.; Polimeni, A.; Ralli, M.; Ceci, F.M.; Ceccanti, M.; Fiore, M. Alcohol as an early life stressor: Epigenetics, metabolic, neuroendocrine and neurobehavioral implications. *Neurosci. Biobehav. Rev.* **2020**, *118*, 654–668. [CrossRef] [PubMed]
66. Carito, V.; Ceccanti, M.; Ferraguti, G.; Coccurello, R.; Ciafrè, S.; Tirassa, P.; Fiore, M. NGF and BDNF Alterations by Prenatal Alcohol Exposure. *Curr. Neuropharmacol.* **2019**, *17*, 308–317. [CrossRef] [PubMed]
67. Boschen, K.E.; Klintsova, A.Y. Neurotrophins in the Brain: Interaction With Alcohol Exposure During Development. *Vitam. Horm.* **2017**, *104*, 197–242. [CrossRef] [PubMed]
68. D'Angelo, A.; Ceccanti, M.; Petrella, C.; Greco, A.; Tirassa, P.; Rosso, P.; Ralli, M.; Ferraguti, G.; Fiore, M.; Messina, M.P. Role of neurotrophins in pregnancy, delivery and postpartum. *Eur. J. Obstet. Gynecol. Reprod. Biol.* **2020**, *247*, 32–41. [CrossRef]

antioxidants

MDPI

Article

Colchicine Protects against Ethanol-Induced Senescence and Senescence-Associated Secretory Phenotype in Endothelial Cells

Huakang Zhou [1], Dilaware Khan [1,*], Norbert Gerdes [2], Carsten Hagenbeck [3], Majeed Rana [4], Jan Frederick Cornelius [1] and Sajjad Muhammad [1,5]

1 Department of Neurosurgery, Medical Faculty and University Hospital Düsseldorf, Heinrich-Heine-University Düsseldorf, 40225 Düsseldorf, Germany
2 Division of Cardiology, Pulmonology and Vascular Medicine, University Hospital and Medical Faculty, Heinrich-Heine-University, 40225 Düsseldorf, Germany
3 Clinic for Gynecology and Obstetrics, University Clinic, 40225 Düsseldorf, Germany
4 Department of Oral, Maxillofacial and Facial Plastic Surgery, University Hospital Düsseldorf, Moorenstrasse 5, 40225 Düsseldorf, Germany
5 Department of Neurosurgery, University Hospital Helsinki, Topeliuksenkatu 5, 00260 Helsinki, Finland
* Correspondence: dilaware.khan@med.uni-duesseldorf.de; Tel.: +0049-21181-08782

Abstract: Inflammaging is a potential risk factor for cardiovascular diseases. It results in the development of thrombosis and atherosclerosis. The accumulation of senescent cells in vessels causes vascular inflammaging and contributes to plaque formation and rupture. In addition to being an acquired risk factor for cardiovascular diseases, ethanol can induce inflammation and senescence, both of which have been implicated in cardiovascular diseases. In the current study, we used colchicine to abate the cellular damaging effects of ethanol on endothelial cells. Colchicine prevented senescence and averted oxidative stress in endothelial cells exposed to ethanol. It lowered the relative protein expression of aging and senescence marker P21 and restored expression of the DNA repair proteins KU70/KU80. Colchicine inhibited the activation of nuclear factor kappa B (NFκ-B) and mitogen activated protein kinases (MAPKs) in ethanol-treated endothelial cells. It reduced ethanol-induced senescence-associated secretory phenotype. In summary, we show that colchicine ameliorated the ethanol-caused molecular events, resulting in attenuated senescence and senescence-associated secretory phenotype in endothelial cells.

Keywords: ethanol; HUVECs; cellular senescence; SASP; inflammation; colchicine; NFκ-B; MAPKs

check for updates

Citation: Zhou, H.; Khan, D.; Gerdes, N.; Hagenbeck, C.; Rana, M.; Cornelius, J.F.; Muhammad, S. Colchicine Protects against Ethanol-Induced Senescence and Senescence-Associated Secretory Phenotype in Endothelial Cells. *Antioxidants* **2023**, *12*, 960. https://doi.org/10.3390/antiox12040960

Academic Editor: Marco Fiore

Received: 13 March 2023
Revised: 11 April 2023
Accepted: 17 April 2023
Published: 19 April 2023

1. Introduction

Inflammaging develops in older individuals and is characterized by elevated levels of pro-inflammatory markers in the serum and different tissues of healthy individuals [1]. Inflammaging is considered a causal risk factor for cardiovascular diseases [1,2]. Throughout life, senescent cells accumulate in vascular tissue, resulting in the development of inflammaging, which consequently leads to cardiovascular diseases [1–4]. Senescent cells have been observed in atherosclerotic tissue, and by eliminating senescent cells, the progression of atherosclerosis could be prevented [1,2,5]. The abolition of senescent cells has been shown to increase lifespan and provide protection against cardiovascular and age-related diseases in experimental animal studies [3,6].

Primary cells grow to a certain limit, after which they stop proliferating and reach growth arrest, a phenomenon termed the Hayflick limit [4,6]. These cells are labeled replicative senescent [4,6,7]. The senescent cells acquire a pro-inflammatory phenotype, called senescence-associated secretory phenotype (SASP) [1,3,8]. These cells increase the expression of cytokines, such as interleukin-1β (IL-β), IL-6, and tumor necrosis factor-α (TNF-α); chemokines, such as monocyte chemoattractant protein-1 (MCP-1) and IL-8;

cell adhesion molecules, such as endothelial selectin (E-selectin), intercellular adhesion molecule-1 (ICAM-1), and vascular cell adhesion molecule-1 (VCAM-1); and matrix metalloproteinase (MMP) such as MMP-2 [1,3]. The SASP-associated molecules have been implicated in several cardiovascular diseases, including atherosclerosis, stroke, and myocardial infarction, and aneurysm formation and rupture [9–15]. The lack or inhibition of these molecules has been shown to reduce atherosclerosis formation [11–14] and decrease aneurysm formation and rupture [9,10,15] in different animal models. The expression of SASP-associated molecules is regulated by P38 via nuclear factor kappa B (NF-κB) transcriptional activity [3,16,17]. Previous studies have shown that inhibiting the activation of P38 and NF-κB delayed cellular senescence and attenuated SASP [18–20].

In addition to replicative senescence, other factors such as oxidative stress, DNA damage, oncogene activation or inactivation, epigenetic alterations, mitochondrial dysfunction, and exposure to damage-associated molecular patterns (DAMPs) released by stressed cells can also induce senescence in cells [1,6,7], contributing to hypertension, arterial stiffness, and atherosclerosis, and thereupon leading to cardiovascular diseases [2,4,5,21]. In addition to age, smoking, alcohol abuse, and hypertension are potential acquired risk factors for cardiovascular diseases. Previously, ethanol has been shown to induce senescence in endothelial [22] and other cell types [23,24], increase the expression of SASP molecules [22], and activate NF-κB and mitogen-activated protein kinases (MAPKs) [23,25,26]. Colchicine is an alkaloid known to impede inflammation and extenuate the expression of pro-inflammatory molecules. Colchicine dampens inflammation and attenuates extracellular remodeling by inhibiting endothelial dysfunction, platelet activation, and platelet aggregation; inhibiting the interaction between endothelial cells and platelets, inflammatory cells and endothelial cells, and platelets and inflammatory cells; and blocking NF-κB activation, resulting in the reduced expression of inflammatory and extracellular remodeling molecules [27]. Colchicine has been used to treat gout flares, familial mediterranean fever, calcium pyrophosphate disease, Adamantiades–Behcet's syndrome, and pericarditis [28,29]. The results from published trials have shown that colchicine has provided benefits against cardiovascular diseases [27,28].

Here, we show that colchicine can reduce ethanol-induced cellular senescence and SASP by inhibiting NFκ-B and MAPKs activation.

2. Methods

2.1. Cell Culture

In the current study, we used human umbilical vein endothelial cells (HUVECs) purchased from Promocell (Heidelberg, Germany). The endothelial cell medium (C-22010, Promocell, Heidelberg, Germany) was added to the endothelial cell growth factors (C-39215, Promocell, Heidelberg, Germany) and used to maintain HUVECs. The cells were kept at 37 degrees Celsius (°C) in a 95% humidified atmosphere containing 5% CO_2. After thawing, the cells were seeded in a T75 cell culture flask. The cells were passaged when they reached 90% confluency. For passaging, the cells were trypsinized at 37 °C for 4 min. A total of 5000 cells/cm^2 were seeded in 10 cm culture plates for protein analysis and 6-well plates for mRNA analysis. For all experiments, endothelial cells at passage 7 were treated with endothelial cell medium containing either 400 millimolar (mM) ethanol (EtOH), 50 nanomolar (nM) colchicine, or 400 mM ethanol combined with 50 nM colchicine. The endothelial cells were treated with a higher concentration of ethanol (400 mM) to induce senescence and SASP over a short period. Controls were treated with endothelial cell medium only. For all experiments, the endothelial cells were treated with different conditions for 24 h, with the exception of immunofluorescence staining, for which the duration of treatment was 2 h. Colchicine was purchased from Sigma-Aldrich, St. Louis, MO, USA (C3915).

2.2. β-Gal Staining

β-Galactosidase (β-Gal) Reporter Gene Staining Kit (Sigma-Aldrich, St. Louis, MO, USA) was used for the detection of β-Gal expression. The manufacturer's instructions were followed to stain the endothelial cells treated for 24 h with different conditions, as described in the Section 2.1. The fixation buffer provided with β-Galactosidase Reporter Gene Staining Kit was used to fix the treated cells. The fixed cells were incubated with the freshly prepared staining solution at 37 °C for 7 h. After that, the staining solution was removed, and the 70% glycerol solution was used to overlay the cells for storage at 4 °C. Using optical microscope, the images were captured. To count the stained cells, ImageJ 1.53c (National Institute of Health, Bethesda, MD, USA) was used. For the experiment, biological triplicates were used.

2.3. Immunofluorescence Staining

Immunofluorescence staining was performed as previously described [22]. After three washing steps with phosphate-buffered saline (PBS) (Thermo Fisher, Waltham, MA, USA), the endothelial cells were fixed with 4% paraformaldehyde (Thermo Fisher, Waltham, MA, USA) for 15 min. For permeabilization, the cells were incubated with 0.2% Triton™ X-100 (Sigma-Aldrich, St. Louis, MO, USA) for 10 min. Subsequently, the cells were treated with 5% bovine serum albumin (BSA) (VWR, Langenfeld, Germany) blocking solution for 1 h at room temperature (RT). After that, the cells were incubated with primary antibody 8-Hydroxydesoxyguanosin (8-OHDG) (1:500, Cat. No.: BSS-BS-1278R, BIOSS, Woburn, MA, USA) at 4 °C overnight. On the following day, the cells were washed three times with PBS and incubated with the secondary antibody (1:1000, Alexa Fluor 488, excitation 495 nm, emission 519 nm, Cat. No.: A48269, Thermo Fisher, Waltham, MA, USA) for 1 h at RT. SlowFade® Gold Antifade Mountant with DAPI (Cat. No.: S36938, Thermo Fisher, Waltham, MA, USA) was used for nuclear staining. The images were captured at 20× magnification. The images were analyzed with ImageJ.

2.4. Western Blot (WB)

For protein analysis, the endothelial cells were treated with different conditions, as described in the Section 2.1 for 24 h. Radioimmunoprecipitation assay buffer was used for total protein extraction. DC protein Assay Kit (500-0116, Bio-Rad, Hercules, CA, USA) was used according to the manufacturer's instructions to measure total protein concentration with the Paradigm micro-plate reader (Beckman Coulter, Krefeld, Germany). An amount of 25 microgram (μg) total protein in reducing conditions was loaded on 12% sodium dodecyl sulfate-polyacrylamide gel. The running conditions were 60 Volts for 20 min, followed by 110 Volts for 30–60 min. The polyvinylidene difluoride membranes were used for transfer at 250 milliampere for 120 min. The membranes were blocked with 5% BSA (0.05% tris buffered saline with tween (TBST)) for 1 h. Subsequently, membranes were incubated overnight at 4 °C with primary antibodies (Supplementary Table S1) and diluted in 5% BSA (0.05% TBST) on a shaking platform. On the next day, membranes were washed three times with TBST for 10 min. The membranes were incubated with secondary antibodies (Supplementary Table S1) at RT for 1 h. α-Tubulin was used as a loading control for all proteins except phosphorylated c-Jun N-terminal kinase (p-JNK), for which β-actin was used as the loading control. α-tubulin and β-actin were probed on different membranes. The probed WB membranes were scanned using Odyssey CLx Imaging system (LI-COR Biosciences, Lincoln, NE, USA). The untrimmed images of membranes have been provided in the Supplementary Material. ImageJ was used to calculate densitometry.

2.5. Quantitative Polymerase Chain Reaction (qPCR)

qPCR was performed as previously described [22]. Tri Reagent (T9424, Sigma-Aldrich, St. Louis, MO, USA) was used to extract total RNA. M-MLV Reverse Transcriptase kit (M1701, Promega, Walldorf, Germany) mixed with RiboLock RNase Inhibitor (EO0384, Thermo Fisher, Waltham, MA, USA) and Random Hexamer Primers (48190011, Thermo

Fisher, Waltham, MA, USA) was used to reverse transcribe 1.2 µg total RNA. AceQ SYBR qPCR Master Mix (Q111-03, Vayzme, Nanjing, China) was used to perform qPCR on BIO-Rad CFX Connect Real-Time PCR System. The primer sequences are provided in Supplementary Table S2. The qPCR protocol was an initial denaturation of 95 °C for 8 min, followed by 45 cycles of 95 °C for 15 s, 58.9 °C for 30 s, and 72 °C for 30 s, which was then followed by a melting curve. β-actin was used for normalizing to calculate relative mRNA expression. To quantify relative mRNA expression, comparative C_T method was used [30]. For the experiment, biological triplicates were used. For each biological replicate, we used three technical triplicates.

2.6. Statistical Analysis

For statistical analysis, we performed one-way analysis of variance (ANOVA) followed by Tukey's test. The level of significance was set to less than 0.05 (* $p < 0.05$).

3. Results

3.1. Colchicine Inhibited Ethanol-Induced Senescence

Ethanol is a known risk factor for cardiovascular diseases. We have already reported that ethanol increases cellular senescence in endothelial cells [22]. To investigate the effects of colchicine, we treated endothelial cells with either an endothelial cell medium alone (control) or an endothelial cell medium combined with either 400 mM ethanol, 50 nM colchicine, or 400 mM ethanol combined with 50 nM colchicine. After 24 h, we performed β-gal staining. Colchicine inhibited ethanol-induced cellular senescence in endothelial cells and lowered the percentage of β-gal positive cells (control = 11.01 ± 1.772%, EtOH = 37.21 ± 1.761%, colchicine = 10.76 ± 2.820%, EtOH + colchicine = 16.22 ± 2.630%, **** $p < 0.0001$, $n = 3$; Figure 1A,B). Next, we investigated the effect of colchicine on aging-associated biomarkers. Colchicine attenuated the relative protein expression of aging-associated biomarker P21 (control = 1.003 ± 0.018, EtOH = 2.398 ± 0.068, colchicine = 0.830 ± 0.060, EtOH + colchicine = 1.468 ± 0.075, **** $p < 0.0001$, $n = 3$; Figure 1C,D). P21 is a cyclin-dependent kinase (CDK) inhibitor, which establishes indefinite cell cycle arrest via the inhibition of CDK-2/4 [21]. The CDK-2/4 inhibition results in the active hypo-phosphorylated form of retinoblastoma protein, which, in turn, mediates cell cycle arrest and senescence phenotypes [21]. Colchicine also recovered the relative protein expression of DNA repair proteins KU70 (control = 1.000 ± 0.066, EtOH = 0.856 ± 0.060, colchicine = 0.995 ± 0.040, EtOH + colchicine = 0.995 ± 0.040, * $p < 0.05$, ** $p < 0.01$, $n = 3$; Figure 1C,E) and KU80 (control = 1.000 ± 0.072, EtOH = 0.838 ± 0.069, colchicine = 1.065 ± 0.004, EtOH + colchicine = 1.097 ± 0.058, * $p < 0.05$, $n = 3$, Figure 1C,F). The reduced expression of KU70 and KU80 has been observed in senescent cells compared to young cells [31]. KU70 and KU80 form a heterodimer and repair DNA double-strand breaks via a nonhomologous end-joining pathway [32]. KU70 and KU80 maintain telomere length, and the inactivation of KU70 and KU80 results in the shortening of telomere length in various primary cell types [33], leading to cellular senescence. Furthermore, KU80 has been shown to hinder oxidative stress-induced DNA damage [34].

Colchicine inhibited ethanol-induced endothelial senescence, ameliorated the relative protein expression of aging-associated biomarker P21, and restored the relative protein expression of the DNA repair proteins KU70 and KU80.

3.2. Colchicine Averted Ethanol-Induced Oxidative Stress in Endothelial Cells

The metabolism of ethanol produces reactive oxygen species (ROS) and reactive nitrogen species, resulting in increased oxidative stress [35]. Colchicine inhibited ethanol-induced oxidative stress in endothelial cells and lowered the percentage of oxidative stress-associated biomarker 8-OHDG positive cells (control = 8.569 ± 4.573%, EtOH = 48.62 ± 10.74%, colchicine = 8.553 ± 3.115%, EtOH + colchicine = 13.60 ± 6.001%, *** $p = 0.001$, $n = 3$; Figure 2). Colchicine obviated oxidative stress in ethanol-treated endothelial cells, which is in accordance with the previously reported findings [36,37].

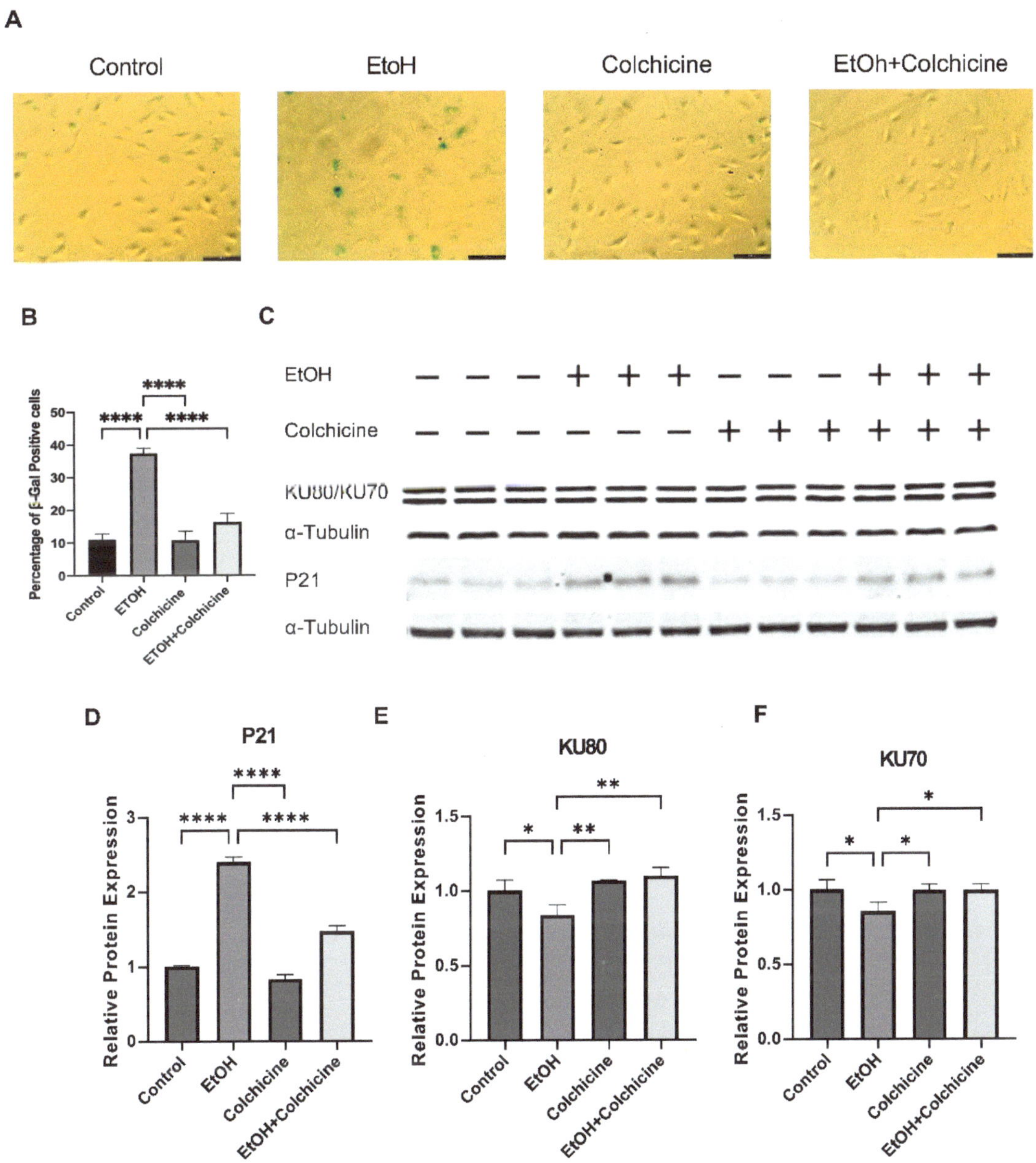

Figure 1. Colchicine inhibited endothelial senescence and restored the relative protein expression of aging-associated biomarkers. (**A**) Senescence in endothelial cells after 24 h of treatment with different conditions. (**B**) Colchicine subdued ethanol (EtOH)-induced senescence in endothelial cells. (**C**) Western blot showing protein expression of P21, KU80, and KU70. (**D**) Colchicine attenuated the relative protein expression of P21 in endothelial cells after ethanol exposure. (**E**,**F**) Colchicine recovered the relative protein expression of KU70 and KU80 in endothelial cells treated with ethanol. α-Tubulin was used as a loading control. Data are the mean of independent biological triplicates. The data was analyzed by performing one-way ANOVA followed by Tukey's test. Scale bar = 100 μm; error bars represent the SD (**** $p < 0.0001$, ** $p < 0.01$, and * $p < 0.05$).

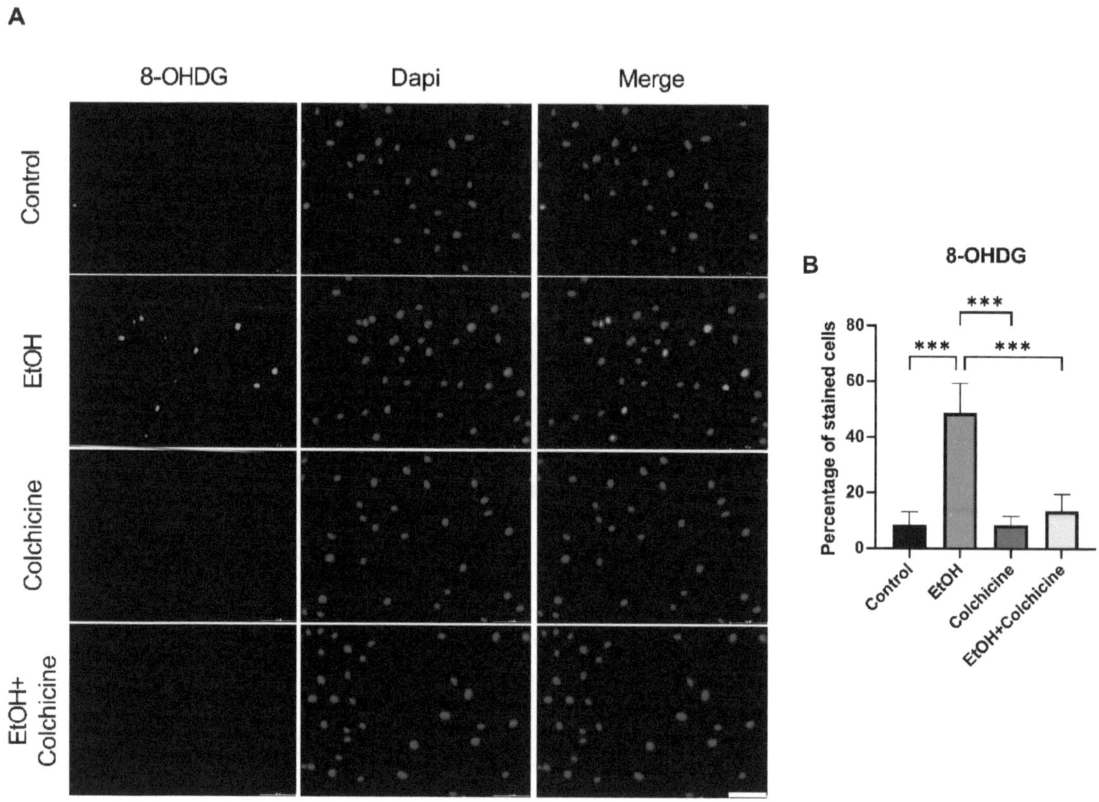

Figure 2. Colchicine restrained ethanol-induced oxidative stress in endothelial cells. (**A**) Immunofluorescence staining for oxidative stress marker 8-Hydroxydesoxyguanosin (8-OHDG) in endothelial cells 2 h after treatment with ethanol, colchicine, and ethanol combined with colchicine. Endothelial medium alone was used for untreated control. (**B**) Colchicine averted the expression of 8-OHDG in ethanol-treated cells. The experiment was performed with independent biological triplicates. The data was analyzed by performing one-way ANOVA followed by Tukey's test. Scale bar = 100 μm; error bars represent the SD (*** $p < 0.001$).

3.3. Colchicine Suppressed the Activation of NF-κB and MAPKs in Ethanol-Treated Endothelial Cells

To investigate the pathways of interest, we performed a protein analysis. NFκ-B has been suggested to play an important role in inflammation, aging, and cellular senescence [8,19]. Our protein analysis showed that colchicine inhibited the ethanol-induced relative protein expression of NF-κB subunit P65 (control = 1.000 ± 0.038, EtOH = 1.172 ± 0.049, colchicine = 1.019 ± 0.034, EtOH + colchicine = 1.053 ± 0.019, * $p < 0.05$, ** $p < 0.01$, $n = 3$; Figure 3A,B). Colchicine also abated the NF-κB activation, and reduced the relative protein expression of NF-κB subunit p-P65 (control = 0.9999 ± 0.083, EtOH = 2.580 ± 0.1417, colchicine = 1.310 ± 0.0488, EtOH + colchicine = 1.506 ± 0.0444, **** $p < 0.0001$, $n = 3$; Figure 3A,C) and *p*-P65/P65 (control = 1.002 ± 0.107, EtOH = 2.208 ± 0.208, colchicine = 1.285 ± 0.014, EtOH + colchicine = 1.431 ± 0.037, *** $p < 0.001$, **** $p < 0.0001$, $n = 3$; Figure 3A,D) was significantly reduced after colchicine treatment in ethanol-treated endothelial cells.

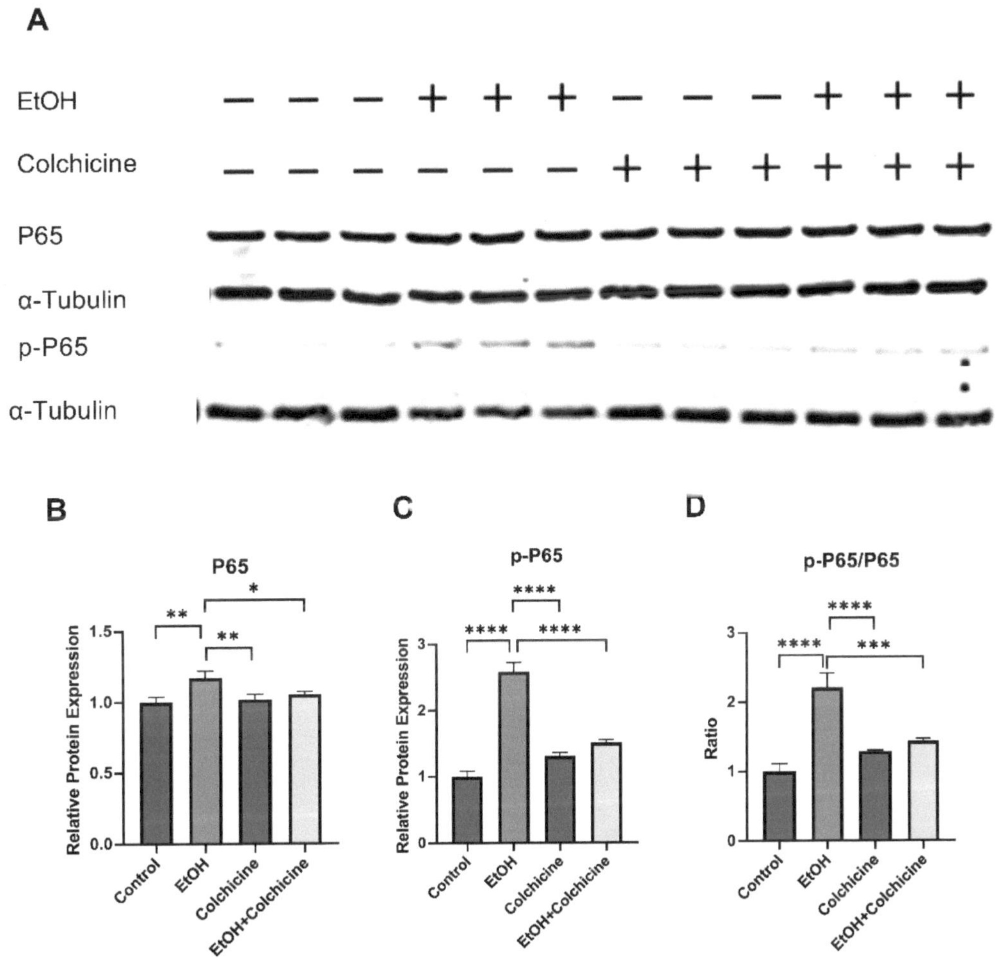

Figure 3. Colchicine inhibited nuclear factor kappa B (NF-κB) activation: (**A**) Western blots for proteins of P65 and p-P65. Colchicine obviated the relative protein expression of (**B**) P65 and (**C**) p-P65 in ethanol-treated endothelial cells. (**D**) The ratio of p-P65/P65 was significantly decreased in ethanol-treated endothelial cells exposed to colchicine. α-Tubulin was used as loading control. Biological triplicates were used for the experiment. Data was analyzed by performing one-way ANOVA followed by Tukey's test. Error bars represent the SD (**** $p < 0.0001$, *** $p < 0.001$, ** $p < 0.01$, and * $p < 0.05$).

It has previously been reported that ethanol activates MAPKs [25] and MAPKs have been implicated in cellular senescence. The protein analysis showed that colchicine impeded the ethanol-induced activation of MAPKs: the relative protein expression of p-P38 (control = 0.9988 ± 0.050, EtOH = 1.553 ± 0.098, colchicine = 0.814 ± 0.103, EtOH + colchicine = 1.070 ± 0.051, *** $p <= 0.001$, **** $p < 0.0001$, $n = 3$; Figure 4A,B), phosphorylated extracellular signal-regulated protein kinase (p-ERK) (control = 1.001 ± 0.1834, ETOH = 2.667 ± 0.4533, colchicine = 0.7198 ± 0.0671, EtOH + colchicine = 1.566 ± 0.1465, ** $p < 0.01$, *** $p < 0.001$, **** $p < 0.0001$, $n = 3$; Figure 4A,C), and p-JNK (control = 1.000 ± 0.039, EtOH = 1.088 ± 0.035, colchicine = 0.899 ± 0.005, EtOH + colchicine = 0.976 ± 0.017, * $p < 0.05$, ** $p < 0.01$, *** $p < 0.001$, $n = 3$; Figure 4A,D).

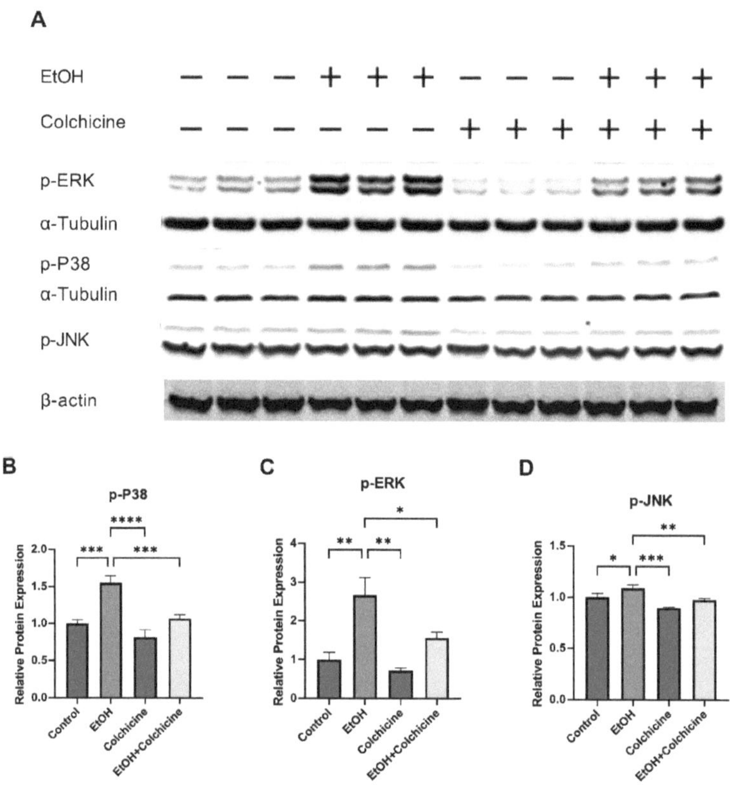

Figure 4. Colchicine inhibited mitogen activated protein kinases (MAPKs) activation: (**A**) Western blots for proteins of p-P38, phosphorylated extracellular signal-regulated protein kinase (p-ERK), and phosphorylated c-Jun N-terminal kinase (p-JNK). Colchicine averted relative protein expression of (**B**) p-P38, (**C**) p-ERK, and (**D**) p-JNK in ethanol-treated endothelial cells. α-Tubulin was used as a loading control for all proteins expression except p-JNK. β-actin was used as a loading control for only p-JNK on a separate membrane. The experiment was performed with independent biological triplicates. The data was analyzed by performing one-way ANOVA followed by Tukey's test. Error bars represent the SD (**** $p < 0.0001$, *** $p < 0.001$, ** $p < 0.01$, and * $p < 0.05$).

3.4. Colchicine Ameliorated Ethanol-Induced SASP in Endothelial Cells

Senescent cells acquire SASP, which is characterized by the increased expression and release of inflammatory cytokines, chemokines, proteases, and growth factors [1,3,8]. Because colchicine inhibited senescence and senescence-associated pathways in ethanol-treated endothelial cells (Figures 1, 3 and 4), we investigated the relative mRNA expression of SASP-associated cytokines, chemokines, and cell adhesion molecules. Colchicine curtailed the relative mRNA expression of the cytokines IL-1β (control = 1.067 ± 0.4931, EtOH = 4.146 ± 0.9146, colchicine = 0.7295 ± 0.1732, EtOH + colchicine = 0.3203 ± 0.09465, *** $p < 0.001$, **** $p < 0.0001$, $n = 3$; Figure 5A) and TNF-α (control = 1.023 ± 0.2783, EtOH = 2.089 ± 0.2934, colchicine = 0.5135 ± 0.0654, EtOH + colchicine = 0.6711 ± 0.2300, ** $p < 0.01$, *** $p < 0.001$, $n = 3$; Figure 5C). It also reduced the relative mRNA expression of the chemokines IL-8 (control = 1.004 ± 0.1091, EtOH = 2.398 ± 0.1235, colchicine = 0.9490 ± 0.0410, EtOH + colchicine = 0.9802 ± 0.0586, **** $p < 0.0001$, $n = 3$; Figure 5D) and MCP-1 (control = 1.013 ± 0.2085, EtOH = 3.341 ± 0.5194, colchicine = 0.7364 ± 0.1053, EtOH + colchicine = 0.4826 ± 0.1187, **** $p < 0.0001$, $n = 3$; Figure 5E). Finally, it decreased the relative mRNA expression of the cell adhesion molecules

ICAM-1 (control = 1.009 ± 0.1649, EtOH = 3.871 ± 0.2693, colchicine = 0.5126 ± 0.1318, EtOH + colchicine = 1.767 ± 1.133, * $p < 0.05$, ** $p < 0.01$, *** $p < 0.001$, $n = 3$; Figure 5F) and E-Selectin (control = 1.012 ± 0.1944, EtOH = 2.645 ± 0.3812, colchicine = 0.5963 ± 0.0850, EtOH + colchicine = 1.366 ± 0.3571, ** $p < 0.01$ *** $p < 0.001$, **** $p < 0.0001$, $n = 3$; Figure 5H). Ethanol did not significantly increase the relative mRNA expression of IL-6 (control = 1.024 ± 0.2673, EtOH = 3.092 ± 0.8467, colchicine = 0.4912 ± 0.1902, EtOH + colchicine = 1.756 ± 1.660, * $p < 0.05$, $n = 3$; Figure 5B) or VCAM-1 (control = 1.028 ± 0.3069, EtOH = 2.108 ± 0.7257, colchicine = 1.004 ± 0.2258, EtOH + colchicine = 0.6622 ± 0.1701, * $p < 0.05$, $n = 3$; Figure 5G). In conclusion, colchicine reduced the relative mRNA expression of SASP-associated cytokines, chemokines, and cell adhesion molecules in endothelial cells exposed to ethanol.

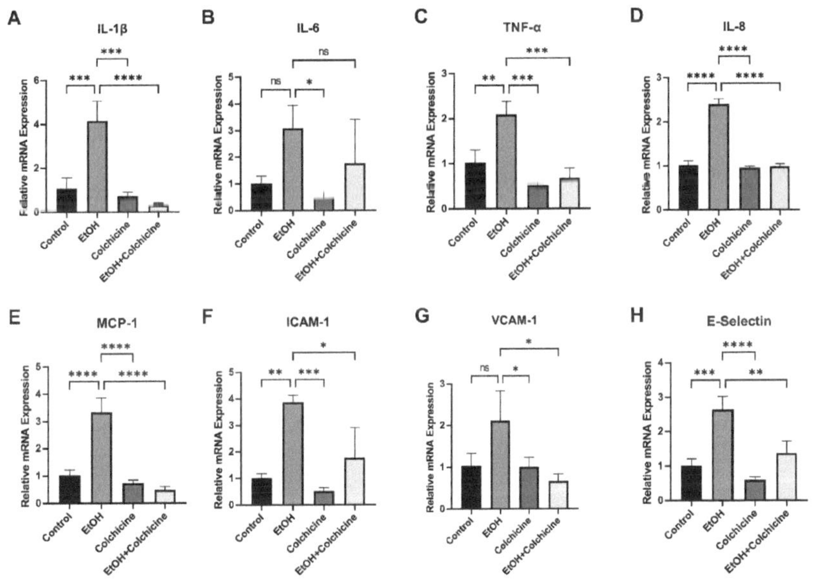

Figure 5. Colchicine attenuated the relative mRNA expression of SASP-associated inflammatory markers. Colchicine reduced the relative mRNA expression of (**A**) interleukin (IL)-1β, (**C**) tumor necrosis factor-α (TNF-α), (**D**) IL-8, (**E**) monocyte chemoattractant protein-1 (MCP-1), (**F**) intercellular adhesion molecule-1 (ICAM-1), and (**H**) endothelial selectin (E-Selectin) in ethanol-treated endothelial cells. Ethanol did not significantly increase the relative mRNA expression of (**B**) IL-6 or (**G**) vascular cell adhesion molecule-1 (VCAM-1). Colchicine significantly lowered the relative mRNA expression of VCAM-1 in ethanol-treated endothelial cells. β-actin was used as a loading control. qPCR data are the mean of three independent technical replicates. The data was analyzed by performing one-way ANOVA followed by Tukey's test. Error bars represent the SD (**** $p < 0.0001$, *** $p < 0.001$, ** $p < 0.01$, and * $p < 0.05$), ns: not significant.

3.5. Colchicine Mitigated Ethanol-Induced Relative mRNA and Relative Protein Expression of MMP-2

Colchicine did not decrease the ethanol-induced relative mRNA expression of MMP-1 (control = 1.000 ± 0.0285, ETOH = 2.617 ± 0.6381, colchicine = 1.847 ± 0.3830, EtOH + colchicine = 3.069 ± 0.9726, * $p < 0.05$, $n = 3$; Figure 6A) or MMP-11 (control = 1.003 ± 0.0929, ETOH = 3.795 ± 1.136, colchicine = 0.5490 ± 0.2415, EtOH + colchicine = 2.214 ± 1.214, * $p < 0.05$, ** $p < 0.01$, $n = 3$; Figure 6D). Both ethanol and colchicine alone or in combination did not significantly affect the relative mRNA expression of MMP-10 (control = 1.003 ± 0.0932, EtOH = 1.686 ± 0.0314, colchicine = 2.585 ± 0.8815, EtOH + colchicine = 1.693 ± 0.3770, $n = 3$; Figure 6C) or TIMP2 (control = 1.004 ± 0.1296, ETOH = 1.775 ± 0.3543, colchicine = 0.8472 ± 0.3676, EtOH + colchicine = 1.165 ± 0.7765, $n = 3$; Figure 6F).

Colchicine inhibited the ethanol-induced relative mRNA expression of TIMP1 (control = 1.001 ± 0.04815, EtOH = 1.773 ± 0.1194, colchicine = 0.5182 ± 0.0901, EtOH + colchicine = 0.7562 ± 0.02403, **** $p < 0.0001$, $n = 3$; Figure 6E) and MMP-2 (control = 1.000 ± 0.0212, EtOH = 2.761 ± 0.3196, colchicine = 0.6408 ± 0.0545, EtOH + colchicine = 1.591 ± 0.3091, *** $p < 0.001$, **** $p < 0.0001$, $n = 3$; Figure 6B). Colchicine also attenuated the relative protein expression of MMP-2 (control = 1.00 ± 0.127, EtOH = 5.161 ± 0.356, colchicine = 2.102 ± 0.381, EtOH + colchicine = 3.102 ± 0.177, **** $p < 0.0001$, $n = 3$; Figure 6G,H).

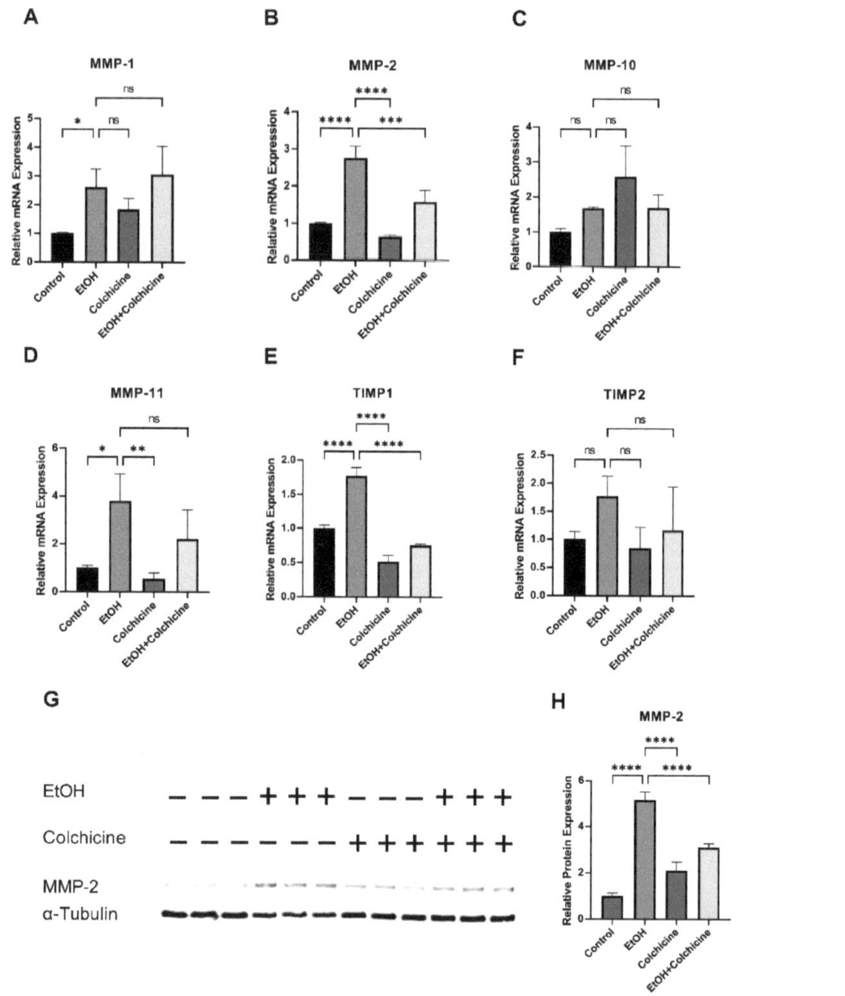

Figure 6. Colchicine abated the relative mRNA and relative protein expression of matrix metalloproteinase (MMP)2. Colchicine did not ameliorate the relative mRNA expression of (**A**) MMP1 or (**D**) MMP-11. Ethanol and colchicine did not alter the relative mRNA expression of (**C**) MMP-10 or (**F**) tissue inhibitor of metalloproteinase (TIMP)2. Colchicine attenuated relative mRNA expression of (**E**) TIMP1 and (**B**) MMP-2 and relative protein expression of (**G**,**H**) MMP-2 in ethanol-treated cells. β-actin for qPCR and α-tubulin for WB were used as a loading control. qPCR data are the mean of three independent technical replicates and WB data are the mean of the independent biological triplicates. The data was analyzed by performing one-way ANOVA followed by Tukey's test. Error bars represent the SD (**** $p < 0.0001$, *** $p < 0.001$, ** $p < 0.01$, and * $p < 0.05$), ns: not significant.

4. Discussion

We showed that colchicine averted cellular senescence and SASP in ethanol-treated endothelial cells. The pathway analysis showed that colchicine inhibited the activation of NF-κB, P38, ERK, and JNK pathways in endothelial cells exposed to ethanol. Ethanol is a potential risk factor for cardiovascular diseases. We, in addition to other researchers, have previously reported that alcohol causes cellular senescence [22,23]. Cellular senescence contributes to cardiovascular diseases via the increase in inflammaging in the vascular endothelium [1–4]. In the current study, we investigated the effects of colchicine on ethanol-treated endothelial cells.

The ethanol treatment induced cellular senescence in endothelial cells (Figure 1) [22–24]. Colchicine inhibited cellular senescence and attenuated oxidative stress in ethanol-treated endothelial cells (Figures 1 and 2). Previously, colchicine has been shown to exert anti-oxidative effects by upregulating the expression of anti-oxidant enzymes such as CAT, GPx-1, and SOD2 in platelets [38]. Ethanol metabolism results in the formation of ROS, which leads to an increase in oxidative stress [35]. The over-accumulation of ROS causes DNA damage and triggers cellular senescence. Colchicine, by inhibiting ROS generation in vivo and in vitro [36–38], can decrease oxidative stress and DNA damage. DNA damage and oxidative stress can activate NF-κB and MAPKs [8,39]. In addition to this, ethanol can activate NF-κB and MAPKs via TLR4/Type I IL-1 receptor signaling [25]. The pathway analysis showed that ethanol activated NF-κB, P38, JNK, and ERK pathways in HUVECs (Figures 3 and 4), which is in agreement with the previously reported findings [22,23,25,26]. These pathways modulate inflammation, contribute to cellular senescence [8,20,39–41], and increase the transcription and expression of cell cycle inhibitor protein P21 via different mechanisms [39,42–46]. The NF-κB activation enhanced the expression of cell cycle inhibitor P21 in response to DNA damage that was independent of the P53 pathway [42]. P38 increases P21 expression by enhancing the expression, stabilization, and promoter activity of P53 [43,44]. Moreover, P38 phosphorylates HuR, which, in turn, increases the cytoplasmic accumulation of HuR and the binding of HuR to P21, consequently improving the stability of P21 mRNA and, thus, enhancing P21 protein levels [45]. ERK1/2 promotes the transcription of P21 via the ELK1, SP1 and SMAD proteins [39,46]. The inhibition of NF-κB and MAPKs activation delayed cellular senescence [18–20], suggesting that inflammation can induce premature senescence in endothelial cells. Colchicine inhibited ethanol-induced senescence by attenuating oxidative stress, recovering the protein expression of KU70 and KU80, ameliorating P21 protein expression, and inhibiting the NF-κB, P38, ERK, and JNK pathways in endothelial cells.

Because Colchicine attenuated ethanol-induced senescence and inhibited the ethanol-induced activation of pro-inflammatory pathways, we investigated the effect of colchicine on SASP-associated cytokines (IL-1β, IL-6, and TNF-α), chemokines (IL-8 and MCP-1), and cell adhesion molecules (ICAM-1, VCAM-1, and E-Selectin) [1,3,8]. Colchicine inhibited the expression of these SASP-associated pro-inflammatory molecules (Figure 5) [38,47,48]. This pro-inflammatory response in senescent and dysfunctional endothelial cells is regulated by the NF-κB complex [3,8]. MAPKs have been suggested to be the upstream regulators of NF-κB [16]. P38 controls NF-κB activity in senescent cells and it induces SASP by increasing the mRNA expression of SASP molecules primarily via NF-κB transcriptional activity [3,16,17].

Senescent endothelial cells promote atherosclerosis and thrombosis by the increased expression and release of SASP-associated inflammatory factors and chemokines [3,8]. MCP-1, IL-8, and cell adhesion molecules E-selectin, VCAM-1, and ICAM-1 promote the extravasation of inflammatory cells from the blood stream across the endothelium [13,49–51]. These infiltrated inflammatory cells create a pro-inflammatory microenvironment, exacerbate inflammation, and lead to atherosclerotic plaque formation [50]. Colchicine reduced the adhesion of monocytes to HUVECs by inhibiting the expression of adhesion molecules VCAM-1 and ICAM-1 [38]. Colchicine decreased the recruitment of monocytes and neutrophils into the atherosclerotic plaque of mice aorta [47]. In addition to their role in the tissue infiltration of inflammatory cells, the SASP-associated molecules have the potential to activate the inflammatory cells [49,50]. MCP-1 activates immune cells, such as monocytes, and regulates the polarization of T-cells and the differentiation of monocytes into dendritic cells [50]. IL-1β and TNF-α released from senescent endothelial cells can activate inflammatory cells, neighboring endothelial and smooth muscle cells [52,53]. Both IL-1β and TNF-α have been shown to decrease collagen synthesis and increase the mRNA expression and activity of MMPs [54], which can consequently result in tissue remodeling [55,56]. The TNF-α-induced phenotype switch in smooth muscle cells can impair vasorelaxation [52]. TNF-α can cause endothelial cell apoptosis and smooth muscle cell proliferation and migration [53,57], leading to the initiation and progression of cardiovascular diseases. TNF-α increased the expression of E-selectin, VCAM-1, and ICAM-1 in HUVECs [58]. Colchicine reduced TNF-α, IL-1β, MCP-1, and ICAM-1 expression at mRNA and protein levels in vivo and in vitro [38,47,48]. Cerebral aneurysm formation and rupture were significantly reduced in MCP1-, TNF-α-, and TNF-α-R1 deficient mice [9,10,15]. The lack of MCP-1, MCP-1 receptor CCR2 inhibition, IL-1β deficiency, and TNF-α inhibition have been shown to decrease atherosclerosis formation [11–14]. The rupture of atherosclerotic plaque leads to thrombus formation. The aggregation, activation, and inflammatory response of platelets is known to play a key role in atherosclerosis and thrombus formation [59]. Colchicine blocked platelet–platelet aggregation in both whole blood and platelet-rich plasma, platelet activation (ROS generation), and procoagulant platelet formation [37,38,60]. The addition of colchicine in whole blood, in vitro, and in oral administration in healthy subjects, in vivo, decreased monocyte-platelet and neutrophil-platelet aggregation [60]. Colchicine has been shown to reduce NETs formation [61], which has been suggested to contribute to thrombosis by accumulating prothrombotic factors such as fibrinogen and von Willebrand factor, and by promoting platelet adhesion, activation, and aggregation [62]. In mice, colchicine inhibited carrageenan-induced thrombosis and ameliorated platelet activation [38]. The authors also showed that colchicine dampened human platelet activation by inhibiting the activation of AKT pathway, which subsequently blocked ERK1/2 activation [38]. These findings indicate that colchicine, by inhibiting senescence and curtailing SASP-induced sterile inflammation in the endothelium, can be potentially useful against cardiovascular diseases [27,28].

Previous studies have shown that senescent cells increase the expression of MMPs [1,3,8,21]. In addition, ethanol has been shown to increase the expression of MMPs in endothelial [22] and other cell types [63]. Alcohol consumption has been reported to increase serum levels of MMPs in alcohol abusers [64] and, in animal studies, alcohol elevated MMP expression in different tissues [65–67]. In the current study, ethanol increased the expression of MMPs (Figure 6). Colchicine inhibited the expression of MMP-2 at mRNA and protein levels (Figure 6). MMPs are known to contribute to cardiovascular diseases via tissue remodeling and scar formation, facilitating the migration and proliferation of smooth muscle cells, the infiltration of inflammatory cells such as monocytes and neutrophils into the endothelium, and promoting inflammation via their protease function on cytokines and chemokines [55,56]. These findings suggest that, by inhibiting MMP-2 mRNA and protein expression, colchicine can potentially suppress tissue remodeling and MMP-2-mediated inflammation in endothelial cells.

5. Conclusions

Alcohol is a potential risk factor for cardiovascular diseases. In the current study, we showed that ethanol induced premature senescence and SASP. Colchicine reduced ethanol-induced inflammaging in HUVECs possibly by inhibiting the activation of the NF-kB and MAPKs pathways. Thus, colchicine could be a potential pharmacological target for cardiovascular diseases.

6. Limitations

Our study has some limitations. The exposure of the endothelial cells to ethanol, for example, was acute and not chronic. The concentration of ethanol (400 mM) used was much higher and the ethanol evaporation from culture media was not prevented. This higher concentration of ethanol was used to induce senescence and SASP in endothelial cells over a short period. Moreover, the study was conducted using endothelial cells in vitro. In addition, we used HUVECs as endothelial cells. The data should be carefully interpreted.

Supplementary Materials: The following supporting information can be downloaded at: https://www.mdpi.com/article/10.3390/antiox12040960/s1, Supplementary Table S1. The list of antibodies used in the study. Supplementary Table S2. The list of primers.

Author Contributions: Conceptualization, D.K. and S.M.; Investigation, H.Z. and D.K.; Methodology, H.Z. and D.K.; Project administration, D.K. and S.M.; Resources, S.M. and N.G.; Writing—original draft, H.Z. and D.K.; Writing—review and editing, N.G., C.H., M.R., J.F.C. and S.M. All authors have read and agreed to the published version of the manuscript.

Funding: The current study was supported by Forschungskommission HHU Düsseldorf, Stiftung Neurochirurgische Forschung (DGNC), EANS Research Funds, BMBF, to S. Muhammad and by the Deutsche Forschungsgemeinschaft (DFG, German Research Foundation) Grant No. 397484323–CRC/TRR259; project A05 to N. Gerdes.

Institutional Review Board Statement: Not applicable.

Informed Consent Statement: Not applicable. The study did not involve human.

Data Availability Statement: All data generated or analyzed during this study are included in this published article.

Conflicts of Interest: The authors declare no conflict of interest.

References

1. Ferrucci, L.; Fabbri, E. Inflammageing: Chronic inflammation in ageing, cardiovascular disease, and frailty. *Nat. Rev. Cardiol.* **2018**, *15*, 505–522. [CrossRef]
2. Ungvari, Z.; Tarantini, S.; Donato, A.J.; Galvan, V.; Csiszar, A. Mechanisms of Vascular Aging. *Circ. Res.* **2018**, *123*, 849–867. [CrossRef]
3. Sun, Y.; Wang, X.; Liu, T.; Zhu, X.; Pan, X. The multifaceted role of the SASP in atherosclerosis: From mechanisms to therapeutic opportunities. *Cell Biosci.* **2022**, *12*, 74. [CrossRef]
4. Katsuumi, G.; Shimizu, I.; Yoshida, Y.; Minamino, T. Vascular Senescence in Cardiovascular and Metabolic Diseases. *Front. Cardiovasc. Med.* **2018**, *5*, 18. [CrossRef]
5. Owens, W.A.; Walaszczyk, A.; Spyridopoulos, I.; Dookun, E.; Richardson, G.D. Senescence and senolytics in cardiovascular disease: Promise and potential pitfalls. *Mech. Ageing Dev.* **2021**, *198*, 111540. [CrossRef]
6. Song, S.; Tchkonia, T.; Jiang, J.; Kirkland, J.L.; Sun, Y. Targeting Senescent Cells for a Healthier Aging: Challenges and Opportunities. *Adv. Sci. Weinh.* **2020**, *7*, 2002611. [CrossRef]
7. Frej, F.; Peter, M.N. Chapter 20—Telomere Biology and Vascular Aging. In *Early Vascular Aging (EVA)*; Nilsson, P.M., Olsen, M.H., Laurent, S., Eds.; Academic Press: Boston, MA, USA, 2015; pp. 201–211.
8. Haga, M.; Okada, M. Systems approaches to investigate the role of NF-kappaB signaling in aging. *Biochem. J.* **2022**, *479*, 161–183. [CrossRef]
9. Aoki, T.; Fukuda, M.; Nishimura, M.; Nozaki, K.; Narumiya, S. Critical role of TNF-alpha-TNFR1 signaling in intracranial aneurysm formation. *Acta Neuropathol. Commun.* **2014**, *2*, 34. [CrossRef]
10. Aoki, T.; Kataoka, H.; Ishibashi, R.; Nozaki, K.; Egashira, K.; Hashimoto, N. Impact of monocyte chemoattractant protein-1 deficiency on cerebral aneurysm formation. *Stroke* **2009**, *40*, 942–951. [CrossRef]
11. Bot, I.; Ortiz Zacarias, N.V.; de Witte, W.E.; de Vries, H.; van Santbrink, P.J.; van der Velden, D.; Kroner, M.J.; van der Berg, D.J.; Stamos, D.; de Lange, E.C.; et al. A novel CCR2 antagonist inhibits atherogenesis in apoE deficient mice by achieving high receptor occupancy. *Sci. Rep.* **2017**, *7*, 52. [CrossRef]
12. Branen, L.; Hovgaard, L.; Nitulescu, M.; Bengtsson, E.; Nilsson, J.; Jovinge, S. Inhibition of tumor necrosis factor-alpha reduces atherosclerosis in apolipoprotein E knockout mice. *Arterioscler. Thromb. Vasc. Biol.* **2004**, *24*, 2137–2142. [CrossRef]
13. Gu, L.; Okada, Y.; Clinton, S.K.; Gerard, C.; Sukhova, G.K.; Libby, P.; Rollins, B.J. Absence of monocyte chemoattractant protein-1 reduces atherosclerosis in low density lipoprotein receptor-deficient mice. *Mol. Cell* **1998**, *2*, 275–281. [CrossRef]
14. Kirii, H.; Niwa, T.; Yamada, Y.; Wada, H.; Saito, K.; Iwakura, Y.; Asano, M.; Moriwaki, H.; Seishima, M. Lack of interleukin-1beta decreases the severity of atherosclerosis in ApoE-deficient mice. *Arterioscler. Thromb. Vasc. Biol.* **2003**, *23*, 656–660. [CrossRef]

15. Starke, R.M.; Chalouhi, N.; Jabbour, P.M.; Tjoumakaris, S.I.; Gonzalez, L.F.; Rosenwasser, R.H.; Wada, K.; Shimada, K.; Hasan, D.M.; Greig, N.H.; et al. Critical role of TNF-alpha in cerebral aneurysm formation and progression to rupture. *J. Neuroinflamm.* **2014**, *11*, 77. [CrossRef]
16. Saha, R.N.; Jana, M.; Pahan, K. MAPK p38 regulates transcriptional activity of NF-kappaB in primary human astrocytes via acetylation of p65. *J. Immunol.* **2007**, *179*, 7101–7109. [CrossRef]
17. Freund, A.; Patil, C.K.; Campisi, J. p38MAPK is a novel DNA damage response-independent regulator of the senescence-associated secretory phenotype. *EMBO J.* **2011**, *30*, 1536–1548. [CrossRef]
18. Tilstra, J.S.; Robinson, A.R.; Wang, J.; Gregg, S.Q.; Clauson, C.L.; Reay, D.P.; Nasto, L.A.; St Croix, C.M.; Usas, A.; Vo, N.; et al. NF-kappaB inhibition delays DNA damage-induced senescence and aging in mice. *J. Clin. Investig.* **2012**, *122*, 2601–2612. [CrossRef]
19. Garcia-Garcia, V.A.; Alameda, J.P.; Page, A.; Casanova, M.L. Role of NF-kappaB in Ageing and Age-Related Diseases: Lessons from Genetically Modified Mouse Models. *Cells* **2021**, *10*, 1906. [CrossRef]
20. Hongo, A.; Okumura, N.; Nakahara, M.; Kay, E.P.; Koizumi, N. The Effect of a p38 Mitogen-Activated Protein Kinase Inhibitor on Cellular Senescence of Cultivated Human Corneal Endothelial Cells. *Investig. Ophthalmol. Vis. Sci.* **2017**, *58*, 3325–3334. [CrossRef]
21. Childs, B.G.; Durik, M.; Baker, D.J.; van Deursen, J.M. Cellular senescence in aging and age-related disease: From mechanisms to therapy. *Nat. Med.* **2015**, *21*, 1424–1435. [CrossRef]
22. Li, X.; Khan, D.; Rana, M.; Hänggi, D.; Muhammad, S. Doxycycline Attenuated Ethanol-Induced Inflammaging in Endothelial Cells: Implications in Alcohol-Mediated Vascular Diseases. *Antioxidants* **2022**, *11*, 2413. [CrossRef]
23. Chen, X.; Li, M.; Yan, J.; Liu, T.; Pan, G.; Yang, H.; Pei, M.; He, F. Alcohol Induces Cellular Senescence and Impairs Osteogenic Potential in Bone Marrow-Derived Mesenchymal Stem Cells. *Alcohol Alcohol.* **2017**, *52*, 289–297. [CrossRef]
24. Chen, J.R.; Lazarenko, O.P.; Haley, R.L.; Blackburn, M.L.; Badger, T.M.; Ronis, M.J. Ethanol impairs estrogen receptor signaling resulting in accelerated activation of senescence pathways, whereas estradiol attenuates the effects of ethanol in osteoblasts. *J. Bone Miner. Res.* **2009**, *24*, 221–230. [CrossRef]
25. Blanco, A.M.; Valles, S.L.; Pascual, M.; Guerri, C. Involvement of TLR4/type I IL-1 receptor signaling in the induction of inflammatory mediators and cell death induced by ethanol in cultured astrocytes. *J. Immunol.* **2005**, *175*, 6893–6899. [CrossRef]
26. Ku, B.M.; Lee, Y.K.; Jeong, J.Y.; Mun, J.; Han, J.Y.; Roh, G.S.; Kim, H.J.; Cho, G.J.; Choi, W.S.; Yi, G.S.; et al. Ethanol-induced oxidative stress is mediated by p38 MAPK pathway in mouse hippocampal cells. *Neurosci. Lett.* **2007**, *419*, 64–67. [CrossRef]
27. Zhang, F.S.; He, Q.Z.; Qin, C.H.; Little, P.J.; Weng, J.P.; Xu, S.W. Therapeutic potential of colchicine in cardiovascular medicine: A pharmacological review. *Acta Pharmacol. Sin.* **2022**, *43*, 2173–2190. [CrossRef]
28. Deftereos, S.G.; Beerkens, F.J.; Shah, B.; Giannopoulos, G.; Vrachatis, D.A.; Giotaki, S.G.; Siasos, G.; Nicolas, J.; Arnott, C.; Patel, S.; et al. Colchicine in Cardiovascular Disease: In-Depth Review. *Circulation* **2022**, *145*, 61–78. [CrossRef] [PubMed]
29. Portincasa, P. Colchicine, Biologic Agents and More for the Treatment of Familial Mediterranean Fever. The Old, the New, and the Rare. *Curr. Med. Chem.* **2016**, *23*, 60–86. [CrossRef]
30. Schmittgen, T.D.; Livak, K.J. Analyzing real-time PCR data by the comparative C(T) method. *Nat. Protoc.* **2008**, *3*, 1101–1108. [CrossRef]
31. Seluanov, A.; Danek, J.; Hause, N.; Gorbunova, V. Changes in the level and distribution of Ku proteins during cellular senescence. *DNA Repair. Amst.* **2007**, *6*, 1740–1748. [CrossRef]
32. Liang, F.; Romanienko, P.J.; Weaver, D.T.; Jeggo, P.A.; Jasin, M. Chromosomal double-strand break repair in Ku80-deficient cells. *Proc. Natl. Acad. Sci. USA* **1996**, *93*, 8929–8933. [CrossRef]
33. d'Adda di Fagagna, F.; Hande, M.P.; Tong, W.M.; Roth, D.; Lansdorp, P.M.; Wang, Z.Q.; Jackson, S.P. Effects of DNA nonhomologous end-joining factors on telomere length and chromosomal stability in mammalian cells. *Curr. Biol.* **2001**, *11*, 1192–1196. [CrossRef]
34. Smith, A.J.; Ball, S.S.; Manzar, K.; Bowater, R.P.; Wormstone, I.M. Ku80 Counters Oxidative Stress-Induced DNA Damage and Cataract Formation in the Human Lens. *Investig. Ophthalmol. Vis. Sci.* **2015**, *56*, 7868–7874. [CrossRef]
35. Das, S.K.; Vasudevan, D.M. Alcohol-induced oxidative stress. *Life Sci.* **2007**, *81*, 177–187. [CrossRef]
36. Zalar, D.M.; Pop, C.; Buzdugan, E.; Kiss, B.; Stefan, M.G.; Ghibu, S.; Crisan, D.; Buruiana-Simic, A.; Grozav, A.; Borda, I.M.; et al. Effects of Colchicine in a Rat Model of Diet-Induced Hyperlipidemia. *Antioxidants* **2022**, *11*, 230. [CrossRef]
37. Pennings, G.J.; Reddel, C.J.; Traini, M.; Campbell, H.; Chen, V.; Kritharides, L. Colchicine inhibits ROS generation in response to glycoprotein VI stimulation. *Sci. Rep.* **2021**, *11*, 11965. [CrossRef]
38. Zhang, B.; Huang, R.; Yang, D.; Chen, G.; Chen, Y.; Han, J.; Zhang, S.; Ma, L.; Yang, X. Combination of Colchicine and Ticagrelor Inhibits Carrageenan-Induced Thrombi in Mice. *Oxid. Med. Cell. Longev.* **2022**, *2022*, 3087198. [CrossRef]
39. Anerillas, C.; Abdelmohsen, K.; Gorospe, M. Regulation of senescence traits by MAPKs. *Geroscience* **2020**, *42*, 397–408. [CrossRef]
40. Chen, Z.; Yao, L.; Liu, Y.; Pan, Z.; Peng, S.; Wan, G.; Cheng, J.; Wang, J.; Cao, W. Astragaloside IV regulates NF-kappaB-mediated cellular senescence and apoptosis of hepatic stellate cells to suppress PDGF-BB-induced activation. *Exp. Ther. Med.* **2019**, *18*, 3741–3750. [CrossRef]
41. Rovillain, E.; Mansfield, L.; Caetano, C.; Alvarez-Fernandez, M.; Caballero, O.L.; Medema, R.H.; Hummerich, H.; Jat, P.S. Activation of nuclear factor-kappa B signalling promotes cellular senescence. *Oncogene* **2011**, *30*, 2356–2366. [CrossRef]

42. Nicolae, C.M.; O'Connor, M.J.; Constantin, D.; Moldovan, G.L. NFkappaB regulates p21 expression and controls DNA damage-induced leukemic differentiation. *Oncogene* **2018**, *37*, 3647–3656. [CrossRef]
43. Saha, K.; Adhikary, G.; Kanade, S.R.; Rorke, E.A.; Eckert, R.L. p38delta regulates p53 to control p21Cip1 expression in human epidermal keratinocytes. *J. Biol. Chem.* **2014**, *289*, 11443–11453. [CrossRef]
44. Bulavin, D.V.; Saito, S.; Hollander, M.C.; Sakaguchi, K.; Anderson, C.W.; Appella, E.; Fornace, A.J., Jr. Phosphorylation of human p53 by p38 kinase coordinates N-terminal phosphorylation and apoptosis in response to UV radiation. *EMBO J.* **1999**, *18*, 6845–6854. [CrossRef]
45. Lafarga, V.; Cuadrado, A.; Lopez de Silanes, I.; Bengoechea, R.; Fernandez-Capetillo, O.; Nebreda, A.R. p38 Mitogen-activated protein kinase- and HuR-dependent stabilization of p21(Cip1) mRNA mediates the G(1)/S checkpoint. *Mol. Cell. Biol.* **2009**, *29*, 4341–4351. [CrossRef]
46. Shin, S.Y.; Kim, C.G.; Lim, Y.; Lee, Y.H. The ETS family transcription factor ELK-1 regulates induction of the cell cycle-regulatory gene p21(Waf1/Cip1) and the BAX gene in sodium arsenite-exposed human keratinocyte HaCaT cells. *J. Biol. Chem.* **2011**, *286*, 26860–26872. [CrossRef]
47. Meyer-Lindemann, U.; Mauersberger, C.; Schmidt, A.C.; Moggio, A.; Hinterdobler, J.; Li, X.; Khangholi, D.; Hettwer, J.; Grasser, C.; Dutsch, A.; et al. Colchicine Impacts Leukocyte Trafficking in Atherosclerosis and Reduces Vascular Inflammation. *Front. Immunol.* **2022**, *13*, 898690. [CrossRef]
48. Li, J.J.; Lee, S.H.; Kim, D.K.; Jin, R.; Jung, D.S.; Kwak, S.J.; Kim, S.H.; Han, S.H.; Lee, J.E.; Moon, S.J.; et al. Colchicine attenuates inflammatory cell infiltration and extracellular matrix accumulation in diabetic nephropathy. *Am. J. Physiol. Renal. Physiol.* **2009**, *297*, F200–F209. [CrossRef]
49. Gschwandtner, M.; Derler, R.; Midwood, K.S. More Than Just Attractive: How CCL2 Influences Myeloid Cell Behavior Beyond Chemotaxis. *Front. Immunol.* **2019**, *10*, 2759. [CrossRef]
50. Singh, S.; Anshita, D.; Ravichandiran, V. MCP-1: Function, regulation, and involvement in disease. *Int. Immunopharmacol.* **2021**, *101*, 107598. [CrossRef]
51. Muller, W.A. Getting leukocytes to the site of inflammation. *Vet. Pathol.* **2013**, *50*, 7–22. [CrossRef]
52. Choi, S.; Park, M.; Kim, J.; Park, W.; Kim, S.; Lee, D.K.; Hwang, J.Y.; Choe, J.; Won, M.H.; Ryoo, S.; et al. TNF-alpha elicits phenotypic and functional alterations of vascular smooth muscle cells by miR-155-5p-dependent down-regulation of cGMP-dependent kinase 1. *J. Biol. Chem.* **2018**, *293*, 14812–14822. [CrossRef]
53. Chen, T.; Zhang, X.; Zhu, G.; Liu, H.; Chen, J.; Wang, Y.; He, X. Quercetin inhibits TNF-alpha induced HUVECs apoptosis and inflammation via downregulating NF-kB and AP-1 signaling pathway in vitro. *Med. Baltim.* **2020**, *99*, e22241. [CrossRef] [PubMed]
54. Siwik, D.A.; Chang, D.L.; Colucci, W.S. Interleukin-1beta and tumor necrosis factor-alpha decrease collagen synthesis and increase matrix metalloproteinase activity in cardiac fibroblasts in vitro. *Circ. Res.* **2000**, *86*, 1259–1265. [CrossRef] [PubMed]
55. Cabral-Pacheco, G.A.; Garza-Veloz, I.; Castruita-De la Rosa, C.; Ramirez-Acuna, J.M.; Perez-Romero, B.A.; Guerrero-Rodriguez, J.F.; Martinez-Avila, N.; Martinez-Fierro, M.L. The Roles of Matrix Metalloproteinases and Their Inhibitors in Human Diseases. *Int. J. Mol. Sci.* **2020**, *21*, 9739. [CrossRef]
56. Young, D.; Das, N.; Anowai, A.; Dufour, A. Matrix Metalloproteases as Influencers of the Cells' Social Media. *Int. J. Mol. Sci.* **2019**, *20*, 3847. [CrossRef]
57. Rastogi, S.; Rizwani, W.; Joshi, B.; Kunigal, S.; Chellappan, S.P. TNF-alpha response of vascular endothelial and vascular smooth muscle cells involve differential utilization of ASK1 kinase and p73. *Cell Death Differ.* **2012**, *19*, 274–283. [CrossRef]
58. Kjaergaard, A.G.; Dige, A.; Krog, J.; Tonnesen, E.; Wogensen, L. Soluble adhesion molecules correlate with surface expression in an in vitro model of endothelial activation. *Basic Clin. Pharmacol. Toxicol.* **2013**, *113*, 273–279. [CrossRef]
59. Wang, L.; Tang, C. Targeting Platelet in Atherosclerosis Plaque Formation: Current Knowledge and Future Perspectives. *Int. J. Mol. Sci.* **2020**, *21*, 9760. [CrossRef]
60. Shah, B.; Allen, N.; Harchandani, B.; Pillinger, M.; Katz, S.; Sedlis, S.P.; Echagarruga, C.; Samuels, S.K.; Morina, P.; Singh, P.; et al. Effect of Colchicine on Platelet-Platelet and Platelet-Leukocyte Interactions: A Pilot Study in Healthy Subjects. *Inflammation* **2016**, *39*, 182–189. [CrossRef]
61. Vaidya, K.; Tucker, B.; Kurup, R.; Khandkar, C.; Pandzic, E.; Barraclough, J.; Machet, J.; Misra, A.; Kavurma, M.; Martinez, G.; et al. Colchicine Inhibits Neutrophil Extracellular Trap Formation in Patients With Acute Coronary Syndrome After Percutaneous Coronary Intervention. *J. Am. Heart Assoc.* **2021**, *10*, e018993. [CrossRef]
62. Zhou, Y.; Xu, Z.; Liu, Z. Impact of Neutrophil Extracellular Traps on Thrombosis Formation: New Findings and Future Perspective. *Front. Cell. Infect. Microbiol.* **2022**, *12*, 910908. [CrossRef] [PubMed]
63. Kim, M.J.; Nepal, S.; Lee, E.S.; Jeong, T.C.; Kim, S.H.; Park, P.H. Ethanol increases matrix metalloproteinase-12 expression via NADPH oxidase-dependent ROS production in macrophages. *Toxicol. Appl. Pharmacol.* **2013**, *273*, 77–89. [CrossRef] [PubMed]
64. Sillanaukee, P.; Kalela, A.; Seppa, K.; Hoyhtya, M.; Nikkari, S.T. Matrix metalloproteinase-9 is elevated in serum of alcohol abusers. *Eur. J. Clin. Investig.* **2002**, *32*, 225–229. [CrossRef] [PubMed]
65. Koken, T.; Gursoy, F.; Kahraman, A. Long-term alcohol consumption increases pro-matrix metalloproteinase-9 levels via oxidative stress. *J. Med. Toxicol.* **2010**, *6*, 126–130. [CrossRef]

66. Wang, J.; Liu, Y.; Zhang, L.; Ji, J.; Wang, B.; Jin, W.; Zhang, C.; Chu, H. Effects of increased matrix metalloproteinase-9 expression on skeletal muscle fibrosis in prolonged alcoholic myopathies of rats. *Mol. Med. Rep.* **2012**, *5*, 60–65. [CrossRef]
67. Yin, L.; Li, F.; Li, J.; Yang, X.; Xie, X.; Xue, L.; Li, Y.; Zhang, C. Chronic Intermittent Ethanol Exposure Induces Upregulation of Matrix Metalloproteinase-9 in the Rat Medial Prefrontal Cortex and Hippocampus. *Neurochem. Res.* **2019**, *44*, 1593–1601. [CrossRef]

MDPI AG
Grosspeteranlage 5
4052 Basel
Switzerland
Tel.: +41 61 683 77 34

Antioxidants Editorial Office
E-mail: antioxidants@mdpi.com
www.mdpi.com/journal/antioxidants

www.ingramcontent.com/pod-product-compliance
Lightning Source LLC
Chambersburg PA
CBHW041356210926
51850CB00013B/3097